New Research Trends for Textiles, a Bright Future

New Research Trends for Textiles, a Bright Future

Laurent Dufossé

Basel • Beijing • Wuhan • Barcelona • Belgrade • Novi Sad • Cluj • Manchester

Editor
Laurent Dufossé
CHEMBIOPRO Laboratory
University of Réunion Island
Sainte-Clotilde
France

Editorial Office
MDPI AG
Grosspeteranlage 5
4052 Basel, Switzerland

This is a reprint of articles from the Special Issue published online in the open access journal *Textiles* (ISSN 2673-7248) (available at: www.mdpi.com/journal/textiles/special_issues/GD2Z85X58V).

For citation purposes, cite each article independently as indicated on the article page online and as indicated below:

Lastname, A.A.; Lastname, B.B. Article Title. *Journal Name* **Year**, *Volume Number*, Page Range.

ISBN 978-3-7258-2590-5 (Hbk)
ISBN 978-3-7258-2589-9 (PDF)
doi.org/10.3390/books978-3-7258-2589-9

© 2024 by the authors. Articles in this book are Open Access and distributed under the Creative Commons Attribution (CC BY) license. The book as a whole is distributed by MDPI under the terms and conditions of the Creative Commons Attribution-NonCommercial-NoDerivs (CC BY-NC-ND) license.

Contents

About the Editor vii

Laurent Dufossé
New Research Trends for Textiles, a Bright Future
Reprinted from: *Textiles* 2024, 4, 356-358, doi:10.3390/textiles4030021 1

Fulga Tanasa, Carmen-Alice Teaca, Marioara Nechifor, Maurusa Ignat, Ioana Alexandra Duceac and Leonard Ignat
Highly Specialized Textiles with Antimicrobial Functionality—Advances and Challenges
Reprinted from: *Textiles* 2023, 3, 219-245, doi:10.3390/textiles3020015 4

Arjunsing Girase, Donald Thompson and Robert Bryan Ormond
Comparative Analysis of the Liquid CO_2 Washing with Conventional Wash on Firefighters' Personal Protective Equipment (PPE)
Reprinted from: *Textiles* 2022, 2, 624-632, doi:10.3390/textiles2040036 31

Josephine Taiye Bolaji and Patricia I. Dolez
Supportive, Fitted, and Comfortable Bras for Individuals with Atypical Breast Shape/Size: Review of the Challenges and Proposed Roadmap
Reprinted from: *Textiles* 2022, 2, 560-578, doi:10.3390/textiles2040032 40

Jully Schmidt Pinto Filippi, Angelo Oliveira Silva, Cintia Marangoni, Jeferson Correia, José Alexandre Borges Valle and Rita de Cassia Siqueira Curto Valle
Turkey Red Oil as a Renewable Leveling and Dispersant Option for Polyester Dyeing with Dispersed Dyes
Reprinted from: *Textiles* 2023, 3, 163-181, doi:10.3390/textiles3020012 59

Christian Dils, Sebastian Hohner and Martin Schneider-Ramelow
Use of Rotary Ultrasonic Plastic Welding as a Continuous Interconnection Technology for Large-Area e-Textiles
Reprinted from: *Textiles* 2023, 3, 66-87, doi:10.3390/textiles3010006 78

Sara A. Ebrahim, Hanan A. Othman, Mohamed M. Mosaad and Ahmed G. Hassabo
Eco-Friendly Natural Thickener (Pectin) Extracted from Fruit Peels for Valuable Utilization in Textile Printing as a Thickening Agent
Reprinted from: *Textiles* 2023, 3, 26-49, doi:10.3390/textiles3010003 100

Waleed Hassan Akhtar, Chihiro Watanabe, Yuji Tou and Pekka Neittaanmäki
A New Perspective on the Textile and Apparel Industry in the Digital Transformation Era
Reprinted from: *Textiles* 2022, 2, 633-656, doi:10.3390/textiles2040037 124

Abdullah Al Mamun, Koushik Kumar Bormon, Mst Nigar Sultana Rasu, Amit Talukder, Charles Freeman and Reuben Burch et al.
An Assessment of Energy and Groundwater Consumption of Textile Dyeing Mills in Bangladesh and Minimization of Environmental Impacts via Long-Term Key Performance Indicators (KPI) Baseline
Reprinted from: *Textiles* 2022, 2, 511-523, doi:10.3390/textiles2040029 148

Yu Wang, Xuejiao Li, Junbo Xie, Ning Wu, Yanan Jiao and Peng Wang
Numerical and Experimental Investigation on Bending Behavior for High-Performance Fiber Yarns Considering Probability Distribution of Fiber Strength
Reprinted from: *Textiles* 2023, 3, 129-141, doi:10.3390/textiles3010010 161

Raphael Romao Santos, Masumi Nakanishi and Sachiko Sukigara
Tactile Perception of Woven Fabrics by a Sliding Index Finger with Emphasis on Individual Differences
Reprinted from: *Textiles* **2023**, 3, 115-128, doi:10.3390/textiles3010009 . . . . **174**

Meenakshi Ahirwar and B. K. Behera
Prediction of Shrinkage Behavior of Stretch Fabrics Using Machine-Learning Based Artificial Neural Network
Reprinted from: *Textiles* **2023**, 3, 88-97, doi:10.3390/textiles3010007 . . . . **188**

Ryoga Miyauchi, Xiaoxiao Zhou and Yuki Inoue
Design Elements That Increase the Willingness to Pay for Denim Fabric Products
Reprinted from: *Textiles* **2023**, 3, 11-25, doi:10.3390/textiles3010002 . . . . **198**

About the Editor

Laurent Dufossé

Laurent Dufossé has held the position of Professor of Food Science and Biotechnology since 2006 at Reunion Island University, which is located on a volcanic island in the Indian Ocean, near Madagascar and Mauritius. The island is one of France's overseas territories, with almost one million inhabitants, and the university has 19,000 students. Previously, Professor Dufossé was a researcher and senior lecturer at the Université de Bretagne Occidentale, Quimper, Brittany, France. He attended the University of Burgundy, Dijon, France, where he received his PhD in Food Science in 1993, and he has been involved in the field of the biotechnology of food ingredients for more than 30 years.

His main research interests over the last 20 years have mainly been the microbial production of pigments and the study of aryl carotenoids, such as isorenieratene, C50 carotenoids, azaphilones, and anthraquinones. These studies have relevance for applications in science and technology in areas such as the supply of sustainable components in many industrial sectors and the development of biobased pigments and dyeing agents for the textile industry.

Editorial

New Research Trends for Textiles, a Bright Future

Laurent Dufossé

CHEMBIOPRO Laboratoire de Chimie et Biotechnologie des Produits Naturels, ESIROI Agroalimentaire, Université de La Réunion, 15 Avenue René Cassin, F-97400 Saint-Denis, Ile de La Réunion, France; laurent.dufosse@univ-reunion.fr; Tel.: +262-692402400

Citation: Dufossé, L. New Research Trends for Textiles, a Bright Future. *Textiles* **2024**, *4*, 356–358. https://doi.org/10.3390/textiles4030021

Received: 14 August 2024
Accepted: 16 August 2024
Published: 19 August 2024

Copyright: © 2024 by the author. Licensee MDPI, Basel, Switzerland. This article is an open access article distributed under the terms and conditions of the Creative Commons Attribution (CC BY) license (https://creativecommons.org/licenses/by/4.0/).

The *Textiles* journal is a peer-reviewed, open access journal, officially launched in 2021. It concerns research and innovation in the field of textile materials. This field is very broad and covers many topics. Textile materials composed of fibers linked by weaving, braiding, knitting, or sewing constitute a wide range of materials and are essential for many applications. They are ancestral materials used since antiquity and yet find themselves utilized in advanced applications, such as in composites in aeronautics or the medical industry [1].

Textiles, an open access international journal by MDPI (Basel, Switzerland), focuses on the broad field of textile materials and topics including, but not limited to, the following: fibers and yarns for textiles, properties, and microstructures; advances in weaving, braiding, and knitting technologies; 3D textiles; nonwovens; the structure and properties of high-performance textiles; the characterization and testing of textiles; fatigue, damage, and the failure of textiles; friction in textile materials; simulations in textiles; textile and clothing science; sustainable fibers and textiles; dyeing textiles; microbial [2,3] and plant pigments for the textile industry; microbial enzymes in the textile industry; bio-polishing; the bioconversion of waste fabric; microbial wastewater treatment; microbial silk; bacterial cellulose; recycling in textiles; fashion and apparel design; textile composite; preform and prepreg draping; medical textile materials; textile materials for civil engineering applications; geotextiles; smart textiles; protective and thermal protective textiles; textile history and archaeology [1,4].

To date, *Textiles* has published 115 articles that can be found here: https://www.mdpi.com/search?q=&journal=textiles&sort=pubdate&page_count=50 (accessed on 14 August 2024).

In order to demonstrate the huge impact of textile research and technology on the world, the Publisher, the Editorial Board, and myself decided to invite contributions and feature papers from key world-class researchers which were collected in a single Special Issue entitled 'New research trends for textiles, a bright future', from September 2022 to May 2023.

In total, 17 papers were submitted and 12 published. These 12 papers can be roughly subdivided into four parts: functional textiles; process and modeling; control; consumers and behavior.

1. Functional Textiles

Tanasa, F.; Teaca, C.; Nechifor, M.; Ignat, M.; Duceac, I.; Ignat, L. Highly Specialized Textiles with Antimicrobial Functionality—Advances and Challenges. Textiles 2023, 3(2), 219–245; https://doi.org/10.3390/textiles3020015. https://www.mdpi.com/2673-7248/3/2/15 (accessed on 14 August 2024).

Girase, A.; Thompson, D.; Ormond, R. Comparative Analysis of the Liquid CO_2 Washing with Conventional Wash on Firefighters' Personal Protective Equipment (PPE). Textiles 2022, 2(4), 624–632; https://doi.org/10.3390/textiles2040036. https://www.mdpi.com/2673-7248/2/4/36 (accessed on 14 August 2024).

Bolaji, J.; Dolez, P. Supportive, Fitted, and Comfortable Bras for Individuals with Atypical Breast Shape/Size: Review of the Challenges and Proposed Roadmap. Textiles

2022, 2(4), 560–578; https://doi.org/10.3390/textiles2040032. https://www.mdpi.com/2673-7248/2/4/32 (accessed on 14 August 2024).

2. Process and Modelling

Filippi, J.; Silva, A.; Marangoni, C.; Correia, J.; Valle, J.; Valle, R. Turkey Red Oil as a Renewable Leveling and Dispersant Option for Polyester Dyeing with Dispersed Dyes. Textiles 2023, 3(2), 163–181; https://doi.org/10.3390/textiles3020012. https://www.mdpi.com/2673-7248/3/2/12 (accessed on 14 August 2024).

Dils, C.; Hohner, S.; Schneider-Ramelow, M. Use of Rotary Ultrasonic Plastic Welding as a Continuous Interconnection Technology for Large-Area e-Textiles. Textiles 2023, 3(1), 66–87; https://doi.org/10.3390/textiles3010006. https://www.mdpi.com/2673-7248/3/1/6 (accessed on 14 August 2024).

Ebrahim, S.; Othman, H.; Mosaad, M.; Hassabo, A. Eco-Friendly Natural Thickener (Pectin) Extracted from Fruit Peels for Valuable Utilization in Textile Printing as a Thickening Agent. Textiles 2023, 3(1), 26–49; https://doi.org/10.3390/textiles3010003. https://www.mdpi.com/2673-7248/3/1/3 (accessed on 14 August 2024).

Akhtar, W.; Watanabe, C.; Tou, Y.; Neittaanmäki, P. A New Perspective on the Textile and Apparel Industry in the Digital Transformation Era. Textiles 2022, 2(4), 633–656; https://doi.org/10.3390/textiles2040037. https://www.mdpi.com/2673-7248/2/4/37 (accessed on 14 August 2024).

Mamun, A.; Bormon, K.; Rasu, M.; Talukder, A.; Freeman, C.; Burch, R.; Chander, H. An Assessment of Energy and Groundwater Consumption of Textile Dyeing Mills in Bangladesh and Minimization of Environmental Impacts via Long-Term Key Performance Indicators (KPI) Baseline. Textiles 2022, 2(4), 511–523; https://doi.org/10.3390/textiles2040029. https://www.mdpi.com/2673-7248/2/4/29 (accessed on 14 August 2024).

3. Control

Wang, Y.; Li, X.; Xie, J.; Wu, N.; Jiao, Y.; Wang, P. Numerical and Experimental Investigation on Bending Behavior for High-Performance Fiber Yarns Considering Probability Distribution of Fiber Strength. Textiles 2023, 3(1), 129–141; https://doi.org/10.3390/textiles3010010. https://www.mdpi.com/2673-7248/3/1/10 (accessed on 14 August 2024).

Romao Santos, R.; Nakanishi, M.; Sukigara, S. Tactile Perception of Woven Fabrics by a Sliding Index Finger with Emphasis on Individual Differences. Textiles 2023, 3(1), 115–128; https://doi.org/10.3390/textiles3010009. https://www.mdpi.com/2673-7248/3/1/9 (accessed on 14 August 2024).

Ahirwar, M.; Behera, B. Prediction of Shrinkage Behavior of Stretch Fabrics Using Machine-Learning Based Artificial Neural Network. Textiles 2023, 3(1), 88–97; https://doi.org/10.3390/textiles3010007. https://www.mdpi.com/2673-7248/3/1/7 (accessed on 14 August 2024).

4. Consumers and Behavior

Miyauchi, R.; Zhou, X.; Inoue, Y. Design Elements That Increase the Willingness to Pay for Denim Fabric Products. Textiles 2023, 3(1), 11–25; https://doi.org/10.3390/textiles3010002. https://www.mdpi.com/2673-7248/3/1/2 (accessed on 14 August 2024).

I, as the Guest Editor, trust all readers of this Special Issue to enjoy the contents, and I would like to deeply thank all the wonderful authors who contributed, Prof. Philippe Boisse, Editor-in-Chief of *Textiles*, the numerous reviewers, and the whole team at MDPI (editing, production, website, etc.).

Conflicts of Interest: The author declares no conflict of interest.

References

1. Boisse, P. Textiles: Multidisciplinary Open Access Journal in Research and Innovation of Textiles. *Textiles* **2021**, *1*, 1–3. [CrossRef]
2. Venil, C.K.; Velmurugan, P.; Dufossé, L.; Devi, P.R.; Ravi, A.V. Fungal pigments: Potential coloring compounds for wide ranging applications in textile dyeing. *J. Fungi* **2020**, *6*, 68. [CrossRef] [PubMed]

3. Venil, C.K.; Dufossé, L.; Velmurugan, P.; Malathi, M.; Lakshmanaperumalsamy, P. Extraction and Application of Pigment from *Serratia marcescens* SB08, an Insect Enteric Gut Bacterium, for Textile Dyeing. *Textiles* **2021**, *1*, 21–36. [CrossRef]
4. Dufossé, L. *New Research Trends for Textiles*; Print & Ebook; MDPI Publisher: Basel, Switzerland, 2022; 506p, ISBN 978-3-0365-6082-3 (Hardback)/978-3-0365-6081-6(PDF). [CrossRef]

Disclaimer/Publisher's Note: The statements, opinions and data contained in all publications are solely those of the individual author(s) and contributor(s) and not of MDPI and/or the editor(s). MDPI and/or the editor(s) disclaim responsibility for any injury to people or property resulting from any ideas, methods, instructions or products referred to in the content.

Review

Highly Specialized Textiles with Antimicrobial Functionality—Advances and Challenges

Fulga Tanasa [1,*], Carmen-Alice Teaca [2], Marioara Nechifor [1], Maurusa Ignat [2], Ioana Alexandra Duceac [1] and Leonard Ignat [2]

[1] Polyaddition and Photochemistry Department, "Petru Poni" Institute of Macromolecular Chemistry, 41A Grigore Ghica-Vodă Alley, 700487 Iaşi, Romania
[2] Center for Advanced Research in Bionanoconjugates and Biopolymers, "Petru Poni" Institute of Macromolecular Chemistry, 41A Grigore Ghica-Vodă Alley, 700487 Iaşi, Romania
* Correspondence: ftanasa@icmpp.ro

Abstract: Textiles with antimicrobial functionality have been intensively and extensively investigated in the recent decades, mostly because they are present in everyday life in various applications: medicine and healthcare, sportswear, clothing and footwear, furniture and upholstery, air and water purification systems, food packaging etc. Their ability to kill or limit the growth of the microbial population in a certain context defines their activity against bacteria, fungi, and viruses, and even against the initial formation of the biofilm prior to microorganisms' proliferation. Various classes of antimicrobials have been employed for these highly specialized textiles, namely, organic synthetic reagents and polymers, metals and metal oxides (micro- and nanoparticles), and natural and naturally derived compounds, and their activity and range of applications are critically assessed. At the same time, different modern processing techniques are reviewed in relation to their applications. This paper focuses on some advances and challenges in the field of antimicrobial textiles given their practical importance as it appears from the most recent reports in the literature.

Keywords: antimicrobial textiles; synthetic antimicrobial reagents; polymers; natural antimicrobial compounds; applications

Citation: Tanasa, F.; Teaca, C.-A.; Nechifor, M.; Ignat, M.; Duceac, I.A.; Ignat, L. Highly Specialized Textiles with Antimicrobial Functionality—Advances and Challenges. *Textiles* **2023**, *3*, 219–245. https://doi.org/10.3390/textiles3020015

Academic Editor: Desislava Staneva

Received: 22 March 2023
Revised: 9 May 2023
Accepted: 16 May 2023
Published: 18 May 2023

Copyright: © 2023 by the authors. Licensee MDPI, Basel, Switzerland. This article is an open access article distributed under the terms and conditions of the Creative Commons Attribution (CC BY) license (https://creativecommons.org/licenses/by/4.0/).

1. Introduction

1.1. General Background

Health risks management has been constantly considered in recent decades in all relevant domains in daily life due to the spectacular worldwide increase in number and variety of microbial infestation and proliferation, ranging from local to global, and from aggressive to violent and nonresponsive epidemics/pandemics (plague, SARS, West Nile, SARS-CoV-2, COVID-19, cholera, smallpox, scarlet rash, HIV-AIDS, Marburg, Ebola, Spanish flu, MERS) [1–3]. Thus, the employ of textiles with antimicrobial functionality has expanded up to unexpected rates. This market was estimated at USD 10.7 billion in 2021 and was projected to reach a 50% increase by 2026 at a compound annual growth rate (CAGR) of 6.5% in the same interval [4].

Subsequently, the scientific literature recorded an increase in the number of articles reporting on antimicrobial textiles and their specific finishing, reagents, and processing. A bibliometric analysis indicated in 2021 a number of publications of 534 articles per year [5], but the domain is very active and the rapid progress is abundantly documented by the most recent literature reports, which also illustrate the variety of new features connected to the subject [6–25].

Furthermore, recent surveys confirmed this trend. For example, data from the Web of Science Core Collection confirmed a number of 50 review articles published in the interval 2018–2023 on topics considered relevant for this manuscript. Moreover, a significant number of patents—245—has been reported in the interval 2018–2023 (https://patents.google.com; accessed on 26 April 2023). Some of these data are illustrated in Figures 1 and 2, where the selection criteria are given in the legend.

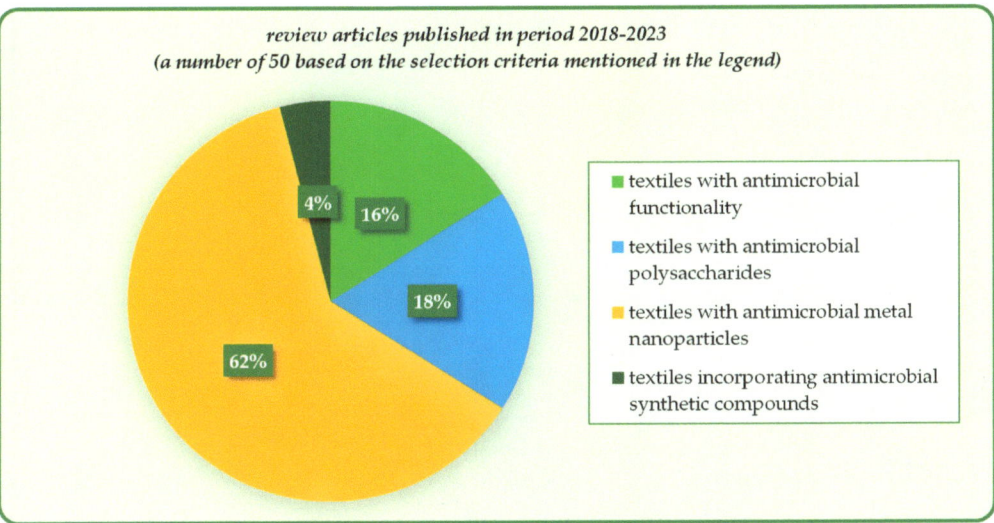

Figure 1. Review articles published in 2018–2023 (data from Web of Science Core Collection).

First and foremost, textiles with antimicrobial finishing have to comply with several requirements: prevent, control, and/or eliminate microbial infestation, growth, and cross-infection over a wide spectrum; reduce odor, prevent staining, and maintain freshness for long intervals; must be stable, safe, durable, and reusable (in certain applications) [26]. Considering their antimicrobial effectiveness and the mechanism of action, as well as their toxicity versus tolerance, nature of fibers, and durability, textiles with antimicrobial functionality may be divided into several classes [18]:

- biostats, biocides (antibacterial, antifungal, antiviral), barriers, and antibiofilm;
- textiles with bound or leaching antimicrobial finishing;
- textiles made of natural (cotton, wool, silk, linen) or synthetic fibers (PP, PE, PES) or blends (cotton/elastane, cotton/PES, wool/acrylic);
- textiles able to release compounds with biologic activity;
- wearable and washing resistant.

Commonly, microorganisms are divided into different classes: bacteria, archaea, protozoa, algae, fungi, viruses, and multicellular animal parasites [27]. They have distinct features; most of them do not negatively interfere with human biota, but some can be or become pathogenic when certain favorable conditions are met. Bacteria are mainly divided into Gram-positive (*Staphylococcus aureus*) and Gram-negative (*Escherichia coli*). Other pathogenic bacteria (*Plasmodium malariae, Mycobacterium tuberculosis, Clostridium tetani, Corynebacterium diphtheriae, Treponema pallidum*), fungi (*Cryptococcus neoformans, Candida auris, Aspergillus fumigatus, Candida albicans, Candida glabrata*), and viruses (Ebola, herpes, hantavirus, papillomavirus, HIV, COVID-19) of particular concern have been used to evaluate the level of performance of antimicrobial textiles.

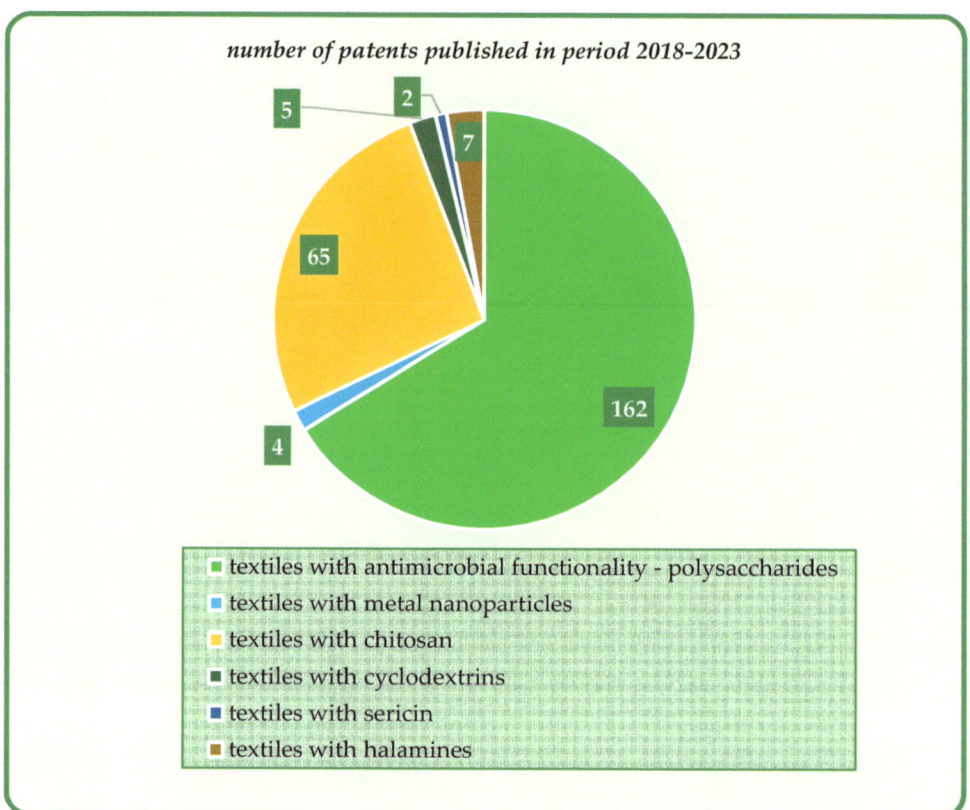

Figure 2. Patents published in the interval 2018–2023 (data acquired from https://patents.google.com; accessed on 26 April 2023).

The present review surveyed some of the most recent and relevant papers in the field of highly specialized textiles with acquired antimicrobial functionality. This allowed the identification not only of new trends and advances, but also of challenges in the field, mainly when it came to the use of high-tech processing methods, variety of applications, employ of complex formulations which include several antimicrobial agents that act in synergy, manufacture of multitask antimicrobial textiles, safety, and environmental risks.

1.2. Processing Techniques

Textiles with antimicrobial functionality are materials of high interest; therefore, their processing is a key factor in their activity and stability. Padding, spraying, grafting, and cross-linking are some of the most relevant techniques. However, the development of biocide/biostatic textiles made of synthetic fibers has allowed new methods, such as compounding extrusion and melt blending [28,29]. At the same time, the employ of colloidal solutions, plasma treatments, magnetron sputtering, sol–gel processes, microencapsulation techniques, or even *in situ* formation/growth of different antimicrobials onto textile supports are modern processing methods that grant textiles enhanced activity and stability [28].

Coating is one of the most popular procedures and is suitable for both yarns and fabrics, natural and synthetic fibers, knitted, woven, and nonwoven textiles. Direct coating can be achieved by knife, roller, or calendaring, and the finishing must be viscous in order to form a satisfactory coating. The spray-coating technique uses an airbrush and the finishing solution must be less viscous. The method may be applied to nanoparticles deposition as well [30].

The exhaust method was "imported" from dyeing processing and comprises the transfer of the active reagent from a batch to the textile substrate, sometimes in the presence of a binder, and a curing stage is required to stabilize the coating. Thiazole-derived reagents have been successfully applied by this method to textiles which subsequently exhibited high effectiveness against Gram-positive and Gram-negative bacteria [31].

The pad–dry–cure approach, also known as the mechanical thermal fixation or padding, is suitable for micro- and nanoparticulate coating materials with low or no affinity toward the textile substrate. The thermal treatment must be short (1–5 min) and at high temperatures (100–150 °C) in order to reach an appropriate cross-linking degree (thermal fixation). The method is simple and effective [28].

Textile substrates may be submitted to different methods of surface modification in order to achieve better compatibility with the antimicrobial finishing reagents. Plasma techniques, microencapsulation and ultrasound methods are among the most employed.

Plasma treatments are highly effective and environmentally friendly, despite their drawbacks (high-energy-consuming process, expensive equipment), and are used to clean/etch or create new functional groups onto textile surfaces, to deposit thin films of nanometric thickness, or even grow nanoparticles *in situ*. The possibility to limit the in-depth alteration of the support is considered the main advantage of this method because it prevents the alteration of the bulk properties of the textile [21,32]. Plasma grafting and polymerization can be applied to a wide range of antimicrobial finishing reagents (quaternary ammonium salts derivatives, dichlorophenol, triclosan, chitosan, guanidine-based compounds, metal and metal oxides nanoparticles) when natural, synthetic, or blended textiles are used as support [21]. Plasma and magnetron sputtering were preferred for metals and metal oxides nanoparticles deposition (Ag, Ti, Cu) onto different substrates when stable coatings were obtained [9,33,34]. Moreover, it was recently reported that the emergence of highly effective antiviral textiles for personal protective equipment was favored by the employ of plasma processing [35,36].

The microencapsulation technique is a modern method used to manufacture antimicrobial textiles, having the advantage that the core is protected and thus the degradation under the action of external factors is prevented. Moreover, the microcapsules are stable and safe to handle and apply to the textile support [37,38]. The approach is preferred when natural and naturally derived compounds are used as antimicrobial finishing reagents. It can be achieved by chemical (*in situ* polymerization in oil-in-water emulsion; interfacial polymerization) and physico–chemical (coacervation, molecular inclusion complexes) methods, and the obtained coatings are resistant to friction, sunlight, washing, and wet/dry cleaning [39].

Nanotechnology is also employed in the manufacture of antimicrobial textiles in various manners. The sol–gel method is a wet chemical procedure and uses colloidal solutions of monomers as precursor to form an interpenetrated network with the textile support or to deposit particles onto the textile surface [28,40]. Metals and metal oxides can also be applied onto textiles by this method, as in the case of titanium dioxide and zinc oxide nanoparticles used for coating fabrics able to prevent the spreading of nosocomial infections [41] or for textiles with antibacterial activity and self-cleaning properties [42]. Cotton, wool, and silk fabrics are suitable for this method and a wise selection of reagents for the sol phase can impart in the end a multiple functionality to the textiles, alongside their biocide activity [28].

In situ synthesis of nanoparticles has the advantage of nanoparticles deposited directly onto the textile support, rather homogeneously, without binders or stabilizers, thus significantly reducing the waste and pollution (and the safety and environmental risks, respectively) and increasing the stability of deposition. Metals and metal oxides (Ag, ZnO, Fe, Au) are mostly used for this technique applied to natural or synthetic textiles [1,28,43].

Highly specialized fibers with antimicrobial activity have been successfully obtained by electrospinning, a modern technique that allows materials made of biopolymers or synthetic polymers, with fibrous/porous morphology, and having tailored biocide properties [44,45].

In the following, some new trends and advances in the field of highly specialized textiles with antimicrobial functionality are presented, as illustrated by recent reports.

2. Synthetic Antimicrobial Agents for Textile Finishing

Antimicrobials encompass a large variety of chemical compounds and physical agents that act on microbes (bacteria, fungi, viruses, protozoa) in general. They are used to kill bacteria or to prevent their development. However, many of them exhibit some serious drawbacks that restrict or prohibit their use, such as the emergence of resistance developed shortly after their introduction, and undesired side effects. At the same time, chemical biocides are potentially harmful substances for the environment and human health if not handled or processed properly.

N-halamine compounds are organic biocides capable of killing microorganisms without releasing free oxidative halogen until they come into contact with microorganisms. They present efficiency against a broad spectrum of microorganisms, long-term stability, non–toxicity to humans, regenerability upon exposure to aqueous free chlorine solutions, and excellent biocompatibility. In addition, microorganisms do not develop resistance to this class of antimicrobials. The surface of the materials influences the antibacterial mechanism of *N*-halamines and has an important role in their antibacterial effectiveness. A large number of places of contact with bacteria increases the inactivation rate and is favored by a larger surface area.

N-halamine biocides have been used in different applications such as water filtration systems, disinfectants in pools, textiles, and medical devices [46]. *N*-halamines and some other synthetic compounds, such as quaternary ammonium compounds, polyhexamethylene biguanide, and triclosan, have been applied for antimicrobial treatment of textiles. Antimicrobial fabrics have found different applications in pharmaceutical, medical, engineering, agricultural, and food industries [47,48]. The *N*-halamine-treated fabrics can be rendered as having excellent antimicrobial activity through a bleaching process and can inactivate a broad spectrum of microorganisms, including Gram-negative and Gram-positive bacteria, in relatively short contact times. When the oxidative halogen is consumed, textiles modified with *N*-halamines regain their antimicrobial properties by exposing them to diluted household bleach. However, the practical application of *N*-halamines involves some disadvantages. For example, the cost of the treatment increases in the case of the use of organic solvents necessary to dissolve some *N*-halamine derivatives, which also presents safety risks.

As surfactants, quaternary ammonium compounds concentrate at the interface between the lipid-containing bacterial cell membrane and the surrounding aqueous environment. There are two types of interaction between quaternary ammonium salts and microbes: a polar interaction, occurring by cationic nitrogen, and a non–polar one, attributed to the hydrophobic chain. The cationic ammonium group can interact with the negatively charged cell membrane of bacteria. This attraction force induces the generation of a surfactant–microbe complex which can interrupt the activity of proteins, including all of the important functions in the cell membrane and even bacterial DNA. Furthermore, hydrophobic groups can penetrate into the microorganism and cancel all of the key cell functions. Increasing the length of the alkyl chain results in increasing the antibacterial activity of quaternary ammonium salts [49].

Quaternary ammonium compounds have no effectiveness against difficult-to-kill non–enveloped viruses. Among the extremely effective disinfectants with a wide spectrum and short contact times (3–5 min), we can count the formulations with low alcohol content used against bacteria, enveloped viruses, pathogenic fungi, and mycobacteria. Disinfectants based on quaternary ammonium salts with the addition of alcohol or solvents bring about a much faster drying of the products on the applied surface, which results in an ineffective or incomplete disinfection. In addition, quaternary ammonium compounds kill algae and are used in industrial water systems to counteract unwanted biological growth. Cetrimide (alkyltrimethylammonium bromide) and benzalkonium chloride have antibacterial, antifungal, and antiviral (enveloped viruses) properties and can be applied to the skin or mucous membranes to avoid or minimize the risk of infection. Hard water, anionic detergents, and organic matter reduce the activity of these disinfectants based on quaternary ammonium salts, which is a disadvantage. Moreover, *Pseudomonas* can metabolize cetrimide, using it as a carbon, nitrogen, and energy source.

Triclosan has antiseptic and disinfectant properties and a significant action against Gram-negative and Gram-positive bacteria. The acaricide benzyl benzoate in its structure accounts for protection against mites and it is used in acaricide (spray or powder) formulas, and for the treatment of scabies as a solution (25% concentration). Triclosan has been widely used in a large number of consumer products, such as cosmetics, toothpastes, deodorants, soaps, toys, and surgical cleaning treatments, based on its non-toxicity and antibacterial properties. Although triclosan is not considered to be as toxic as other pollutants, its occurrence in wastewaters, biosolids, and aquatic and terrestrial environments remains a concern. Furthermore, triclosan exhibits certain physicochemical characteristics that make it difficult to remove from the environment. There are studies that attribute some harmful health effects to triclosan, such as skin irritation, hormonal disruption, interference with muscle function, and contribution to antibacterial resistance [50].

Chlorhexidine has a cationic molecular component that attaches to negatively charged cell membrane area and causes cell lysis. As an antiseptic, chlorhexidine is used as a mouth rinse and endodontic irrigant due to long-lasting antimicrobial effect attributed to its binding to hydroxyapatite. It is commonly held that chlorhexidine would be less caustic than sodium hypochlorite. Similar to sodium hypochlorite, heating chlorhexidine in low concentration increases its local efficacy in the root canal system and maintains low systemic toxicity. Chlorhexidine presents drawbacks, such as its incapacity to dissolve necrotic tissue remnants and chemically clean the canal system, and lower effectiveness on Gram-negative than on Gram-positive bacteria [51].

Common antimicrobial agents are prepared from natural or low-molecular-weight compounds. Due to biocidal diffusion, they present toxicity to the human body. In addition, they are easily susceptible to resistance and can lead to environmental contamination. Antimicrobial polymeric materials can overcome these problems by promoting antimicrobial efficiency and reducing residual toxicity. Moreover, antimicrobial polymers exhibit chemical stability, non–volatility, and long-term activity. Polymers containing covalently linked antimicrobial moieties avoid the penetration of low-molecular-weight biocides from the polymer matrices, unlike antimicrobial polymers obtained by physical methods (trapping or coating of organic and/or inorganic active agents during or after processing). These antimicrobial polymers are environmentally friendly and show durability over time. The most studied antimicrobial polymeric materials, and probably the most used, are those based on quaternary ammonium and/or phosphonium salts [52]. In addition, polymeric N-halamines with or without reactive functional groups were used to coat different fabrics by various approaches [49].

During the last two decades, synthetic (co)polymers have been designed to mimic the prominent physio–chemical characteristics of host defense peptides. Although these polymers have revealed a broad-spectrum antimicrobial activity, rapid bactericidal kinetics, and a very low propensity to induce resistance, none of them has been currently in clinical trials [53]. The schematic reaction mechanism of passive and active action of the antimicrobial polymers is presented in Scheme 1.

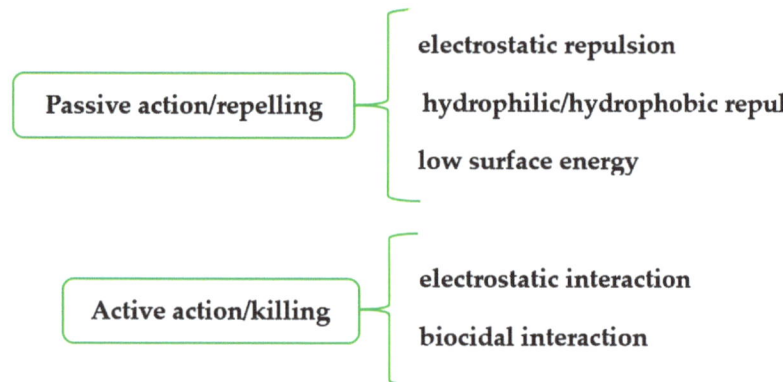

Scheme 1. The mechanism of action of antimicrobial polymers.

Concerning the conducting polymers, namely, polyaniline, polypyrrole, and polythiophene, their biomedical applications have not been well studied even though they have good antimicrobial activity. This limitation may be dampered by the preparation of polymer blends and nanocomposites with different (bio)polymers and nanomaterials, respectively, to achieve the desirable biocompatibility and physicochemical properties [54]. Table 1 summarizes the most relevant antimicrobial agents presented above, their applications, and mechanism of action, and Scheme 2 illustrates the chemical structures of the most important antibacterial compounds.

Table 1. Synthetic antimicrobial products, their applications, and mode of action.

Antimicrobial Agent	Properties and Applications	Antimicrobial Mechanism	Ref.
Quaternary ammonium compounds Polymeric materials having onium salts (quaternary ammonium and/or phosphonium salts) Quaternary ammonium polyethylenimine	- Healthcare, household products, surface preservation, food industry, pharmaceutical/cosmetic (preservation) - Highly effective as antimicrobial agents in orthodontic cements to introduce antibacterial activity toward *S. mutants* and *L. casei*	The long, lipophilic alkyl chain of the quaternary ammonium compounds perforates cell membranes, and produces the release of cytoplasmic components, autolysis and cell death of the microbial strain	[52,55–59]
Halogenated phenols Triclosan	- Antiseptic, disinfectant, fungicide, pesticide, antimicrobial, antiseptic, preservative - Antimicrobial activity against many types of Gram-positive and Gram-negative non-spore-forming bacteria, some fungi - Clinical settings, consumer products (cosmetics, cleaning products, paint, plastic materials, toys) - Durable antifungal finishing of cotton fabrics	Inhibits the active site of enoyl-acyl carrier protein reductase enzyme, which is essential to the fatty acids synthesis of bacteria and the building of the cell membrane	[10,58,60,61]

Table 1. Cont.

Antimicrobial Agent	Properties and Applications	Antimicrobial Mechanism	Ref.
Chlorhexidine Hexametaphosphate salt of chlorhexidine (as nanoparticles) Polyhexamethylene biguanide (PHMB)	- Preoperative skin cleansing preparations, hand disinfectants, and oral mouth rinses - Efficient antimicrobial agent against gram-negative and -positive bacteria and yeasts. - Biomedical materials and consumer products - Antimicrobial efficacy against MRSA and *P. aeruginosa*, in both planktonic and biofilm growth conditions - Healthcare uniforms - Nonspecific antimicrobial properties and remained efficient (>99% against *S. aureus* and *K. pneumoniae*) after use for 5 months	Chlorhexidine inhibits membrane-bound ATPase, based on cell membrane disruption and leakage of intracellular constituents, a rapid process with most damage occurring within 20 s of exposure. The positively charged biguanidines bind to negatively charged phosphate group of the bacterial cell wall or virus envelope, breaking the membrane integrity, which leads to cell lysis and subsequent cell death	[25,62–64]
N-halamines	- Antimicrobial activity against a broad spectrum of microorganisms, rechargeability, nontoxicity to humans - Medical devices, water purification, hospitals, antibacterial modification of cotton fabrics - Antimicrobial activity against aerosolized bacteria - Air filtration technology - Biocidal properties against *S. aureus* and *E. coli* - Food packaging and biomedical applications	The direct transfer of oxidative halogens to a cell after contact resulting in oxidation of the amino acids in the cell membrane and inactivation the microorganism	[46,49,65–68]
5,5-dimethylhydantoin	Cotton fabric with regenerable antibacterial properties against *S. aureus*	Coating dimethylhydantoin on cotton fabric (by pad–dry–plasma–cure process) followed by chlorination inhibits the bacteria	[69]
Cinnamic acid derivatives	Pharmacological, antifungal, and antibacterial action	Plasma membrane disruption, nucleic acid and protein damage, and the induction of intracellular reactive oxygen species	[70–72]
Polyaniline and its derivatives	- Bacteria-resistant surfaces against *S. aureus* and *E. coli* - Wall and room-door coatings in hospitals	Different oxidation states of polyaniline and presence of functional groups	[73]
Polypyrrole (nanoparticles)	Antimicrobial treatment against *S. aureus* and *E. coli* of polyester fabrics	The positive charges (=NH+) in the polypyrrole backbone that are created by dopant compounds	[74,75]
Polythiophenes	Antimicrobial compounds able to kill bacteria selectively by damaging negatively charged cell envelopes	Cationic charges with capacity to create huge amounts of singlet oxygen that interact with organism	[76]

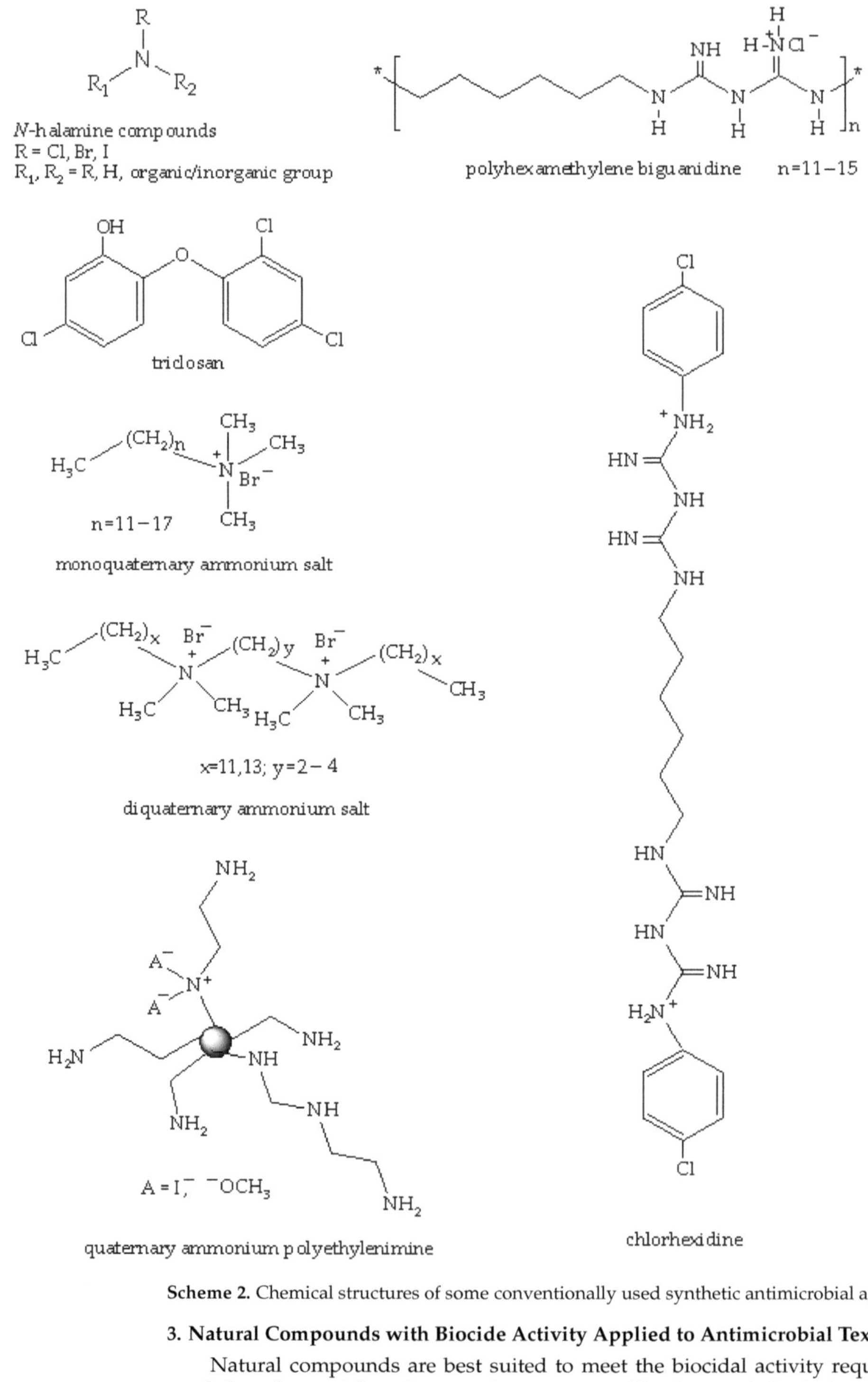

Scheme 2. Chemical structures of some conventionally used synthetic antimicrobial agents.

3. Natural Compounds with Biocide Activity Applied to Antimicrobial Textiles

Natural compounds are best suited to meet the biocidal activity requirements of textile-based materials and present important specific characteristics, being non–harmful in relation to the toxicity issues, environmentally friendly, and renewable. This biocidal property is manifested towards microorganisms' inherent presence, namely, bacteria and fungi, which may cause microbiological destruction of the textile materials. Such issue is of real significance when applications relying on the use of textile materials derived from natural fibers are considered. The most sensitive components of the textile materials to the microbiological action are the cellulose fibers. Some effective biocidal formulations applied to impart antimicrobial properties to textile materials were recently reviewed [77], with focus mainly on the natural compounds such as pectin and lignin, which exhibit important biocidal peculiarities, and the methods which can be employed in order to confer increased resistance as biocidal effect in relation to textile materials' applications. Methods employed in order to apply natural compounds having antimicrobial activity on textile materials are presented in Scheme 3.

Natural compounds with biocide activity applied for textiles protection are referred to as biopolymer matrices (such as chitosan, lignin, starch, cyclodextrins, zein, gelatin) and biological active components extracted from plants (such as essential oils) [18,78–81]. Cellulose-based fibrous scaffolds produced by electrospinning have effectively encapsulated cinnamon, lemongrass, and peppermint essential oils and could be very useful for topical treatments even at low concentration levels due to their significant biocidal resistance against a Gram-negative bacilli, namely *Escherichia coli* [82].

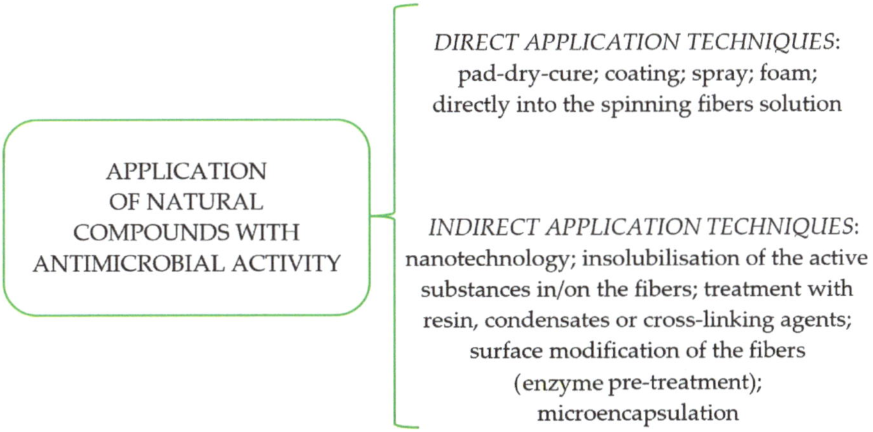

APPLICATION OF NATURAL COMPOUNDS WITH ANTIMICROBIAL ACTIVITY

DIRECT APPLICATION TECHNIQUES: pad-dry-cure; coating; spray; foam; directly into the spinning fibers solution

INDIRECT APPLICATION TECHNIQUES: nanotechnology; insolubilisation of the active substances in/on the fibers; treatment with resin, condensates or cross-linking agents; surface modification of the fibers (enzyme pre-treatment); microencapsulation

Scheme 3. Some of the usual methods employed for application of natural compounds with antimicrobial activity to textile materials [83].

Natural-fibers-based fabrics present valuable enhanced properties through application of natural compounds for their functional finishing, including [18,77,82,84]:

- UV protection properties (conferred by using lignin extracts, natural dye extracts);
- Antioxidant properties (conferred by using natural dye extracts);
- Antimicrobial properties (conferred by using chitosan, lignin, cyclodextrins, essential oils).

Generally, the natural compounds, polysaccharides and oligosaccharides, employed for the antimicrobial finishing of textiles (chitosan, starch, cyclodextrins), as well as lignin, are largely abundant as environmentally friendly waste products [85].

Chitosan modified with hinokitiol (a natural monoterpenoid, namely, a tropolone derivative, found in the wood of trees in the family *Cupressaceae*) is a natural product with very good prospective as antibacterial agent for textiles. The treated cotton fabric exhibited good antibacterial properties while maintaining its initial properties such as hydrophilicity, handle, and strength [86]. Significant antibacterial properties were also conferred to cotton fabrics when using *Aloe vera* gel for finishing, with the bacterial growth being strongly inhibited [87].

In the following, aspects referring to some biopolymer matrices usually applied for textiles finishing and protection are considered.

3.1. Chitosan

Chitosan, a cationic polysaccharide originating from crustaceans and fungi, is obtained by alkaline deacetylation of chitin. Its valuable advantages for adding functionalities to the textile surfaces finishing include biocompatibility, biodegradability, and properties such as antimicrobial, antistatic, nontoxicity, chelating ability, deodorizing, film-forming ability, reactivity in chemical media, presence of ionizable groups, dyeing enhancement, efficacy of cost, thickening ability, and wound alleviation [85,88]. Application of chitosan under hydrogel form on cellulosic fabric conferred antibacterial resistance against bacteria strains such as *Staphylococcus aureus*, *Escherichia coli*, and *Listeria monocytogenes* [89].

The poor binding ability of chitosan with the fibers from textile materials is usually addressed by employment of various cross-linking agents which grant an improved antimicrobial activity. Mostly used and safer agents are:

- 1,2,3,4-butanetetracarboxylic acid (BTCA) and citric acid (CA), when cellulose fibers are considered;
- organic anhydrides, such as succinic and phthalic ones, for grafting chitosan on wool fabrics;
- citric acid in combination with oxidizing agents having reduced toxicity, such as potassium permanganate and sodium hypophosphite, for an effective cross-linking between chitosan and textile substrates—cotton cellulose, wool fabrics).

The application of chitosan on textiles by UV radical curing is also a feasible innocuous methodology for yielding fabrics with finishes having lasting microbial resistance [90].

3.2. Lignin

Lignin, a dark-colored phenolic compound provides resistance against microbial attack in lignocellulose resources (plants and trees). It is generally separated during the processing (delignification or pulping process) when cellulose fibers are obtained. Lignocellulose resources mainly comprise biopolymers with resistance against microorganisms, cellulose and lignin; therefore, these materials can have antimicrobial potential [91].

A coating formulation using lignin extracts derived from sugarcane bagasse was proved to impart good antibacterial activity against *Staphylococcus epidermidis* to the textile support, and the effect was manifest by the reduction of the inherent formation of bacteria onto the textile sample [92,93].

3.3. Cyclodextrins

Cyclodextrins (CDs) are a family of water-soluble cyclic oligosaccharides having two components, one hydrophilic (outer surface) and one lipophilic (central cavity). They are produced during enzymatic conversion of the starch by the enzyme, namely cyclodextrin glycosyltransferase. CDs are composed of alpha-1,4-linked glucopyranosides subunits, and the most commonly available types are α-CD (6 moieties), β-CD (8 moieties)—the most used in research studies, and γ-CD (10 moieties).

The main advantages of using CDs in different applications [85] include eco-friendly character, ability to form inclusion complexes, insecticides carrier ability, fragrances slow-releasing ability, solubilization ability, facile production, efficacy of cost, ability for chelation, and drug-delivery ability. Application of cyclodextrins in textile functional finishing can effectively aid properties such as antimicrobial, fragrance, and dyeing (CDs act as encapsulating, dispersing, and leveling agents) [94].

Feasible interactions between β-CD and some textile fibers include ionic interactions (for wool fibers), covalent bonds, cross-linking agents, and graft polymerization (for both cotton and wool fibers). CDs can impart better UV protection and odor reduction through complexing and controlled releasing of different fragrances (perfumes, aromas), substances with therapeutical effects or "skincare-active" compounds (vitamins, caffeine, menthol), as well as bioactive agents (biocides, insecticides—mosquito repellents).

A significant application of CDs for various textile materials finishing is represented by water and soil remediation and catalysis (e.g., adsorption of small pollutants from waste waters and polluted soil) when such fabrics act as effective selective filters—so-called "preparation of textile nanosponges" [95]. Last, but not the least, CDs have an essential contribution as guest molecules employed in antimicrobial textile modifications by grafting using citric acid as cross-linker in the presence of sodium hypophosphite when a most efficient, lasting antibacterial textile having a pleasurable fragrance was obtained [96].

Improvement in the grafting yield of the cyclodextrin derivative monochlorotriazinyl-β-cyclodextrin (MCT-β-CD) on organic cotton was attained by previously applying a biopolishing procedure, a cellulase enzyme treatment of the textile substrate [97]. An enhanced antibacterial activity and improved durability (upon repeated washing process) for the MCT-β-CD grafted enzymatic treated organic cotton were imparted through incorporation of thymol.

A recent report [98] presented the ability of β-CDs to form complexes with essential oils, and the application of β-CD nano/microcapsules to produce aromatic textiles with focus on the various assembly methods of these aromatic β-CD nano/microcapsules by incorporation of essential oils, as well as on the large range of methodologies employed for the production of such textiles with aromatic character.

3.4. Sericin

Sericin is a natural protein derived from silk worm, *Bombyx mori*, with important characteristics being biocompatible, biodegradable, UV-resistant, oxidative-resistant, good moisture retention receptor, antibacterial, prone to gelling, and adherent [85]. The action against microbes in testing resistance of cotton fabric against bacterial strains, namely *Escherichia coli* and *Staphylococcus aureus*, was enhanced after applying a sericin-based coating [99].

4. Metal and Metal Oxide Nanoparticles

The associations of fibers and textile materials with metal stripes, wires, or plates made of gold, silver, copper, or their alloys have been used in artworks and luxury objects since ancient times [100]. Later on, the progress in both metal and textile processing also led to practical uses, starting with protective/strengthened items and, more recently, to multilayered and composite textiles with an extended range of engineered functionalities, from stimuli-responsive clothes and devices to medicine and electronics [101–103].

This evolution was highly enhanced in the last decades by the significant advances made in the field of nanotechnologies, polymer nanocomposites, and nanosized inorganic particles. In this regard, a major breakthrough in healthcare and medical tools was the successful integration of metal and metal oxide nanoparticles within a large spectrum of natural and synthetic fibers, yarns, and fabrics, otherwise prone to microbial colonization and conveyance, to impart their antibiotic and even antiviral properties. Additional benefits consist of increased resilience at discoloration, decay, and odor formation [104–107]. Unlike other inorganics, such as clays, graphene, or carbon nanotubes, which, rather, passivate the textile host, metal-based nanoparticles also act as biocides through the active release of metal ions that compromise the cell membrane and subsequently the cytoplasmic metabolism in a cascade of events driven by enhanced free radical formation and biomolecule conjugation [108]. A generally accepted mechanism of action is depicted in Figure 3.

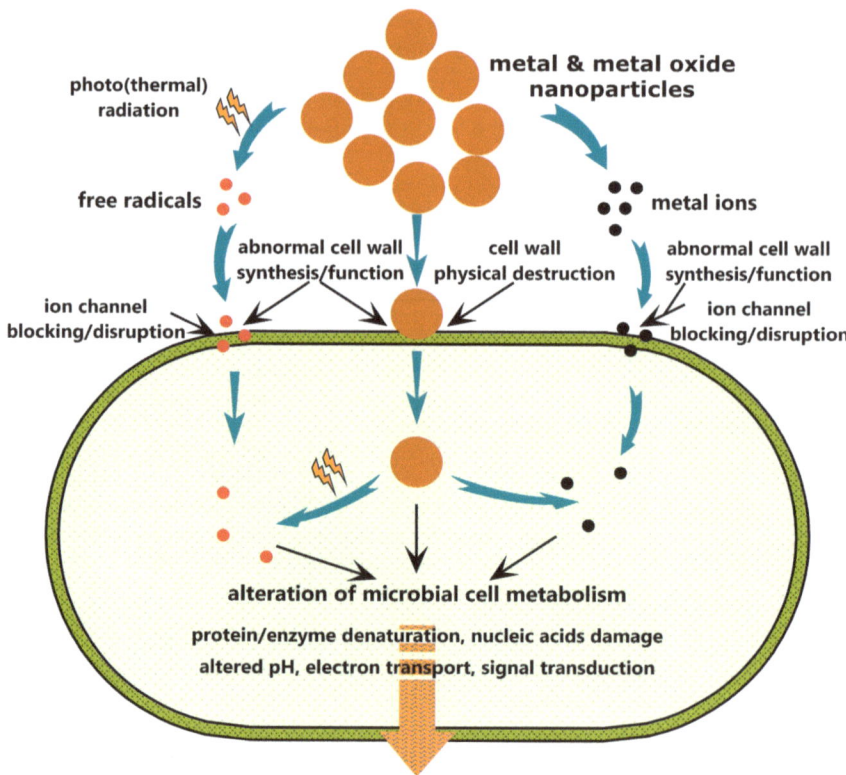

Figure 3. The main mechanisms of action exhibited by metal and metal oxide nanoparticles as antimicrobial active agents.

However, despite the fact that various effects against a plethora of microbial and viral types and strains are frequently reported and reviewed, specific mechanisms, targets, and taxonomies are still far from a complete elucidation [109–116].

The most studied and used to date for textile modification are silver and copper oxide nanoparticles, which are considered to be the most effective antimicrobial agents, followed by zinc oxide and titanium dioxide (Tables 2–5). Other metals and metal oxides are also applied; however, to a lesser extent [117]. The application of other potential metal-based nanoparticles may be limited either by price (gold) or facile surface oxidation (copper), or is prohibited due to their higher toxicity to humans and environment, as in the case of chromium and nickel. It must be mentioned that a high number of heavy metal species, including copper, zinc, and titanium salts and complexes, could be present in traces to sizeable amounts within the unmodified textile materials, originating from raw materials and processing, but especially from the dyeing steps, which may interfere with the further added nanoparticles [118,119].

Table 2. Examples of textile modification with silver nanoparticles (AgNPs).

Fiber/Textile Type	Preparation	Morphology/Content	Microbial Strains	Applicative Characteristics	Refs.
Plain weave 100% bleached organic cotton fabric	Dip and dry coating	49.23 mg/kg, 73.28 mg/kg; eventually embedded in alginate matrix	Gram-positive *Staphylococcus aureus* ATCC 6538/Gram-negative *Escherichia coli* ATCC 873937	Antibacterial and UV protection; superior coloration effect; leakage of about 2.04 mg/kg/cycle during first fifteenth washing cycle	[120]
Brown cotton fiber	*In situ* one-step process under heating	8–21 nm spherical particles; 12.8 µg/kg weight fraction formation of Ag NPs	Gram-positive *Staphylococcus aureus* ATCC 6538/Gram-negative *Pseudomonas aeruginosa* ATCC 9027	Stable antibacterial activity for 50 cycles of laundering; good dispersion; potential applications in sportswear, underwear, and medical textiles	[121]
Commercial polyamide 6,6 fabric	PVP-AgNP dispersions deposited on PA66 with/without DBD plasma pretreatment	20 nm PVP-AgNP colloids	Gram-positive *Staphylococcus aureus* ATCC 6538/Gram-negative *Escherichia coli* ATCC 25922	Plasma-treated polyamide fabric maintains antimicrobial activity even at very low Ag concentration after five washing cycles	[122]
Cellulosic/cotton fabrics	Photochemical reduction in Na–CMC solutions	2–8 nm/5–35 nm spherical polydisperse/monodisperse nanoparticles	*Staphylococcus epidermidis*/*Candida albicans*	Antifungal effect; prevents odor formation	[123]
Polyamide 6 fibers	Electroless plating method; fibers pretreated with a dopamine/CuSO$_4$/H$_2$O$_2$ system	Average particle size of 223 nm; surface continuous and compact silver layer	*Escherichia coli* AATCC 11229/*Staphylococcus aureus* AATCC 6538	Antimicrobial efficiency of 99.9% and 100% against *E. coli* and *S. aureus* decrease to 83.5% and 87.9%, respectively after 1 h/2 h of ultrasonic washing; potential use as antibacterial/conductive textiles	[124]
Commercial prewashed PES fabric	Spray coating of PES with layers of chitosan or HMDSO before and after AgNP deposition	Quasi-spherical particles of 20–30 nm with relative uniform distribution	*Staphylococcus aureus*/*Escherichia coli*	Fast and cost-effective method; controlled release of silver; antimicrobial effect reduced by washing; applications in medical textiles	[23]
Reusable and single use face masks	Testing of commercial face masks from a safe-by-design perspective	Silver detected in both the external and the internal layer under both ionic and nanoparticulate form; mostly near-spherical particles of 13 to 155 nm	viral pathogens	Evaluation of content, type and *in situ* localization of silver-based biocides face; safety of silver-containing masks	[125]

Table 2. Cont.

Fiber/Textile Type	Preparation	Morphology/Content	Microbial Strains	Applicative Characteristics	Refs.
Scoured and bleached 100% cotton fabric of plain weave structure	In situ deposition of Ag nanoparticles on cotton fabrics premodified with dopamine	Medium size of 33–43 nm	Staphylococcus aureus/Escherichia coli	Dopamine is effective in nanoparticles immobilization; ~98% remanent activity after twenty wash cycles	[126]

Na–CMC: sodium–carboxymethylcellulose. DBD: dielectric barrier discharge. PVP: poly(N-vinylpyrrolidone).

Table 3. Examples of textile modification with copper oxides nanoparticles (CuO NPs).

Fiber/Textile Type	Preparation	Morphology/Content	Microbial Strains	Applicative Characteristics	Refs.
Bleached and mercerized cotton fabric (100%)	Pure and hybrid CuO/colloidal chitosan nanosol sonochemically prepared; cotton coating by pad–dry–cure method	Spherical morphology with irregular formation; medium size of 58 nm	Staphylococcus aureus/Escherichia coli	Improved antibacterial activity for hybrid coatings after ten wash cycles	[127]
Fine–medium-weight 100% cotton woven fabric	Ex situ by wet chemical method/pad–dry–cure method	Spherical shape; size of 60–75 nm	Staphylococcus aureus/Escherichia coli	Antimicrobial activity decreases at laundering (S. aureus: 74.36%/12.05% after 10/20 cycles; E. coli: 69.54%/9.85% after 10/20 cycles washes; potential healthcare and hygiene uses	[128]
Cotton fabrics	Green synthesis with R. tuberosa leaf extract	Polydisperse nanorods ranging from 20 to 100 nm	Staphylococcus aureus; Escherichia coli; Klebsiella pneumoniae	Prevention of fabrics microbial damage; bioremediation of industrial dyes	[129]
Polyester/cotton 65/35 blend fabric (PES/CO)	In situ impregnation by the pad–dry/pad–dry/pad–thermofix process	Sizes of about 3 nm and 20 nm	antiviral species: SARS-CoV-2_COV2019 ITALY/INMI1 and Human Corona Virus 229E strain ATCC VR-740; Escherichia coli ATCC 25922 strains	99.93%; 99.96% inactivation efficiency (30; 60 min exposure) against SARS-CoV-2; 99% efficiency on E. coli growth after 20 wash cycles; reusable face masks with antiviral/antibacterial properties and reduced environmental contamination	[130]
70% cotton and 30% polyester mixed textiles	In situ and ex situ green and chemical syntheses	Green route: spherical morphology with sizes of 2.4 ± 0.5 nm; chemical route: no defined geometry with average size of 75 ± 28 nm	E. coli ATCC No. 8739/S. aureus ATCC No. 6538 bacteria; Aspergillus brasiliensis ATCC No. 16 404 fungus	In situ method and 734 ppm Cu$_2$O gives better antifungal effects; high potential against aspergillosis	[131]

Table 3. Cont.

Fiber/Textile Type	Preparation	Morphology/Content	Microbial Strains	Applicative Characteristics	Refs.
Rolls of cotton, plain unbleached woven cotton	*In situ* sonochemical method; "throwing the stones" (TTS) technology with preformed colloids and ultrasound impregnation	Homogeneous layer of ~40 nm nanoparticles on cotton fibers (0.9% w/w CuO)	HDF/HepG2 cells	Low toxicity (>95% HDF cell viability); nanoparticles do not penetrate the skin barrier; potential uses as antimicrobial fabrics for bed sheets, curtains, and laboratory coats	[132]
100% cotton fabric	*In situ* by exhaust dyeing method	Small nanoparticles of different sizes and shapes randomly distributed on fiber surfaces	*Escherichia coli*	Still efficient after 20 washes, could be an economic alternative for antimicrobial textiles	[133]
Fabric samples	CuO biosynthesis with *Aspergillus terreus* strain AF-1; *ex situ* pad–dry–cure method	Homogeneous distributions of spherical, 11–47 nm nanoparticles; 6.1% elemental composition	*Bacillus subtilis* ATCC 6633, *Staphylococcus aureus* ATCC 6538, *Escherichia coli* ATCC 8739, and *Pseudomonas aeruginosa* ATCC 9027	Green synthesis; potential uses in healthcare and hygiene products	[134]
Cotton fabric	Plasma pretreated cotton fabric; *ex situ* coating by pad–dry–cure method	Fabric roughness gradually rises with increases in plasma treatment time; 40 nm sized CuO nanoparticles	*Bacillus subtilis*, *Staphylococcus aureus*, *Salmonella typhimurium*, *Klebsiella pneumoniae*	Potential uses in various biomedical applications	[135]

Table 4. Examples of textile modification with zinc oxide nanoparticles (ZnONPs).

Fiber/Textile Type	Preparation	Morphology/Content	Microbial Strains	Applicative Characteristics	Refs.
Bleached woven cotton fabric (100%; 144 g/m^2)	Single-step sonoenzymatic process	30–120 nm Zn nanoparticles	*Staphylococcus aureus*; *Escherichia coli*	Nanoparticles agglomeration regardless the enzyme used; 33.4% Zn retained on fabrics after ten washing cycles; potential antibacterial medical textiles	[136]
100% plain woven cotton fabrics	Plasma pretreated cotton woven fabric; *in situ*, sonochemically	Spherical shape with 20–90 nm diameter	*Staphylococcus aureus*	Stability improves by cotton fabric prefunctionalization with plasma; Zn content goes from 5.63% to 5.41% after five washing cycles	[137]
Polyamide 6 (PA), polyethylene terephthalate (PET) and polypropylene (PP) textiles	Chemical bath deposition; washing; thermal stabilization; hydrothermal formation of nano/microrods	Irregular needles, flower-like agglomerates and nano/microrods	*Escherichia coli*; *Staphylococcus aureus*	Significant antibacterial activity, particularly in the case of PA/ZnO and Gram-negative bacteria; potential uses in everyday life applications	[138]

Table 4. Cont.

Fiber/Textile Type	Preparation	Morphology/Content	Microbial Strains	Applicative Characteristics	Refs.
100% cotton yarns and polyester/cotton (67/33) blend yarns	Exhaust–dry–cure method	Sizes ranging between 30 and 90 nm	Staphylococcus aureus; Escherichia coli	Antimicrobial efficacy of samples increases at blends, higher yarn twists and lower particle sizes	[139]
100% cellulose cotton	Pad–dry–cure method assisted by open-air plasma modification; green sonochemically nanoparticle synthesis with Psidium guajava Linn (guava) plant extract	Hexagonal nanoparticles of about 41 nm agglomerated into larger clusters	Staphylococcus aureus; Escherichia coli	Open-air plasma treatment enhances nanoparticle adsorption; self-cleaning activity of 94% after five washing cycles	[140]
Gray cotton fabric (100% cotton) of plain weave structure	Pad–dry–cure method and thermo-fixation with sonochemically synthesized ZnO nanoparticles	Nearly spherical nanoparticles with an average size of 40–100 nm; 0.5%, 1%, and 2% ZnO content	Staphylococcus aureus; Escherichia coli	86% reduction of microorganisms after 15 washes; uses as multifunctional textiles with antimicrobial, self-cleaning, and UV protective properties	[141]

Table 5. Examples of textile modification with titanium dioxide nanoparticles (TiO$_2$NPs).

Fiber/Textile Type	Preparation	Morphology/Content	Microbial Strains	Applicative Characteristics	Refs.
Polyamide 66 cloth in plain weave	Pad–dry–cure process	700 nm particles	Aspergillus niger NRRL-A326 (fungus)/Staphylococcus aureus ATCC 6538-P (G+)/Escherichia coli ATCC 25933 (G)/Candida albicans ATCC 10231 (yeast)	Hydrophobic; photocatalytic self-cleaning activity; UV protection activity; potential applications in air filters, outdoor textiles, furniture, and medical textiles	[142]
Nylon 66 knitted fabrics	Synthesized by sol–gel method; subsequently applied by layer by layer (LBL) technique	Medium size of 40–60 nm; tendency to aggregation	Staphylococcus aureus (NCTC 3750)/Escherichia coli (AATCC-10148)	Potential applications in optics, biosensing, separation membranes and technical textile	[143]
Cotton–polyester twill fabric (70–30%)	In situ coating	Average diameter size of 98 nm	-	UV protective properties	[144]
Cotton fabric	Immersion in a mixture of perfluorodecyl triethoxysilane and TiO$_2$NPs solution	Medium size of 50 nm; uniform coating	E. coli	Water repellency; self-cleaning; oil–water separation; stain resistance; antibacterial properties	[145]

Table 5. Cont.

Fiber/Textile Type	Preparation	Morphology/Content	Microbial Strains	Applicative Characteristics	Refs.
Plain woven cotton mercerized fabric	Pad–dry–cure method; functionalization with trimethyl[3-trimethoxysilyl) propyl] ammonium chloride to enhance the affinity	Particles of 30 nm; 4% dried TiO_2 NPs by weight	S. aureus / E. coli	Dye degradation; antibacterial properties; multifunctional cotton fabric for outdoor, industrial and medical applications	[146]
Cotton fabrics lab coat and indiolino fabrics	In situ impregnation; sonochemically, hydrothermal and solvothermal synthesis of TiO_2 particles	Homogeneous distribution on the cotton fabric surface	Escherichia coli / Bacillus pumilus	Sonosynthesis with Ti isopropoxide as precursor enhances the bactericidal activity; self-cleaning properties; potential use for face masks	[147]

Despite their proven efficiency and specific advantages, the application of metallic and metal oxide nanoparticles as antimicrobial additives for textile materials should always take into account their toxicity and environmental impact by leaching and disposal [148–150]. Leaching, furthermore, limits the type and number of uses for a given item, but also exhibits higher biocidal activity [18,151].

There are basically two methods of producing antimicrobial textiles based on metal and metal oxides: *ex situ* and *in situ* [13,152–154]. *Ex situ* methods are related to the incorporation of previously synthesized nanoparticles through direct application to the targeted textile matrix, commonly by the pad–dry–cure technique, which involves a chain sequence of immersion into the nanoparticulate colloidal solution followed by pressurization, drying, and curing. The main drawback of this simple technique, given by the poor adhesion to the constitutive fibers, which favors nanoparticle leaks, agglomeration, and inhomogeneity, could be addressed by the addition of carboxylic acids, thiols, or generation of reactive and negatively charged groups onto the initial fabric surface through chemical or physical means, cross-linking, as well as by incorporating macromolecular stabilizing agents in either one or both raw media.

In situ methods suppose the initial adsorption of metal ions at the level of fiber surfaces followed by their conversion to nanoparticles by chemical reduction or radiation, which improves the stability and distribution. Before, during, or after nanoparticles formation, the surface of textile scaffold may be modified in similar ways. As an alternative, nanoparticles could be also synthesized during polymerization or fiber spinning, followed by processing into final textile products.

5. Challenges in Antimicrobial Textiles Manufacturing

Antimicrobial textiles have to meet a series of requirements due to their wide range of applications (hydrophilic/hydrophobic, breathable, safe, nontoxic, resistance to cleaning cycles, etc.) and one of the most relevant for their purpose is the antimicrobial activity.

Tests for antimicrobial activity are standardized by international organizations and can be divided into two classes [7,14,155]:

a. qualitative tests—AATCC TM147, AATCC TM30 (American Association of Textile Chemists and Colorists Test Method), ISO 20645, ISO 11721 (International Organization for Standardization) and SN 195920, SN 195921 (Swiss standard);
b. quantitative tests—AATCC TM100, ISO 20743, SN 195924, JIS L 1902 (Japanese industry standards) and ASTM E 2149 (or its modification) [156].

Qualitative evaluation is fast and simple, based on the formation of an inhibition area around the tested sample. This does not necessarily mean the sample is biocide, but that it is only biostatic. Therefore, it is not possible to compare the activity of different antimicrobial agents or textiles. Quantitative assessment provides information on the level of performance, but it can be also used as criterion for the optimization of the finishing reagent and/or method. It requires more time and is more specific as it relies on the count of microorganisms. The main drawback of these tests is their high susceptibility to be contaminated and compromised. Therefore, they are performed under strictly controlled conditions. At the same time, the lack of a unitary standard, the poor reproducibility, and the effectiveness of the microbial extraction from samples are factors that affect the tests accuracy in a negative way. Complementary tests, such as viability tests, colorimetric analysis, staining, and microscopy, are useful and their results can be corroborated.

Another issue that has to be addressed is the environmental impact of antimicrobial textiles waste. On one hand, there is the problem of the non-biodegradable textile support (synthetic polymers), and the most eloquent example is the massive accumulation of protective masks discarded in nature in the last years. On the other hand, some antimicrobial reagents used as textile finishing may end up in water biotopes and their accumulation will negatively affect the natural balance, as in the case of quaternary ammonium salts and derivatives, and triclosan (half-life in lake water is approximately 10 days and the degradation products, such as methyl triclosan, are more toxic) [7].

Associated with this issue is the problem of nanoparticles released from the antimicrobial textile and which migrate into the human body, where they can accumulate within various tissues as they can easily penetrate the cell wall barrier. This concern arose along with the development of nanotechnology and its applications in medicine, healthcare, pharmaceutics, and cosmetics. For example, clinical studies on the accumulation of silver nanoparticles in living tissues confirmed its toxicity [157].

Other challenges in antimicrobial textiles that have been already tackled are:

- the use of natural plant fibers with intrinsic antimicrobial activity, raw or modified [158];
- employ of biopolymers with intrinsic antimicrobial activity (i.e., chitosan) as both support and antimicrobial finishing, and with multiple functionality [159];
- combining various antimicrobial compounds in order to enhance the effect in the final product; for example, plant extracts and plant-derived molecules with biologic activity have been encapsulated in chitosan particles that were subsequently used as antimicrobial finishing for cotton fabrics [160];
- use of complex antimicrobial formulations including metals, metal oxides, and other nanoparticles (Ag, TiO_2, silica), natural compounds (curcumin, *Aloe vera*), and binders;
- increasing the compatibility between the textile substrate and the antimicrobial finishing in order to achieve materials with enhance stability and wearability;
- a constant concern to maintain the production cost of most of these materials in the affordable range for the public—this can be achieved through an increased funding of research, both from public and private funds, and a more active involvement of the business community in healthcare and environmental protection.

6. Conclusions and Future Trends

The domain of highly specialized textiles with antimicrobial functionality is, without a doubt, a very active field of research, both theoretically and practically, and a continuously expanding market as a result of the societal demand. The multivalent nature of the textile substrate (natural fibers, synthetic fibers, blends of natural and synthetic fibers, biopolymers, natural plant fibers with intrinsic biocide/biostatic activity), the wide variety of antimicrobial finishing materials (organic synthetic compounds, synthetic polymers, natural and naturally-derived compounds, metals and metal oxides, raw or functionalized silica micro- and nanoparticles), the broad range of processing techniques (coating, microencapsulation, grafting and copolymerization, plasma processing, electrospinning, sol–gel methods, etc.) and applications (biocides/biostatics, antibacterial, antifungal, antiviral, water and air filtration media, protective personal clothing and masks, sports- and footwear, upholstery, hospital beddings, wound dressings, etc.) are factors that clearly illustrate the complexity of this domain. At the same time, each and every one of them can become a driving force orienting the research toward new frontiers, as presented in this paper using some of the most recent advances reported in the literature.

New trends have already emerged. One major advance is represented by the highly specialized textiles with antiviral activity which are even more relevant given the viruses' natural ability to evolve by mutations, as substantiated by studies on aggressive epidemics/pandemics (SARS-CoV, MERS-CoV, SARS-CoV-2, Ebola, West Nile, etc.). The development of up-to-date antiviral drugs and vaccines is time-consuming, so antiviral textiles are a realistic alternative and can contribute to significantly limit, or even control, the viruses' proliferation and spreading (textile biosensors). Even more, computational modeling can be considered a valuable tool in order to assess the virus–receptor interaction and factors affecting the binding affinity, and then to model the corresponding counterparts designed to bind and neutralize the virus. Modern processing techniques, such as plasma-assisted methods, are helpful as well.

Combining green processing, such as sonochemical methods, plasma-assisted procedures, sol–gel techniques, *in situ* growth of nanoparticles (i.e., green synthesis of Ag nanoparticles is nontoxic, cost-effective, and accurate), and green antimicrobial reagents (natural and naturally-derived compounds) for textile finishing represent another trend that has already confirmed most expectations. By this approach, antimicrobial textiles with multiple functionality (anti-inflammatory, antibacterial, antifungal, anti-odor, etc.) can be manufactured.

Last, but not least, the increasing awareness of the environmental risk associated with the careless disposal of textiles with antimicrobial finishing must be considered. Medical waste is disposed of in a controlled manner, but the reckless discharge of some antimicrobial textiles from domestic applications has become rapidly a source of concern (i.e., the massive accumulation in nature of personal masks after the SARS-CoV-2 pandemics). The management of non-biodegradable plastic waste, as well as the monitoring and neutralization of toxic reagents accumulated in various biotopes, are valid solutions that must be considered in a general plan for the coherent elimination of antimicrobial textiles, or even the partial recycling of some of them, at least those designed to be wearable and resistant to multiple cycles of washing and wet/dry cleaning.

Highly specialized textiles with antimicrobial functionality are becoming more and more part of our everyday lives. Therefore, regardless of how much we appreciate the advantages, we must not minimize the risks and disadvantages of their use. In order to control and limit them, we need a very active research–development–innovation flow, which has been shown, and the commitment to transfer the solutions offered by research to practice.

Author Contributions: Conception and design: F.T.; data collection: F.T., C.-A.T., M.N., I.A.D., M.I. and L.I.; writing original manuscript: F.T., C.-A.T., M.N., I.A.D., M.I. and L.I.; writing revised manuscript: F.T., C.-A.T., M.N., I.A.D., M.I. and L.I.; supervision, F.T. All authors have read and agreed to the published version of the manuscript.

Funding: This research received no external funding.

Institutional Review Board Statement: Not applicable.

Informed Consent Statement: Not applicable.

Data Availability Statement: Not applicable.

Acknowledgments: This paper is dedicated to our colleague, Madalina Zanoaga, as a sign of appreciation and to honor her scientific activity, as she has initiated this study in our group.

Conflicts of Interest: The authors declare no conflict of interest.

References

1. Novi, V.T.; Gonzalez, A.; Brockgreitens, J.; Abbas, A. Highly Efficient and Durable Antimicrobial Nanocomposite Textiles. *Sci. Rep.* **2022**, *12*, 17332. [CrossRef] [PubMed]
2. Roychoudhury, S.; Das, A.; Sengupta, P.; Dutta, S.; Roychoudhury, S.; Choudhury, A.P.; Fuzayel Ahmed, A.B.; Bhattacharjee, S.; Slama, P. Viral Pandemics of the Last Four Decades: Pathophysiology, Health Impacts and Perspectives. *Int. J. Environ. Res. Public Health* **2020**, *17*, 9411. [CrossRef] [PubMed]
3. Piret, J.; Boivin, G. Pandemics Throughout History. *Front. Microbiol.* **2021**, *11*, 631736. [CrossRef] [PubMed]
4. Antimicrobial Textiles Market, Global Industry Size Forecast. Available online: https://www.marketsandmarkets.com/Market-Reports/antimicrobial-textile-market-254286152.html (accessed on 21 March 2023).
5. Halepoto, H.; Gong, T.; Memon, H. A Bibliometric Analysis of Antibacterial Textiles. *Sustainability* **2022**, *14*, 11424. [CrossRef]
6. Miraftab, M. *High Performance Medical Textiles: An Overview*; Woodhead Publishing Limited: Sawston, UK, 2014; ISBN 9780857099075.
7. Morais, D.S.; Guedes, R.M.; Lopes, M.A. Antimicrobial Approaches for Textiles: From Research to Market. *Materials* **2016**, *9*, 498. [CrossRef] [PubMed]
8. Jiang, C.; Dejarnette, S.; Chen, W.; Scholle, F.; Wang, Q.; Ghiladi, R.A. Color-Variable Dual-Dyed Photodynamic Antimicrobial Polyethylene Terephthalate (PET)/Cotton Blended Fabrics. *Photochem. Photobiol. Sci.* **2023**, 1–18. [CrossRef]
9. Antunes, J.; Matos, K.; Carvalho, S.; Cavaleiro, A.; Cruz, S.M.A.; Ferreira, F. Carbon-Based Coatings in Medical Textiles Surface Functionalisation: An Overview. *Processes* **2021**, *9*, 1997. [CrossRef]
10. Zanoaga, M.; Tanasa, F. Antimicrobial Reagents as Functional Finishing for Textiles Intended for Biomedical Applications. I. Synthetic Organic Compounds. *Chem. J. Mold.* **2017**, *9*, 14–32. [CrossRef]
11. Tanasa, F.; Zanoaga, M. Antimicrobial Reagents as Functional Finishing for Textiles Intended for Biomedical Applications. II. Metals and Metallic Compounds: Silver. In *IFMBE Proceedings*; Springer: Berlin/Heidelberg, Germany, 2016; Volume 55, pp. 305–308. [CrossRef]
12. Zanoaga, M.; Tanasa, F. Antimicrobial Reagents as Functional Finishing for Textiles Intended for Biomedical Applications. III. Other Metals and Metallic Compounds. In *IFMBE Proceedings*; Springer: Berlin/Heidelberg, Germany, 2016; Volume 55, pp. 309–314.
13. Riaz, S.; Ashraf, M. Recent Advances in Development of Antimicrobial Textiles. In *Advances in Functional Finishing of Textiles*; Springer: Singapore, 2020; pp. 129–168. [CrossRef]
14. Ibrahim, A.; Laquerre, J.-É.; Forcier, P.; Deregnaucourt, V.; Decaens, J.; Vermeersch, O.; Ibrahim, A.; Laquerre, J.-É.; Forcier, P.; Deregnaucourt, V.; et al. Antimicrobial Agents for Textiles: Types, Mechanisms and Analysis Standards. In *Textiles for Functional Applications*; IntechOpen: London, UK, 2021. [CrossRef]
15. Jin, L.; Zhou, F.; Wu, S.; Cui, C.; Sun, S.; Li, G.; Chen, S.; Ma, J. Development of Novel Segmented-Pie Microfibers from Copper-Carbon Nanoparticles and Polyamide Composite for Antimicrobial Textiles Application. *Text. Res. J.* **2021**, *92*, 3–14. [CrossRef]
16. Suh, I.-Y.; Kim, Y.-J.; Zhao, P.; Cho, D.S.; Kang, M.; Huo, Z.-Y.; Kim, S.-W. Self-Powered Microbial Blocking Textile Driven by Triboelectric Charges. *Nano Energy* **2023**, *110*, 108343. [CrossRef]
17. Saha, J.; Mondal, M.I.H. Antimicrobial Textiles from Natural Resources: Types, Properties and Processing. In *Antimicrobial Textiles from Natural Resources*; Woodhead Publishing: Sawston, UK, 2021; pp. 1–43. [CrossRef]
18. Gulati, R.; Sharma, S.; Sharma, R.K. Antimicrobial Textile: Recent Developments and Functional Perspective. *Polym. Bull.* **2022**, *79*, 5747–5771. [CrossRef]
19. Favatela, M.F.; Otarola, J.; Ayala-Peña, V.B.; Dolcini, G.; Perez, S.; Torres Nicolini, A.; Alvarez, V.A.; Lassalle, V.L. Development and Characterization of Antimicrobial Textiles from Chitosan-Based Compounds: Possible Biomaterials against SARS-CoV-2 Viruses. *J. Inorg. Organomet. Polym. Mater.* **2022**, *32*, 1473–1486. [CrossRef] [PubMed]
20. Assylbekova, G.; Alotaibi, H.F.; Yegemberdiyeva, S.; Suigenbayeva, A.; Sataev, M.; Koshkarbaeva, S.; Abdurazova, P.; Sakibayeva, S.; Prokopovich, P. Sunlight Induced Synthesis of Silver Nanoparticles on Cellulose for the Preparation of Antimicrobial Textiles. *J. Photochem. Photobiol.* **2022**, *11*, 100134. [CrossRef]
21. Naebe, M.; Haque, A.N.M.A.; Haji, A. Plasma-Assisted Antimicrobial Finishing of Textiles: A Review. *Engineering* **2022**, *12*, 145–163. [CrossRef]
22. Nortjie, E.; Basitere, M.; Moyo, D.; Nyamukamba, P. Extraction Methods, Quantitative and Qualitative Phytochemical Screening of Medicinal Plants for Antimicrobial Textiles: A Review. *Plants* **2022**, *11*, 2011. [CrossRef] [PubMed]
23. Ribeiro, A.I.; Shvalya, V.; Cvelbar, U.; Silva, R.; Marques-Oliveira, R.; Remião, F.; Felgueiras, H.P.; Padrão, J.; Zille, A. Stabilization of Silver Nanoparticles on Polyester Fabric Using Organo-Matrices for Controlled Antimicrobial Performance. *Polymers* **2022**, *14*, 1138. [CrossRef]
24. Liu, Z.; Wang, Z.; Meng, Y.; Song, Y.; Li, L.; Yu, M.; Li, J. Electron Beam Irradiation Grafting of Metal-Organic Frameworks onto Cotton to Prepare Antimicrobial Textiles. *RSC Adv.* **2023**, *13*, 1853–1861. [CrossRef]

25. Yim, S.; Cheung, J.W.; Cheng, I.Y.; Ho, L.W.; Szeto, S.S.; Chan, P.; Lam, Y.; Kan, C. Longitudinal Study on the Antimicrobial Performance of a Polyhexamethylene Biguanide (PHMB)-Treated Textile Fabric in a Hospital Environment. *Polymers* **2023**, *15*, 1203. [CrossRef]
26. Morris, H.; Murray, R. Medical Textiles. In *Textile Progress*; Taylor & Francis: Abingdon, UK, 2020; Volume 52, pp. 1–127.
27. Britannica. Microbiology: Definition, History, & Microorganisms. Available online: https://www.britannica.com/science/microbiology (accessed on 28 April 2023).
28. Tania, I.S.; Ali, M.; Arafat, M.T. Processing Techniques of Antimicrobial Textiles. In *Antimicrobial Textiles from Natural Resources*; Woodhead Publishing: Sawston, UK, 2021; pp. 189–215. [CrossRef]
29. Gao, Y.; Cranston, R. Recent Advances in Antimicrobial Treatments of Textiles. *Text. Res. J.* **2008**, *78*, 60–72. [CrossRef]
30. Sarathi, P.; Thilagavathi, G. Synthesis and characterization of titanium dioxide nano-particles and their applications to textiles for microbe resistance. *J. Text. Appar. Technol. Manag.* **2009**, *6*, 138525674.
31. Mohamed, F.A.; Abd El-Megied, S.A.; Bashandy, M.S.; Ibrahim, H.M. Synthesis, Application and Antibacterial Activity of New Reactive Dyes Based on Thiazole Moiety. *Pigment Resin Technol.* **2018**, *47*, 246–254. [CrossRef]
32. Tanasă, F.; Nechifor, M.; Teacă, C.A.; Stanciu, M.C. Physical Methods for the Modification of the Natural Fibers Surfaces. In *Surface Treatment Methods of Natural Fibres and their Effects on Biocomposites*; Woodhead Publishing: Sawston, UK, 2022; pp. 125–146. ISBN 9780128218631.
33. Yuan, X.; Yin, W.; Ke, H.; Wei, Q.; Huang, Z.; Chen, D. Properties and Application of Multi-Functional and Structurally Colored Textile Prepared by Magnetron Sputtering. *J. Ind. Text.* **2022**, *51*, 1295–1311. [CrossRef]
34. Shahidi, S.; Ghoranneviss, M. Plasma Sputtering for Fabrication of Antibacterial and Ultraviolet Protective Fabric. *Cloth. Text. Res. J.* **2015**, *34*, 37–47. [CrossRef]
35. Ma, C.; Nikiforov, A.; De Geyter, N.; Dai, X.; Morent, R.; Ostrikov, K. Future Antiviral Polymers by Plasma Processing. *Prog. Polym. Sci.* **2021**, *118*, 101410. [CrossRef] [PubMed]
36. Parthasarathi, V.; Thilagavathi, G. Development of Plasma Enhanced Antiviral Surgical Gown for Healthcare Workers. *Fash. Text.* **2015**, *2*, 4. [CrossRef]
37. Singh, N.; Sheikha, J. Microencapsulation and Its Application in Production of Functional Textiles. *Indian J. Fibre Text. Res.* **2020**, *45*, 495–509. [CrossRef]
38. Yip, J.; Luk, M.Y.A. Microencapsulation Technologies for Antimicrobial Textiles. In *Antimicrobial Textiles*; Woodhead Publishing: Sawston, UK, 2016; pp. 19–46. ISBN 9780081005859.
39. Systematic, B.A.; Podgornik, B.B.; Šandri, S. Microencapsulation for Functional Textile Coatings With. *Coatings* **2021**, *11*, 1371.
40. Rivero, P.J.; Goicoechea, J. Sol-Gel Technology for Antimicrobial Textiles. In *Antimicrobial Textiles*; Woodhead Publishing: Sawston, UK, 2016; pp. 47–72. ISBN 9780081005859.
41. Abramova, A.V.; Abramov, V.O.; Bayazitov, V.M.; Voitov, Y.; Straumal, E.A.; Lermontov, S.A.; Cherdyntseva, T.A.; Braeutigam, P.; Weiße, M.; Günther, K. A Sol-Gel Method for Applying Nanosized Antibacterial Particles to the Surface of Textile Materials in an Ultrasonic Field. *Ultrason. Sonochem.* **2020**, *60*, 104788. [CrossRef] [PubMed]
42. Pakdel, E.; Daoud, W.A.; Wang, X. Assimilating the Photo-Induced Functions of TiO_2-Based Compounds in Textiles: Emphasis on the Sol-Gel Process. *Text. Res. J.* **2015**, *85*, 1404–1428. [CrossRef]
43. Bu, Y.; Zhang, S.; Cai, Y.; Yang, Y.; Ma, S.; Huang, J.; Yang, H.; Ye, D.; Zhou, Y.; Xu, W.; et al. Fabrication of Durable Antibacterial and Superhydrophobic Textiles via in Situ Synthesis of Silver Nanoparticle on Tannic Acid-Coated Viscose Textiles. *Cellulose* **2019**, *26*, 2109–2122. [CrossRef]
44. Rodríguez-Tobías, H.; Morales, G.; Grande, D. Comprehensive Review on Electrospinning Techniques as Versatile Approaches toward Antimicrobial Biopolymeric Composite Fibers. *Mater. Sci. Eng. C* **2019**, *101*, 306–322. [CrossRef] [PubMed]
45. Han, D.; Sherman, S.; Filocamo, S.; Steckl, A.J. Long-Term Antimicrobial Effect of Nisin Released from Electrospun Triaxial Fiber Membranes. *Acta Biomater.* **2017**, *53*, 242–249. [CrossRef]
46. Dong, A.; Wang, Y.J.; Gao, Y.; Gao, T.; Gao, G. Chemical Insights into Antibacterial N-Halamines. *Chem. Rev.* **2017**, *117*, 4806–4862. [CrossRef]
47. Zain, N.M.; Akindoyo, J.O.; Beg, M.D.H. Synthetic Antimicrobial Agent and Antimicrobial Fabrics: Progress and Challenges. *IIUM Eng. J.* **2018**, *19*, 10–29. [CrossRef]
48. Li, Z.; Chen, J.; Cao, W.; Wei, D.; Zheng, A.; Guan, Y. Permanent Antimicrobial Cotton Fabrics Obtained by Surface Treatment with Modified Guanidine. *Carbohydr. Polym.* **2018**, *180*, 192–199. [CrossRef] [PubMed]
49. Liu, Y.; Ren, X.; Liang, J. Antimicrobial Modification Review. *BioResources* **2015**, *10*, 1964–1985.
50. Dhillon, G.S.; Kaur, S.; Pulicharla, R.; Brar, S.K.; Cledón, M.; Verma, M.; Surampalli, R.Y. Triclosan: Current Status, Occurrence, Environmental Risks and Bioaccumulation Potential. *Int. J. Environ. Res. Public Health* **2015**, *12*, 5657–5684. [CrossRef]
51. Kumar, S.B. Chlorhexidine Mouthwash—A Review. *J. Pharm. Sci. Res.* **2017**, *9*, 1450–1452.
52. Xue, Y.; Xiao, H.; Zhang, Y. Antimicrobial Polymeric Materials with Quaternary Ammonium and Phosphonium Salts. *Int. J. Mol. Sci.* **2015**, *16*, 3626–3655. [CrossRef] [PubMed]

53. Ergene, C.; Yasuhara, K.; Palermo, E.F. Biomimetic Antimicrobial Polymers: Recent Advances in Molecular Design. *Polym. Chem.* **2018**, *9*, 2407–2427. [CrossRef]
54. Huang, K.S.; Yang, C.H.; Huang, S.L.; Chen, C.Y.; Lu, Y.Y.; Lin, Y.S. Recent Advances in Antimicrobial Polymers: A Mini-Review. *Int. J. Mol. Sci.* **2016**, *17*, 1578. [CrossRef] [PubMed]
55. Gerba, C.P. Quaternary Ammonium Biocides: Efficacy in Application. *Appl. Environ. Microbiol.* **2015**, *81*, 464–469. [CrossRef] [PubMed]
56. Jiao, Y.; Niu, L.; Ma, S.; Li, J.; Tay, F.R.; Chen, J. hua Quaternary Ammonium-Based Biomedical Materials: State-of-the-Art, Toxicological Aspects and Antimicrobial Resistance. *Prog. Polym. Sci.* **2017**, *71*, 53–90. [CrossRef]
57. El-Newehy, M.H.; Meera, M.A.; Aldalbahi, A.K.; Thamer, B.M.; Mahmoud, Y.A.G.; El-Hamshary, H. Biocidal Polymers: Synthesis, Characterization and Antimicrobial Activity of Bis-Quaternary Onium Salts of Poly(Aspartate-Co-Succinimide). *Polymers* **2020**, *13*, 23. [CrossRef]
58. Foksowicz-Flaczyk, J.; Walentowska, J.; Przybylak, M.; Maciejewski, H. Multifunctional Durable Properties of Textile Materials Modified by Biocidal Agents in the Sol-Gel Process. *Surf. Coat. Technol.* **2016**, *304*, 160–166. [CrossRef]
59. Sharon, E.; Sharabi, R.; Eden, A.; Zabrovsky, A.; Ben-Gal, G.; Sharon, E.; Pietrokovski, Y.; Houri-Haddad, Y.; Beyth, N. Antibacterial Activity of Orthodontic Cement Containing Quaternary Ammonium Polyethylenimine Nanoparticles Adjacent to Orthodontic Brackets. *Int. J. Environ. Res. Public Health* **2018**, *15*, 606. [CrossRef]
60. Daoud, F.C.; Coppry, M.; Moore, N.; Rogues, A.M. Do Triclosan Sutures Modify the Microbial Diversity of Surgical Site Infections? A Systematic Review and Meta-Analysis. *Microorganisms* **2022**, *10*, 927. [CrossRef]
61. Ahmed, I.; Boulton, A.J.; Rizvi, S.; Carlos, W.; Dickenson, E.; Smith, N.A.; Reed, M. The Use of Triclosan-Coated Sutures to Prevent Surgical Site Infections: A Systematic Review and Meta-Analysis of the Literature. *BMJ Open* **2019**, *9*, e029727. [CrossRef]
62. Subramani, K.; Seo, H.N.; Dougherty, J.; Chaudhry, K.; Bollu, P.; Rosenthal, K.S.; Zhang, J.F. In Vitro Evaluation of Antimicrobial Activity of Chlorhexidine Hexametaphosphate Nanoparticle Coatings on Orthodontic Elastomeric Chains. *Mater. Res. Express* **2020**, *7*, 075401. [CrossRef]
63. Wood, N.J.; Jenkinson, H.F.; Davis, S.A.; Mann, S.; O'Sullivan, D.J.; Barbour, M.E. Chlorhexidine Hexametaphosphate Nanoparticles as a Novel Antimicrobial Coating for Dental Implants. *J. Mater. Sci. Mater. Med.* **2015**, *26*, 201. [CrossRef] [PubMed]
64. Jones, I.A.; Joshi, L.T. Biocide Use in the Antimicrobial Era: A Review. *Molecules* **2021**, *26*, 2276. [CrossRef]
65. Cheng, X.; Li, R.; Du, J.; Sheng, J.; Ma, K.; Ren, X.; Huang, T.S. Antimicrobial Activity of Hydrophobic Cotton Coated with N-Halamine. *Polym. Adv. Technol.* **2015**, *26*, 99–103. [CrossRef]
66. Demir, B.; Cerkez, I.; Worley, S.D.; Broughton, R.M.; Huang, T.S. N-Halamine-Modified Antimicrobial Polypropylene Nonwoven Fabrics for Use against Airborne Bacteria. *ACS Appl. Mater. Interfaces* **2015**, *7*, 1752–1757. [CrossRef] [PubMed]
67. Ren, T.; Dormitorio, T.V.; Qiao, M.; Huang, T.S.; Weese, J. N-Halamine Incorporated Antimicrobial Nonwoven Fabrics for Use against Avian Influenza Virus. *Vet. Microbiol.* **2018**, *218*, 78–83. [CrossRef]
68. Li, R.; Sheng, J.; Cheng, X.; Li, J.; Ren, X.; Huang, T.S. Biocidal Poly (Vinyl Alcohol) Films Incorporated with N-Halamine Siloxane. *Compos. Commun.* **2018**, *10*, 89–92. [CrossRef]
69. Zhou, C.E.; Kan, C.W.; Matinlinna, J.P.; Tsoi, J.K.H. Regenerable Antibacterial Cotton Fabric by Plasma Treatment with Dimethylhydantoin: Antibacterial Activity against S. Aureus. *Coatings* **2017**, *7*, 11. [CrossRef]
70. de Morais, M.C.; de Oliveira Lima, E.; Perez-Castillo, Y.; de Sousa, D.P. Synthetic Cinnamides and Cinnamates: Antimicrobial Activity, Mechanism of Action, and In Silico Study. *Molecules* **2023**, *28*, 1918. [CrossRef]
71. Imai, M.; Yokoe, H.; Tsubuki, M.; Takahashi, N. Growth Inhibition of Human Breast and Prostate Cancer Cells by Cinnamic Acid Derivatives and Their Mechanism of Action. *Biol. Pharm. Bull.* **2019**, *42*, 1134–1139. [CrossRef]
72. Cai, R.; Miao, M.; Yue, T.; Zhang, Y.; Cui, L.; Wang, Z.; Yuan, Y. Antibacterial Activity and Mechanism of Cinnamic Acid and Chlorogenic Acid against Alicyclobacillus Acidoterrestris Vegetative Cells in Apple Juice. *Int. J. Food Sci. Technol.* **2019**, *54*, 1697–1705. [CrossRef]
73. Robertson, J.; Gizdavic-Nikolaidis, M.; Swift, S. Investigation of Polyaniline and a Functionalised Derivative as Antimicrobial Additives to Create Contamination Resistant Surfaces. *Materials* **2018**, *11*, 436. [CrossRef] [PubMed]
74. Sanchez Ramirez, D.O.; Varesano, A.; Carletto, R.A.; Vineis, C.; Perelshtein, I.; Natan, M.; Perkas, N.; Banin, E.; Gedanken, A. Antibacterial Properties of Polypyrrole-Treated Fabrics by Ultrasound Deposition. *Mater. Sci. Eng. C* **2019**, *102*, 164–170. [CrossRef] [PubMed]
75. Zare, E.N.; Agarwal, T.; Zarepour, A.; Pinelli, F.; Zarrabi, A.; Rossi, F.; Ashrafizadeh, M.; Maleki, A.; Shahbazi, M.A.; Maiti, T.K.; et al. Electroconductive Multi-Functional Polypyrrole Composites for Biomedical Applications. *Appl. Mater. Today* **2021**, *24*, 101117. [CrossRef]
76. Wang, C.Y.; Makvandi, P.; Zare, E.N.; Tay, F.R.; Niu, L. Advances in Antimicrobial Organic and Inorganic Nanocompounds in Biomedicine. *Adv. Ther.* **2020**, *3*, 2000024. [CrossRef]
77. Kachuk, D.S.; Mishchenko, E.V.; Venger, E.A.; Popovych, T.A. Biocidal Protection of Textile Materials. *J. Chem. Technol.* **2022**, *30*, 240–252. [CrossRef]

78. Yıldırım, F.F.; Avinc, O.; Yavas, A.; Sevgisunar, G. *Sustainable Antifungal and Antibacterial Textiles Using Natural Resources*; Springer: Berlin/Heidelberg, Germany, 2020; ISBN 9783030385415.
79. Pawłowska, A.; Stepczyńska, M. Natural Biocidal Compounds of Plant Origin as Biodegradable Materials Modifiers. *J. Polym. Environ.* **2022**, *30*, 1683–1708. [CrossRef]
80. Walentowska, J.; Foksowicz-Flaczyk, J. Thyme Essential Oil for Antimicrobial Protection of Natural Textiles. *Int. Biodeterior. Biodegrad.* **2013**, *84*, 407–411. [CrossRef]
81. Singh, N.; Sahu, O. *Sustainable Cyclodextrin in Textile Applications*; Elsevier Ltd.: Amsterdam, The Netherlands, 2018; ISBN 9780081024911.
82. Liakos, I.; Rizzello, L.; Hajiali, H.; Brunetti, V.; Carzino, R.; Pompa, P.P.; Athanassiou, A.; Mele, E. Fibrous Wound Dressings Encapsulating Essential Oils as Natural Antimicrobial Agents. *J. Mater. Chem. B* **2015**, *3*, 1583–1589. [CrossRef]
83. Tawiah, B.; Badoe, W.; Fu, S. Advances in the Development of Antimicrobial Agents for Textiles: The Quest for Natural Products. Review. *Fibres Text. East. Eur.* **2016**, *24*, 136–149. [CrossRef]
84. Rather, L.J.; Shabbir, M.; Li, Q.; Mohammad, F. Coloration, UV Protective, and Antioxidant Finishing of Wool Fabric Via Natural Dye Extracts: Cleaner Production of Bioactive Textiles. *Environ. Prog. Sustain. Energy* **2019**, *38*, 13187. [CrossRef]
85. Shahid-Ul-Islam; Shahid, M.; Mohammad, F. Green Chemistry Approaches to Develop Antimicrobial Textiles Based on Sustainable Biopolymers—A Review. *Ind. Eng. Chem. Res.* **2013**, *52*, 5245–5260. [CrossRef]
86. Liu, Z.; Luo, Y.; Zhao, X.; Zheng, K.; Wu, M.; Wang, L. A Natural Antibacterial Agent Based on Modified Chitosan by Hinokitiol for Antibacterial Application on Cotton Fabric. *Cellulose* **2022**, *29*, 2731–2742. [CrossRef]
87. Ali, S.W.; Purwar, R.; Joshi, M.; Rajendran, S. Antibacterial Properties of Aloe Vera Gel-Finished Cotton Fabric. *Cellulose* **2014**, *21*, 2063–2072. [CrossRef]
88. Shahid-ul-Islam; Butola, B. S. Recent Advances in Chitosan Polysaccharide and Its Derivatives in Antimicrobial Modification of Textile Materials. *Int. J. Biol. Macromol.* **2019**, *121*, 905–912. [CrossRef]
89. Benltoufa, S.; Miled, W.; Trad, M.; Slama, R.B.; Fayala, F. Chitosan Hydrogel-coated Cellulosic Fabric for Medical End-Use: Antibacterial Properties, Basic Mechanical and Comfort Properties. *Carbohydr. Polym.* **2020**, *227*, 115352. [CrossRef]
90. Ferrero, F.; Periolatto, M. Antimicrobial Finish of Textiles by Chitosan UV-Curing. *J. Nanosci. Nanotechnol.* **2012**, *12*, 4803–4810. [CrossRef]
91. Lobo, F.C.M.; Franco, A.R.; Fernandes, E.M.; Reis, R.L. An Overview of the Antimicrobial Properties of Lignocellulosic Materials. *Molecules* **2021**, *26*, 1749. [CrossRef]
92. Sunthornvarabhas, J.; Liengprayoon, S.; Lerksamran, T.; Buratcharin, C.; Suwonsichon, T.; Vanichsriratana, W.; Sriroth, K. Utilization of Lignin Extracts from Sugarcane Bagasse as Bio-Based Antimicrobial Fabrics. *Sugar Tech* **2019**, *21*, 355–363. [CrossRef]
93. Sunthornvarabhas, J.; Liengprayoon, S.; Suwonsichon, T. Antimicrobial Kinetic Activities of Lignin from Sugarcane Bagasse for Textile Product. *Ind. Crops Prod.* **2017**, *109*, 857–861. [CrossRef]
94. Andreaus, J.; Dalmolin, M.C.; De Oliveira, I.B.; Barcellos, I.O. Aplicação de Ciclodextrinas Em Processos Têxteis. *Quim. Nova* **2010**, *33*, 929–937. [CrossRef]
95. Utzeri, G.; Matias, P.M.C.; Murtinho, D.; Valente, A.J.M. Cyclodextrin-Based Nanosponges: Overview and Opportunities. *Front. Chem.* **2022**, *10*, 859406. [CrossRef]
96. Rukmani, A.; Sundrarajan, M. Inclusion of Antibacterial Agent Thymol on β-Cyclodextrin-Grafted Organic Cotton. *J. Ind. Text.* **2012**, *42*, 132–144. [CrossRef]
97. Sundrarajan, M.; Rukmani, A. Biopolishing and Cyclodextrin Derivative Grafting on Cellulosic Fabric for Incorporation of Antibacterial Agent Thymol. *J. Text. Inst.* **2013**, *104*, 188–196. [CrossRef]
98. Ma, J.; Fan, J.; Xia, Y.; Kou, X.; Ke, Q.; Zhao, Y. Preparation of Aromatic β-Cyclodextrin Nano/Microcapsules and Corresponding Aromatic Textiles: A Review. *Carbohydr. Polym.* **2023**, *308*, 120661. [CrossRef]
99. Rajendran, R.; Balakumar, C.; Sivakumar, R.; Amruta, T.; Devaki, N. Extraction and Application of Natural Silk Protein Sericin from Bombyx Mori as Antimicrobial Finish for Cotton Fabrics. *J. Text. Inst.* **2012**, *103*, 458–462. [CrossRef]
100. Karatzani, A. The Use of Metal Threads in the Decoration of Late and Post-Byzantine Embroidered Church Textiles and Post-Byzantine Embroidered Church Textiles. *Cah. Balk.* **2021**, *48*, 233–253. [CrossRef]
101. Schneegass, S.; Amft, O. (Eds.) *Smart Textiles. Fundamentals, Design, and Interaction*; Springer International Publishing: Berlin/Heidelberg, Germany, 2017; ISBN 9783319501239.
102. Rajendran, S. (Ed.) *Advanced Textiles for Wound Care*, 2nd ed.; Woodhead Publishing: Sawston, UK, 2018; ISBN 9780081021927.
103. Elmogahzy, Y.E. *Engineering Textiles: Integrating the Design and Manufacture of Textile Products*, 2nd ed.; Woodhead Publishing: Sawston, UK, 2020; ISBN 9780081024898.
104. Thilagavathi, G.; Rathinamoorthy, R. (Eds.) *Odour in Textiles. Generation and Control*; CRC Press: Boca Raton, FL, USA, 2022; ISBN 9780367693367.
105. Sabba, D. (Ed.) *Nanotechnology in Smart Textiles*; Arcler Press: Burlington, ON, Canada, 2019; ISBN 9781773616490.
106. Joshi, M. (Ed.) *Nanotechnology in Textiles: Advances and Developments in Polymer Nanocomposites*; Jenny Stanford Publishing: Singapore, 2020; ISBN 9789814800815.

107. Militky, J.; Periyasamy, A.P.; Venkataraman, M. (Eds.) *Textiles and Their Use in Microbial Protection: Focus on COVID-19 and Other Viruses*; CRC Press: Boca Raton, FL, USA, 2021; ISBN 9780367691059.
108. Ul-Islam, S.; Butola, B.S. (Eds.) *Nanomaterials in the Wet Processing of Textiles*; Scrivener Publishing: Beverly, MA, USA, 2018; ISBN 978-1-119-45984-2.
109. Franco, D.; Calabrese, G.; Guglielmino, S.P.P.; Conoci, S. Metal-Based Nanoparticles: Antibacterial Mechanisms and Biomedical Application. *Microorganisms* **2022**, *10*, 1778. [CrossRef]
110. Chakraborty, N.; Jha, D.; Roy, I.; Kumar, P.; Gaurav, S.S. Nanobiotics against Antimicrobial Resistance: Harnessing the Power of Nanoscale Materials and Technologies. *J. Nanobiotechnology* **2022**, *20*, 375. [CrossRef]
111. Frei, A.; Verderosa, A.D.; Elliott, A.G.; Zuegg, J.; Blaskovich, M.A.T. Metals to Combat Antimicrobial Resistance. *Nat. Rev. Chem.* **2023**, *7*, 202–224. [CrossRef]
112. Ma, X.; Zhou, S.; Xu, X.; Du, Q. Copper-Containing Nanoparticles: Mechanism of Antimicrobial Effect and Application in Dentistry-a Narrative Review. *Front. Surg.* **2022**, *9*, 905892. [CrossRef]
113. Mendes, C.R.; Dilarri, G.; Forsan, C.F.; Sapata, V.d.M.R.; Lopes, P.R.M.; de Moraes, P.B.; Montagnolli, R.N.; Ferreira, H.; Bidoia, E.D. Antibacterial Action and Target Mechanisms of Zinc Oxide Nanoparticles against Bacterial Pathogens. *Sci. Rep.* **2022**, *12*, 2658. [CrossRef] [PubMed]
114. Rashid, M.M.; Simon, B.; Tom, B. Recent Advances in TiO2-Functionalized Textile Surfaces. *Surf. Interfaces* **2021**, *22*, 100890. [CrossRef]
115. Guisbiers, G. (Ed.) *Antimicrobial Activity of Nanoparticles*, 1st ed.; Elsevier: Amsterdam, The Netherlands, 2023; ISBN 9780128216378.
116. Fernandes, M.; Padrão, J.; Ribeiro, A.I.; Fernandes, R.D.V.; Melro, L.; Nicolau, T.; Mehravani, B.; Alves, C.; Rodrigues, R.; Zille, A. Polysaccharides and Metal Nanoparticles for Functional Textiles: A Review. *Nanomaterials* **2022**, *12*, 1006. [CrossRef]
117. Andra, S.; Balu, S.K.; Jeevanandam, J.; Muthalagu, M. Emerging nanomaterials for antibacterial textile fabrication. *Naunyn-Schmiedeberg's Arch. Pharmacol.* **2021**, *394*, 1355–1382. [CrossRef] [PubMed]
118. Rujido-Santos, I.; Herbello-Hermelo, P.; Barciela-Alonso, M.C.; Bermejo-Barrera, P.; Moreda-Piñeiro, A. Metal Content in Textile and (Nano)Textile Products. *Int. J. Environ. Res. Public Health* **2022**, *19*, 944. [CrossRef]
119. Dolez, P.I.; Benaddi, H. Toxicity Testing of Textiles. *Adv. Charact. Test. Text.* **2018**, 151–188. [CrossRef]
120. Mahmud, S.; Pervez, N.; Taher, M.A.; Mohiuddin, K.; Liu, H.H. Multifunctional Organic Cotton Fabric Based on Silver Nanoparticles Green Synthesized from Sodium Alginate. *Text. Res. J.* **2020**, *90*, 1224–1236. [CrossRef]
121. Nam, S.; Selling, G.W.; Hillyer, M.B.; Condon, B.D.; Rahman, M.S.; Chang, S.C. Brown Cotton Fibers Self-Produce Ag Nanoparticles for Regenerating Their Antimicrobial Surfaces. *ACS Appl. Nano Mater.* **2021**, *4*, 13112–13122. [CrossRef]
122. Ribeiro, A.I.; Senturk, D.; Silva, K.K.; Modic, M.; Cvelbar, U.; Dinescu, G.; Mitu, B.; Nikiforov, A.; Leys, C.; Kuchakova, I.; et al. Antimicrobial Efficacy of Low Concentration PVP-Silver Nanoparticles Deposited on DBD Plasma-Treated Polyamide 6,6 Fabric. *Coatings* **2019**, *9*, 581. [CrossRef]
123. Khe, Y.; Sv, M.; Aa, S.; Jz, J.; Fm, T.; Ssh, R.; Renat, L. Antibacterial effect of cotton fabric treated with silver nanoparticles of different sizes and shapes. *Int. J. Nanomater. Nanotechnol. Nanomed.* **2019**, *5*, 016–023. [CrossRef]
124. Jiang, L.; Zhou, Y.; Guo, Y.; Jiang, Z.; Chen, S.; Ma, J. Preparation of Silver Nanoparticle Functionalized Polyamide Fibers with Antimicrobial Activity and Electrical Conductivity. *J. Appl. Polym. Sci.* **2019**, *136*, 47584. [CrossRef]
125. Mast, J.; Van Miert, E.; Siciliani, L.; Cheyns, K.; Blaude, M.N.; Wouters, C.; Waegeneers, N.; Bernsen, R.; Vleminckx, C.; Van Loco, J.; et al. Application of Silver-Based Biocides in Face Masks Intended for General Use Requires Regulatory Control. *Sci. Total Environ.* **2023**, *870*, 161889. [CrossRef]
126. Tania, I.S.; Ali, M.; Azam, M.S. Mussel-Inspired Deposition of Ag Nanoparticles on Dopamine-Modified Cotton Fabric and Analysis of Its Functional, Mechanical and Dyeing Properties. *J. Inorg. Organomet. Polym. Mater.* **2021**, *31*, 4065–4076. [CrossRef]
127. Dhineshbabu, N.R.; Rajendran, V. Antibacterial Activity of Hybrid Chitosan-Cupric Oxide Nanoparticles on Cotton Fab. *IET Nanobiotechnology* **2016**, *10*, 13–19. [CrossRef]
128. Hasan, R. Production of Antimicrobial Textiles by Using Copper Oxide Nanoparticles. *Int. J. Contemp. Res. Rev.* **2018**, *9*, 20195–20202. [CrossRef] [PubMed]
129. Vasantharaj, S.; Sathiyavimal, S.; Saravanan, M.; Senthilkumar, P.; Gnanasekaran, K.; Shanmugavel, M.; Manikandan, E.; Pugazhendhi, A. Synthesis of Ecofriendly Copper Oxide Nanoparticles for Fabrication over Textile Fabrics: Characterization of Antibacterial Activity and Dye Degradation Potential. *J. Photochem. Photobiol. B Biol.* **2019**, *191*, 143–149. [CrossRef] [PubMed]
130. Román, L.E.; Villalva, C.; Uribe, C.; Paraguay-Delgado, F.; Sousa, J.; Vigo, J.; Vera, C.M.; Gómez, M.M.; Solís, J.L. Textiles Functionalized with Copper Oxides: A Sustainable Option for Prevention of COVID-19. *Polymers* **2022**, *14*, 3066. [CrossRef]
131. Asmat-Campos, D.; de Oca-Vásquez, G.M.; Rojas-Jaimes, J.; Delfín-Narciso, D.; Juárez-Cortijo, L.; Nazario-Naveda, R.; Batista Meneses, D.; Pereira, R.; de la Cruz, M.S. Cu2O Nanoparticles Synthesized by Green and Chemical Routes, and Evaluation of Their Antibacterial and Antifungal Effect on Functionalized Textiles. *Biotechnol. Rep.* **2023**, *37*, e00785. [CrossRef] [PubMed]
132. Singh, J.; Beddow, J.; Mee, C.; Maryniak, J.; Joyce, E.M.; Mason, T.J. Cytotoxicity Study of Textile Fabrics Impregnated With CuO Nanoparticles in Mammalian Cells. *Int. J. Toxicol.* **2017**, *36*, 478–484. [CrossRef] [PubMed]
133. Román, L.E.; Amézquita, M.J.; Uribe, C.L.; Maurtua, D.J.; Costa, S.A.; Costa, S.M.; Keiski, R.; Solís, J.L.; Gómez, M.M. In Situ Growth of CuO Nanoparticles onto Cotton Textiles. *Adv. Nat. Sci. Nanosci. Nanotechnol.* **2020**, *11*, 25009. [CrossRef]

134. Shaheen, T.I.; Fouda, A.; Salem, S.S. Integration of Cotton Fabrics with Biosynthesized CuO Nanoparticles for Bactericidal Activity in the Terms of Their Cytotoxicity Assessment. *Ind. Eng. Chem. Res.* **2021**, *60*, 1553–1563. [CrossRef]
135. Tharchanaa, S.B.; Anupriyanka, T.; Shanmugavelayutham, G. Ecofriendly Surface Modification of Cotton Fabric to Enhance the Adhesion of CuO Nanoparticles for Antibacterial Activity. *Mater. Technol.* **2022**, *37*, 3222–3230. [CrossRef]
136. Petkova, P.; Francesko, A.; Perelshtein, I.; Gedanken, A.; Tzanov, T. Simultaneous Sonochemical-Enzymatic Coating of Medical Textiles with Antibacterial ZnO Nanoparticles. *Ultrason. Sonochem.* **2016**, *29*, 244–250. [CrossRef] [PubMed]
137. Shahidi, S.; Rezaee, H.; Rashidi, A.; Ghoranneviss, M. In Situ Synthesis of ZnO Nanoparticles on Plasma Treated Cotton Fabric Utilizing Durable Antibacterial Activity. *J. Nat. Fibers* **2018**, *15*, 639–647. [CrossRef]
138. Fiedot-Toboła, M.; Ciesielska, M.; Maliszewska, I.; Rac-Rumijowska, O.; Suchorska-Woźniak, P.; Teterycz, H.; Bryjak, M. Deposition of Zinc Oxide on Different Polymer Textiles and Their Antibacterial Properties. *Materials* **2018**, *11*, 707. [CrossRef]
139. Palaniappan, G. Study on the Antimicrobial Efficacy of Fabrics Finished with Nano Zinc Oxide Particles. *J. Textile Sci. Eng.* **2020**, *10*.
140. Irfan, M.; Naz, M.Y.; Saleem, M.; Tanawush, M.; Głowacz, A.; Glowacz, W.; Rahman, S.; Mahnashi, M.H.; Alqahtani, Y.S.; Alyami, B.A.; et al. Statistical Study of Nonthermal Plasma-Assisted ZnO Coating of Cotton Fabric through Ultrasonic-Assisted Green Synthesis for Improved Self-Cleaning and Antimicrobial Properties. *Materials* **2021**, *14*, 6998. [CrossRef]
141. Tania, I.S.; Ali, M.; Akter, M. Fabrication, Characterization, and Utilization of ZnO Nanoparticles for Stain Release, Bacterial Resistance, and UV Protection on Cotton Fabric. *J. Eng. Fiber. Fabr.* **2022**, *17*, 14238. [CrossRef]
142. Abdel Salam, K.A.; Ibrahim, N.A.; Maamoun, D.; Abdel Salam, S.H.; Fathallah, A.I.; Abdelrahman, M.S.; Mashaly, H.; Hassabo, A.G.; Khattab, T.A. Anti-Microbial Finishing of Polyamide Fabric Using Titanium Dioxide Nanoparticles. *J. Text. Color. Polym. Sci.* **2023**, *20*, 171–174. [CrossRef]
143. Kale, R.D.; Meena, C.R. Synthesis of Titanium Dioxide Nanoparticles and Application on Nylon Fabric Using Layer by Layer Technique for Antimicrobial Property. *Adv. Appl. Sci. Res.* **2012**, *3*, 3073–3080.
144. Rabiei, H.; Farhang Dehghan, S.; Montazer, M.; Khaloo, S.S.; Koozekonan, A.G. UV Protection Properties of Workwear Fabrics Coated with TiO_2 Nanoparticles. *Front. Public Health* **2022**, *10*, 929095. [CrossRef]
145. Tudu, B.K.; Sinhamahapatra, A.; Kumar, A. Surface Modification of Cotton Fabric Using TiO_2 Nanoparticles for Self-Cleaning, Oil-Water Separation, Antistain, Anti-Water Absorption, and Antibacterial Properties. *ACS Omega* **2020**, *5*, 7850–7860. [CrossRef] [PubMed]
146. Riaz, S.; Ashraf, M.; Aziz, H.; Younus, A.; Umair, M.; Salam, A.; Iqbal, K.; Hussain, M.T.; Hussain, T. Cationization of TiO_2 Nanoparticles to Develop Highly Durable Multifunctional Cotton Fabric. *Mater. Chem. Phys.* **2022**, *278*, 125573. [CrossRef]
147. Alvarez-Amparán, M.A.; Martínez-Cornejo, V.; Cedeño-Caero, L.; Hernandez-Hernandez, K.A.; Cadena-Nava, R.D.; Alonso-Núñez, G.; Moyado, S.F. Characterization and Photocatalytic Activity of TiO_2 Nanoparticles on Cotton Fabrics, for Antibacterial Masks. *Appl. Nanosci.* **2022**, *12*, 4019–4032. [CrossRef] [PubMed]
148. Riaz, S.; Ashraf, M.; Hussain, T.; Hussain, M.T.; Rehman, A.; Javid, A.; Iqbal, K.; Basit, A.; Aziz, H. Functional Finishing and Coloration of Textiles with Nanomaterials. *Color. Technol.* **2018**, *134*, 327–346. [CrossRef]
149. Abu-Qdais, H.A.; Abu-Dalo, M.A.; Hajeer, Y.Y. Impacts of Nanosilver-Based Textile Products Using a Life Cycle Assessment. *Sustainability* **2021**, *13*, 3436. [CrossRef]
150. Ramzan, U.; Majeed, W.; Hussain, A.A.; Qurashi, F.; Qamar, S.U.R.; Naeem, M.; Uddin, J.; Khan, A.; Al-Harrasi, A.; Razak, S.I.A.; et al. New Insights for Exploring the Risks of Bioaccumulation, Molecular Mechanisms, and Cellular Toxicities of AgNPs in Aquatic Ecosystem. *Water* **2022**, *14*, 2192. [CrossRef]
151. Owen, L.; Laird, K. Development of a Silver-Based Dual-Function Antimicrobial Laundry Additive and Textile Coating for the Decontamination of Healthcare Laundry. *J. Appl. Microbiol.* **2021**, *130*, 1012–1022. [CrossRef]
152. Shahidi, S.; Jamali, A.; Dalal Sharifi, S.; Ghomi, H. In-Situ Synthesis of CuO Nanoparticles on Cotton Fabrics Using Spark Discharge Method to Fabricate Antibacterial Textile. *J. Nat. Fibers* **2018**, *15*, 870–881. [CrossRef]
153. Shahid-ul-Islam; Butola, B.S.; Kumar, A. Green Chemistry Based In-Situ Synthesis of Silver Nanoparticles for Multifunctional Finishing of Chitosan Polysaccharide Modified Cellulosic Textile Substrate. *Int. J. Biol. Macromol.* **2020**, *152*, 1135–1145. [CrossRef] [PubMed]
154. Huang, C.; Cai, Y.; Chen, X.; Ke, Y. Silver-based nanocomposite for fabricating high performance value-added cotton. *Cellulose* **2021**, *29*, 723–750. [CrossRef] [PubMed]
155. Haase, H.; Jordan, L.; Keitel, L.; Keil, C.; Mahltig, B. Comparison of Methods for Determining the Effectiveness of Antibacterial Functionalized Textiles. *PLoS ONE* **2017**, *12*, e0188304. [CrossRef]
156. Ščasníková, K.; Sibilová, A.; Bánovská, Z. Comparison of quantitative methods for determining the antibacterial effectiveness of non-woven textiles. *Fibres Text.* **2023**, *29*, 38–44. [CrossRef]
157. Boudreau, M.D.; Imam, M.S.; Paredes, A.M.; Bryant, M.S.; Cunningham, C.K.; Felton, R.P.; Jones, M.Y.; Davis, K.J.; Olson, G.R. Differential Effects of Silver Nanoparticles and Silver Ions on Tissue Accumulation, Distribution, and Toxicity in the Sprague Dawley Rat Following Daily Oral Gavage Administration for 13 Weeks. *Toxicol. Sci.* **2016**, *150*, 131–160. [CrossRef] [PubMed]
158. Zamora-Mendoza, L.; Guamba, E.; Miño, K.; Romero, M.P.; Levoyer, A.; Alvarez-Barreto, J.F.; Machado, A.; Alexis, F. Antimicrobial Properties of Plant Fibers. *Molecules* **2022**, *27*, 7999. [CrossRef]

159. Tien, N.D.; Lyngstadaas, S.P.; Mano, J.F.; Blaker, J.J.; Haugen, H.J. Recent Developments in Chitosan-Based Micro/Nanofibers for Sustainable Food Packaging, Smart Textiles, Cosmeceuticals, and Biomedical Applications. *Molecules* **2021**, *26*, 2683. [CrossRef]
160. Antunes, J.C.; Domingues, J.M.; Miranda, C.S.; Silva, A.F.G.; Homem, N.C.; Amorim, M.T.P.; Felgueiras, H.P. Bioactivity of Chitosan-Based Particles Loaded with Plant-Derived Extracts for Biomedical Applications: Emphasis on Antimicrobial Fiber-Based Systems. *Mar. Drugs* **2021**, *19*, 359. [CrossRef]

Disclaimer/Publisher's Note: The statements, opinions and data contained in all publications are solely those of the individual author(s) and contributor(s) and not of MDPI and/or the editor(s). MDPI and/or the editor(s) disclaim responsibility for any injury to people or property resulting from any ideas, methods, instructions or products referred to in the content.

Article

Comparative Analysis of the Liquid CO₂ Washing with Conventional Wash on Firefighters' Personal Protective Equipment (PPE)

Arjunsing Girase *, Donald Thompson and Robert Bryan Ormond *

Textile Protection and Comfort Center, Wilson College of Textiles, North Carolina State University, Raleigh, NC 27695, USA
* Correspondence: aggirase@ncsu.edu (A.G.); rbormond@ncsu.edu (R.B.O.)

Abstract: Firefighters are exposed to several potentially carcinogenic fireground contaminants. The current NFPA 1851 washing procedures are less effective in cleaning due to the limited intensity of the washing conditions that are used. The 2020 edition of NFPA 1851 has added limited specialized cleaning for higher efficacy. The liquid carbon dioxide (CO_2) laundering technique has gained popularity in recent years due to its availability to remove contaminants and its eco-friendliness. The primary aim of this study is to address the firefighter questions regarding the efficacy of cleaning with liquid CO_2 and to compare it with the conventional washing technique. The unused turnout jackets were contaminated with a mixture of fireground contaminants. These turnout jackets were cleaned with conventional NFPA 1851-apvoved aqueous washing and a commercially available liquid CO_2 method. Post-cleaning samples were analyzed for contamination using pressurized solvent extraction and GC-MS. The liquid CO_2 technique demonstrated considerable improvement in washing efficiency compared to the conventional washing.

Keywords: fireground contaminants; liquid CO_2; NFPA 1851; carcinogenic; cleaning efficiency; PAHs; phenols; phthalates

Citation: Girase, A.; Thompson, D.; Ormond, R.B. Comparative Analysis of the Liquid CO₂ Washing with Conventional Wash on Firefighters' Personal Protective Equipment (PPE). *Textiles* **2022**, *2*, 624–632. https://doi.org/10.3390/textiles2040036

Academic Editor: Laurent Dufossé

Received: 18 October 2022
Accepted: 22 November 2022
Published: 25 November 2022

Publisher's Note: MDPI stays neutral with regard to jurisdictional claims in published maps and institutional affiliations.

Copyright: © 2022 by the authors. Licensee MDPI, Basel, Switzerland. This article is an open access article distributed under the terms and conditions of the Creative Commons Attribution (CC BY) license (https:// creativecommons.org/licenses/by/ 4.0/).

1. Introduction

The International Agency for Research on Cancer (IARC) has stated that the profession of firefighting is a known carcinogen to human beings [1]. Firefighters are exposed to several chemicals during fire suppression activities. Polycyclic aromatic hydrocarbons (PAHs) are compounds that are released due to the incomplete combustion of materials. PAHs have toxic and mutagenic properties while some of them are endocrine disruptors. Multiple PAHs, including the known carcinogen benzo[a]pyrene, have been found in firefighters' personal protective equipment (PPE) and on their skin [2]. Among several different compounds, plasticizers are also found on firefighter PPE. When used samples of PPE were analyzed, 20 different PAHs and 6 phthalate esters were found. Phthalates are ubiquitously found in polyvinyl plastic materials that are used abundantly in flooring, wire sheathing, and home furnishings [3].

The NFPA 1851 Standard on Selection, Care, and Maintenance of Protective Ensembles for Structural Fire Fighting and Proximity Fire Fighting has standard guidelines and requirements for inspecting, cleaning, and maintaining firefighter turnout gear. These guidelines include washing at temperatures less than 105 °F (40 °C), G-forces in the extractor should be less than 100 G, and prohibiting the use of bleaching or oxidizing agents [4]. The standard has categorized the decontamination techniques as (1) Preliminary Exposure Reduction (PER), (2) Advanced Cleaning, (3) Specialized Cleaning. The advanced cleaning procedure permits the use of programmable washing machines and detergents. The specialized cleaning is performed by a verified independent service provider (ISP). The standard clearly states to use specialized cleaning when the ensemble is inadequately cleaned by advanced cleaning [4].

A limited number of studies have been conducted that indicate residual contamination after using the standard aqueous wash. Mayer et al., 2019 investigated the impact of routine laundering on firefighter hoods. The study was performed on two sets of hoods that were exposed to the same structural fire. One set was routinely laundered after every fire scenario and in total, was washed four times in a standard washer extractor. The other set was kept unlaundered to assess the contamination. The analysis between the two sets showed that overall, laundered hoods had 81% lower PAH contamination than unlaundered hoods. The pre-wash and post-wash analyses were performed on completely different sets of hoods. The high values of standard deviation in contamination indicated high spatial variability that may have affected washing efficiency results [5]. A study of water-only decontamination of turnout gear used in live-structure burns showed an increase in contamination by 42%; however, this increase could have been attributed to the disparity in sampling sites for pre- and post-washing samples [6]. Thus, the uneven contamination on the gear is a major hindrance in calculating the washing efficiency of the method and prevents from gaining a comprehensive understanding of the process. The above studies indicated a need for a controlled assessment that includes uniform contamination and targeted contaminants as opposed to using the highly variable live-fire scenarios to contaminate the materials.

Dry cleaning is a technique of removing soils and contaminants from textiles using a non-aqueous solvent. In conventional dry cleaning, perchloroethylene (PER) is most commonly used. PER has a toxic effect on the human body. Several alternatives have been looked at for textile dry cleaning applications such as hydrocarbon solvents, Green earth®, acetal silicon-based solvents, and carbon dioxide (CO_2) [7]. CO_2 has distinctive advantages over other solvents such as being non-toxic, non-flammable, non-corrosive, environmentally benign, and economical [8]. Some studies have indicated that the cleaning efficiency of CO_2 for non-particulate soil removal is comparable to that of preliminary exposure reduction, which is conducted on-scene with a brush, soap, and water. The particulate removal for CO_2 dry cleaning was lower [8]. For dry cleaning operations, the liquid state of CO_2 is preferred over the supercritical state since the two-phase gas-liquid interface is beneficial for trapping soil particles. The substantially higher pressure in CO_2 cleaning makes it easy to separate the CO_2 from the detergent formulation and the soil post-cleaning process. Additionally, the spontaneous evaporation of CO_2 from the fabric during depressurization saves the energy of drying [9].

The following study was conducted to evaluate and compare the cleaning efficacies of liquid CO_2 washing and conventional aqueous wash for the application of firefighter protective clothing. The novelty of the experiment is that controlled contamination of the targeted fireground contaminants is used to evaluate two different washing techniques: Conventional wash and liquid CO_2, which is a novel technique to clean the fireground contaminants and has not been used and compared with conventional washThe null hypothesis for the experiment was that there is no significant difference in cleaning efficacies of conventional and liquid CO_2 methods.

2. Materials and Methods

For this study, five new turnout jackets were used to mount the test samples (swatches) for cleaning. On every single jacket, eight hook-and-loop patches were stitched. The hook part was stitched on the jacket and the loop part was stitched to the test samples. The positions of the hook-patches are shown in the schematic in Figure 1. The size of each patch was 5 cm × 5 cm. The test sample swatches (5 cm × 5 cm) were prepared separately using the outer shell material, PBI Max™ Gold (6 oz.), with a fluorinated durable water-repellent finish. These swatches were spiked with targeted fireground contaminants using analytical standards (Table 1). Three analytical standards for phenols, PAHs, and phthalates (2000 ng/μL for each compound in the mix) were used to contaminate the test samples. The solutions were diluted to 1000 ng/μL using n-hexane. Twenty 3-μL drops of each standard mix were applied on the swatch from the stock solution using a repeater pipette. Thus, the amount of each contaminant present on a single test sample was 60,000 ng. All the test samples were allowed to dry for 24 h. After contamination, the test samples were adhered to the positions on the turnout jackets. In addition to the contaminated swatches, blank samples were shipped in a plastic bag along with the samples for liquid CO_2 cleaning to identify any cross-contamination during transportation. Similarly, positive control (contaminated-but-not-washed) fabric samples were prepared before the extraction process.

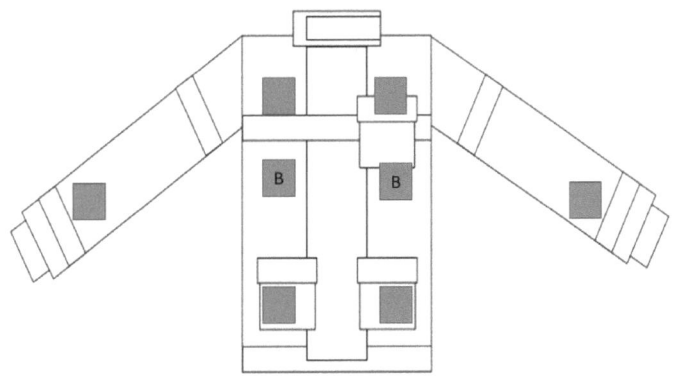

B= Backside

Figure 1. Schematic of the turnout suit.

Table 1. Targeted Contaminants and their relevant properties.

Compound	Boiling Point (°C)	Volatile/ Semi-volatile	K_{OW} [A]	LOD (ng/μL)	LOQ (ng/μL)	RSQ
Phenol	182	Volatile	1.46	0.29	0.90	0.9988
2,4,6-Trichlorophenol (2,4,6-TCP)	246	Volatile	3.69	0.17	0.52	0.9973
Pentachlorophenol (PCP)	310	Semi-volatile	5.12	0.22	0.67	0.9927
Di-butyl phthalate (DBP)	340	Semi-volatile	4.50	0.09	0.26	0.9998
Benzyl butyl phthalate (BBP)	370	Semi-volatile	4.73	0.10	0.30	0.9994
Di-ethylhexyl phthalate (DEHP)	384	Semi-volatile	7.60	0.13	0.38	0.9997
Phenanthrene	340	Semi-volatile	4.46	0.22	0.67	0.9992
Pyrene	404	Semi-volatile	4.88	0.07	0.21	0.9997
Benzo[a] pyrene (BaP)	495	Semi-volatile	6.13	0.06	0.18	0.9995

[A] = values taken from PubChem®.

2.1. Liquid CO_2 Protocol

To conduct the liquid CO_2 cleaning of the test samples, the research team employed Tersus Solutions (Denver, CO, USA). All of the test jackets were shipped to the cleaning facility to be washed with liquid CO_2 utilizing a protocol that is proprietary to the facility. The limited details of the method are given in Table 2. After washing, all the test samples attached to the jackets were sent back for analysis. These test samples were removed from the jackets and stored separately in the plastic bag and in the refrigerator at 4 °C for 24 h after receiving. The analysis was done using the analytical method described later in the article.

Table 2. Details of the liquid CO_2 method.

Step	Details
Duration of cycle	50 min
Wash bath: Single wash	8 min
Rinse: Two cycles	4 min each
Pressure range	600–850 psi
Total load	50 lbs.
Detergent	Proprietary
CO2 grade	Beverage

2.2. Conventional Washing Protocol

For comparative analysis, the test sample preparation process was repeated exactly for the samples to receive conventional aqueous wash using a commercially available detergent (CD-1). The ingredients of CD-1 are shown in Table 3. The UNIMAC® 45 lbs. washing extractor was used in this process. The machine was installed in Wilson College of Textiles, Raleigh, NC. The temperature of the wash was kept at 40 °C (105 °F) and the duration of the wash was 60 min. The conventional method was compliant with the NFPA 1851 requirements. Due to the limited availability of the materials (liquid contaminants, unused jackets, velcro patches), the jacket was patched with five contaminated swatches. The positions where test samples were attached were chosen randomly. The amount of detergent CD-1 used was 120 mL per 45 lbs. load and was calculated according to the manufacturer's recommendations. All test samples were air-dried after washing for 24 h and then extracted and analyzed as described below.

Table 3. Ingredients of CD-1.

CD-1
D-Limonene
Non-ionic surfactant: 4-Nonylphenyl-polyethylene glycol
Mackamide C
Glycol ether

2.3. Extraction

All fabric samples were extracted using a pressurized solvent extractor (BUCHI®, Speed Extractor E-916) with n-hexane (Fisher Scientific, Hampton, NH, USA) as the extraction solvent. Outer shell fabrics (pre-wash or post-wash) of size 5 cm × 5 cm were placed in the 10-mL stainless steel extraction cell. Glass beads (4 mm diameter) were sonicated with n-hexane to remove any prior contamination, and 5 g of glass beads were filled inside each cell to fill the void volume to reduce the excess solvent entering the cell. The cell was capped with top and bottom cellulose filters to prohibit unwanted particulate contamination. Each extraction comprised two full extraction cycles and one flush cycle at the end. Every single extraction cycle consisted of a five-minute heat-up followed by a five-minute hold where the solvent and fabric were in contact with each other. The cycle was held at 100 °C and 100 bar, and the extraction was carried out in the nitrogen atmosphere. The extract passed through a condensing coil and was collected in a 60-mL glass vial. The total run time for the extraction process was 32 min.

After collecting, the extract from each cycle was diluted to 10 mL in a standard 10-mL volumetric flask using n-hexane. A sample of the diluted extract was transferred into the 2-mL amber autosampler vial using a 3-mL syringe with 0.2 µm PTFE filters. These vials were loaded on to GC-MS system and analyzed. All the extraction cycles included a positive control (contaminated-but-not-washed) fabric sample and a negative control (uncontaminated and unwashed) fabric sample. The post-washing concentrations of the washed samples were calculated relative to the positive control during that particular extraction. The negative controls were used to check any compounds were present on the fabrics themselves in the first place. They were used only for qualitative analysis.

2.4. Gas Chromatography/Mass Spectrometry

The analysis of the fireground contaminants was carried out using an Agilent 7890B Gas Chromatograph (GC) system coupled to an Agilent 5977B Mass Spectrometer (MS) equipped with Electron Ionization (EI) capability (Agilent Technologies, Inc., Santa Clara, CA, USA). Chromatographic analysis was conducted in the splitless mode with a purge flow of 100 mL/min at 1.0 min. The Agilent Ultra Inert liner (5190–6168, straight 2 mm ID) was used in the GC inlet, which was maintained at 250 °C. An Agilent EPA 8270D fused silica capillary column (30 m × 0.25 mm × 0.25 µm) was used with a helium flow rate of 1.2 mL/min. The oven program was set to begin at 40 °C, then increased to 280 °C at a rate of 10 °C/min with a 1 min hold, followed by a further increase to 300 °C at 25 °C/min with a final hold of 1 min. The total running time was 30.48 min. The MS transfer line was kept at 280 °C throughout the run. The MS quad temp was maintained at 300 °C, and the ion source temp was kept at 200 °C. The gain factor used was 1.00. The analysis was conducted in scan mode (35–550 amu) using EI with an energy of 70 eV. The calibration solutions were prepared to calibrate the instruments using the mix of the compounds (2000 ng/µL) as the stock solutions. The calibration standards, as shown in Table 4, were prepared by diluting n-hexane (Fisher Scientific—95% purity) in a 10-mL volumetric flask. The calibration curve was obtained by averaging out the responses of three replicates. The limit of detection (LOD) and limit of quantitation (LOQ) values were calculated using Equations (1) and (2), respectively. The lowest concentration was run ten times and the standard deviation of the response (area) σ was calculated. The calibration curve provided the equation of the line which provided the slope m.

$$\text{LOD} = 3\sigma/m \quad (1)$$

$$\text{LOQ} = 10\sigma/m \quad (2)$$

Table 4. Calibration solutions for the chromatography method.

Calibration Standard	Target Concentration (ng/µL)	The Volume Injected from the Stock Solution (µL)	Mass per Unit Fabric Area (ng/cm^2)
1	0.6	3	240
2	1.2	6	480
3	3	15	1200
4	6	18	2400
5	9	45	3600
6	12	60	4800

2.5. Data Analysis

Microsoft Excel was used to calculate the R^2, slope (m) of the calibration curve, standard deviation of the responses (σ), LOD and LOQ. The linearity of the calibration solutions (response vs. concentration) was high, as indicated from the R^2 values in Table 1 for all the compounds. This showed that the proportion of the variation in the response generated for various concentrations was predictable and dependent. To quantify the effectiveness of both decontamination methods, the washing efficiency was calculated using Equation (3). For samples that did not show detectable peaks in the chromatogram, $\frac{1}{2}$ LOQ values were used in the equation. For every compound, the arithmetic mean (average) of the washing efficiencies of the replicates was calculated. For liquid CO_2 washed samples, the average of the washing efficiency for any compound was calculated using 40 replicates and for conventional wash, the values of washing efficiency are the average of the 5 replicates. In a separate analysis, data set from five random samples were taken from the liquid CO_2 set and compared with conventional washed samples to perform a comparative analysis at the equal number of data points and to see if that created any difference. Standard errors were calculated to check the variability across samples of the given population. The statistical analysis was done using JMP Pro® statistical software (15.2.0, SAS Institute Inc., Cary, NC, USA). Shapiro–Wilks test was used to check the normal distribution of the data. A single factor ANOVA was conducted to test the variances at $p < 0.05$, confirming the unequal variances for the one tailed t-test. A singled tail t-test was done at alpha-level = 0.05. Random sample picking from the given data set was done using JMP Pro® (15.2.0, SAS Institute Inc., Cary, NC, USA).

$$\text{Cleaning efficiency (\%)} = \frac{(\text{Original concentration}(Cc) - \text{post washing concentration}(Cw))}{\text{Original concentration}} \times 100 \qquad (3)$$

3. Results

The original target concentration applied to the materials (accounting for analytical sample preparation) for all the samples was 6 ng/µL. The blank samples in the plastic bag did not show any compounds, thus indicating no issues with cross-contamination during transportation. The washing efficiency values for both the methods: liquid CO_2 and conventional wash are presented in Table 5 for the targeted contaminants. The comparative analysis of the washing efficiencies is shown in Figure 2 (Average washing efficiencies in Figure S1). The results for the washing efficiencies were very close together as indicated by the error percentage from Table 5 which indicates low variation. The Shapiro–Wilks test indicated a p-value of 0.24 > 0.05 (for Conventional wash) and a p-value of 0.41 > 0.5 (for liquid CO_2 wash), indicating that the data was normally distributed (Figure S2). The single-tailed t-test indicated that the difference was statistically significant for $p < 0.05$ (Figure S3). Thus, we reject our null hypothesis and conclude that there is a significant difference between the means of the different washing methods and liquid CO_2 is more effective than the conventional washing method. For the equal number of samples where for liquid CO_2, average of 5 random replicates was taken, similar results were found (Average washing efficiencies in Figure S4). The Shapiro–Wilks indicated p-value of 0.27 > 0.05 (Conventional wash) and p-value of 0.46 > 0.05 (liquid CO_2 wash) (Figure S5) and the single-tailed t-test showed (Figure S6) that the difference between the means was statistically significant. This indicated that irrespective of the number of samples, liquid CO_2 removed contaminants effectively.

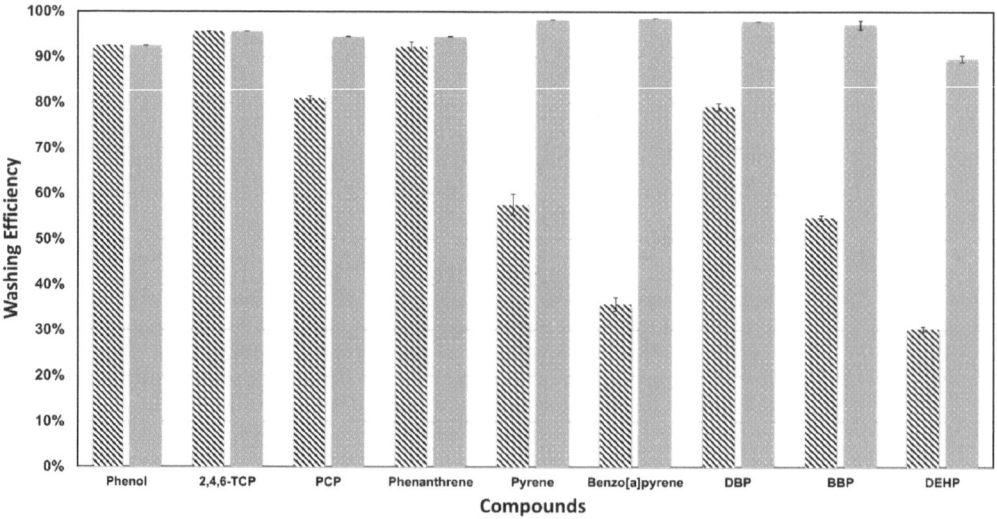

Figure 2. Comparison of washing efficiencies for conventional wash and liquid CO_2.

Table 5. Average washing efficiency of targeted contaminants for conventional and liquid CO_2.

	Compounds	Conventional Wash	Count of ND [a]	Liquid CO_2	Count of ND [a]
Phenols	Phenol	92.46% [†]	5	92.46% [†]	40
	2,4,6-tri-chlorophenol (TCP)	95.59% [†]	5	95.59% [†]	40
	Penta-chloro-phenol (PCP)	80.96%	0	94.43% [†]	40
PAHs	Phenanthrene	92.32%	0	94.44% [†]	40
	Pyrene	57.62%	0	98.22% [†]	40
	Benzo[a]pyrene (BaP)	35.73%	0	98.52% [†]	40
Phthalates	Di-butyl-phthalate (DBP)	79.19%	0	97.80% [†]	40
	Benzyl-butyl-phthalate (BBP)	54.76%	0	97.08%	0
	Di-ethyl-hexyl-phthalate (DEHP)	30.29%	0	89.67%	0

[†] Non-detectable signal—$\frac{1}{2}$ LOQ used for calculation. [a] = number of samples with non-detectable signals.

4. Discussion

For conventional wash, the washing efficiency decreased from phenols to phthalates. The increase in the octanol-water partition coefficient (K_{OW}) values and the decrease in the washing efficiency in a chemical class showed that the relation between the two was evident. The average washing efficiency for conventional wash is shown in Figure 2. The $\frac{1}{2}$ LOQ values from Table 1 were used for calculating phenol and TCP. The conventional wash removed these contaminants below the quantitation limits of the analytical method. The aqueous wash and non-ionic surfactants removed the phenols, phenanthrene and DBP. Phenols are more polar as compared to the other two groups. They are fairly soluble in water; hence, the results were comparable. The detergent contained d-limonene for aqueous washing, a non-polar compound that helped remove the phenanthrene and DBP. For PAHs, an increase in the number of aromatic rings increased hydrophobicity and resulted in decreased removal from the fabric. Thus, a decreasing trend in washing efficiency can be seen in the aqueous washing [10]. A similar trend was observed in phthalates, an increase in alkyl chain length increased the hydrophobicity and thus phthalates were not removed effectively by aqueous washing [11].

From the comparative analysis perspective, Figure 2 demonstrated that conventional wash was not very effective at removing higher molecular weight PAHs and phthalates. This limitation can be attributed to the polar nature of the water and the hydrophobic nature of the compounds.

For liquid CO_2, the $\frac{1}{2}$ LOQ values were used in calculations for all the compounds except for BBP and DEHP, as they were the only compounds that had detectable levels remaining in the fabric. It indicated that the contaminants might be present in trace amounts after washing that cannot be quantified by the analytical method. Even for BBP and DEHP, the average washing efficiency was still greater than 90%.

The results indicate the potency of the liquid CO_2 method in removing the fireground contaminants. The three different chemical groups: phenols, PAHs, and phthalates were all removed effectively using the liquid CO_2 washing method. This may be due to the non-polar nature of liquid CO_2 that aided in solubilizing the more hydrophobic contaminants. The proprietary detergent used has been effective in removing phenols. The high diffusivity and low viscosity helped liquid CO_2 reach the fabric's interstices and remove contamination. The washing system was kept under high pressure, that helped in solubilizing the contaminants from the solvent at a low temperature. This made it a very suitable solvent for removing non-polar contamination.

5. Conclusions

The liquid CO_2 wash was certainly effective in removing the targeted contaminants. The average washing efficiency for liquid CO_2 was 95.36% which was significantly higher than the average washing efficiency of conventional wash: 68.77%. The controlled study included uniform contamination of the garments that helped understand and analyze both methods on the same level. One interesting trend that can be seen here when observing the conventional wash is that the cleaning efficiency and the K_{OW} values have an inverse relation. This makes sense since the octanol-water partition coefficient indicates the ability of the compound to partition in the organic or aqueous phase. So, the higher value shows the reluctance of the compound to partition more towards the water. The results were statistically significant. A limitation of this study design is that the method was tested against liquid contamination and did not account for the particulate contamination that is experienced with smoke and soot in firefighter exposures. Additionally, studies have shown that the lack of mechanical action impedes the removal of particulate contamination for liquid CO_2 [8]. Thus, it will be interesting to evaluate the efficacy of liquid CO_2 when real-world samples are used. Simultaneously, it is important to investigate the redeposition of the contaminants while washing with this technique. Additionally, a further investigation of the impact of liquid CO_2 on the durability of the turnout suit and its accessories is needed along with the operation costs for the method.

Supplementary Materials: The following supporting information can be downloaded at: https://www.mdpi.com/article/10.3390/textiles2040036/s1, Figure S1: Box-plot of average washing efficiencies of all compounds for all samples. Figure S2: *t*-Test analysis (All samples). Figure S3: Normality test when all samples are considered (Left: Conven-tional washing, Right: liquid CO_2 washing) Figure S4: Box-Plot of average washing efficiencies of all compounds (when equal number of samples are considered). Figure S5: Normality test for equal number of samples (Left: Conventional washing, Right: liquid CO_2 washing). Figure S6: *t*-Test analysis (Equal number of samples)

Author Contributions: Conceptualization, D.T. and R.B.O.; Methodology, A.G.; Software, A.G.; Validation, D.T., R.B.O. and A.G.; Writing—original draft preparation, A.G.; Writing—review and editing, A.G. and R.B.O.; Project administration, R.B.O.; Funding acquisition, R.B.O. and D.T. All authors have read and agreed to the published version of the manuscript.

Funding: The following project is funded by the Assistance to Firefighters Grant Program (AFG)-Fire Prevention and Safety (FP&S) Research and Development Grants (EMW-2017-FP-00601).

Institutional Review Board Statement: Not applicable.

Informed Consent Statement: Not applicable.

Data Availability Statement: All the data supporting the findings of the study are available within the article.

Acknowledgments: Our special thanks to Steve Madsen from Tersus solutions and Bill Brooks from UNIMAC for providing access to the cleaning processes.

Conflicts of Interest: The author declares no conflict of interest.

References

1. Demers, P.A.; DeMarini, D.M.; Fent, K.W.; Glass, D.C.; Hansen, J.; Adetona, O.; Andersen, M.H.G.; Freeman, L.E.B.; Caban-Martinez, A.J.; Daniels, R.D.; et al. Carcinogenicity of occupational exposure as a firefighter. *Lancet Oncol.* **2022**, *23*, 985–986. [CrossRef] [PubMed]
2. Abrard, S.; Bertrand, M.; De Valence, T.; Schaupp, T. French firefighters exposure to Benzo[a]pyrene after simulated structure fires. *Int. J. Hyg. Environ. Health* **2019**, *222*, 84–88. [CrossRef] [PubMed]
3. Alexander, B.; Baxter, C.S. Plasticizer Contamination of Firefighter Personal Protective Clothing—A Potential Factor in Increased Health Risks in Firefighters. *J. Occup. Environ. Hyg.* **2014**, *11*, D43–D48. [CrossRef]
4. *NFPA 1851*; Standard on Selection, Care and Maintenance of PPE for Structural and Proximity Firefighting. NFPA: Quincy, MA, USA, 2020.
5. Mayer, A.C.; Fent, K.W.; Bertke, S.; Horn, G.P.; Smith, D.L.; Kerber, S.; La Guardia, M. Firefighter hood contamination: Efficiency of laundering to remove PAHs and FRs. *J. Occup. Environ. Hyg.* **2019**, *16*, 129–140. [CrossRef] [PubMed]
6. Calvillo, A.; Haynes, E.; Burkle, J.; Schroeder, K.; Calvillo, A.; Reese, J.; Reponen, T. Pilot study on the efficiency of water-only decontamination for firefighters' turnout gear. *J. Occup. Environ. Hyg.* **2019**, *16*, 199–205. [CrossRef] [PubMed]
7. Troynikov, O.; Watson, C.; Jadhav, A.; Nawaz, N.; Kettlewell, R. Towards sustainable and safe apparel cleaning methods: A review. *J. Environ. Manag.* **2016**, *182*, 252–264. [CrossRef] [PubMed]
8. Sutanto, S.; van Roosmalen, M.J.E.; Witkamp, G.J. Redeposition in CO_2 textile dry cleaning. *J. Supercrit. Fluids* **2013**, *81*, 183–192. [CrossRef]

9. Banerjee, S.; Sutanto, S.; Kleijn, J.M.; van Roosmalen, M.J.E.; Witkamp, G.-J.; Stuart, M.A.C. Colloidal interactions in liquid CO_2—A dry-cleaning perspective. *Adv. Colloid Interface Sci.* **2012**, *175*, 11–24. [CrossRef] [PubMed]
10. Kim, K.-H.; Jahan, S.A.; Kabir, E.; Brown, R.J.C. A review of airborne polycyclic aromatic hydrocarbons (PAHs) and their human health effects. *Environ. Int.* **2013**, *60*, 71–80. [CrossRef]
11. Staples, C.A. *Phthalate Esters*; Springer: Berlin/Heidelberg, Germany, 2003.

Review

Supportive, Fitted, and Comfortable Bras for Individuals with Atypical Breast Shape/Size: Review of the Challenges and Proposed Roadmap

Josephine Taiye Bolaji and Patricia I. Dolez *

Department of Human Ecology, University of Alberta, Edmonton, AB T6G 2N1, Canada
* Correspondence: pdolez@ualberta.ca

Abstract: Individuals with atypical breast shape/size often find it quite challenging to obtain a comfortable, supportive, and fitted bra off-the-shelf. They include people with very large breasts, who have significant breast asymmetry, and/or have undergone mastectomy or mammoplasty. This paper provides insights in their challenges and attempts to fill the gap in terms of critical review of the current state of knowledge around the topic of bras. Poor and ill fitted bras are associated with breast, chest and shoulder pain, embarrassment, and an overall reduction in quality of life among others. Building upon the advantages and limitations of solutions to improve the fit, support and comfort of bras found in the literature, this paper proposes strategies to solve these challenges. As the problem is multidisciplinary, a human-centered interdisciplinary approach is key to ensure that all aspects are considered at all stages of the process. A modular design allows selecting the fabric characteristics based on the requirements of each bra part. In terms of materials, stretch woven fabrics offer a large potential in the production of bras to enhance the support provided by areas such as the under band and back panels. Bespoke manufacturing takes into account the specificities of each individual. The road map proposed here will contribute to enhance the quality of life of individuals with atypical breast shape/size.

Keywords: bra; atypical breast shape/size; modular design; stretch fabrics; bespoke manufacturing

Citation: Bolaji, J.T.; Dolez, P.I. Supportive, Fitted, and Comfortable Bras for Individuals with Atypical Breast Shape/Size: Review of the Challenges and Proposed Roadmap. *Textiles* **2022**, *2*, 560–578. https://doi.org/10.3390/textiles2040032

Academic Editor: Laurent Dufossé

Received: 30 August 2022
Accepted: 20 October 2022
Published: 24 October 2022

Publisher's Note: MDPI stays neutral with regard to jurisdictional claims in published maps and institutional affiliations.

Copyright: © 2022 by the authors. Licensee MDPI, Basel, Switzerland. This article is an open access article distributed under the terms and conditions of the Creative Commons Attribution (CC BY) license (https://creativecommons.org/licenses/by/4.0/).

1. Introduction

The invention of bras can be dated back to the 14th century [1], and has since evolved with tipping points in the early 19th century when the first bra patents were filed [2]. Since then, there have been landmarks, such as the introduction of large-scale production in 1930 [3] and cup and band sizing in 1932 [4]. Today, there are varieties of bra styles and sizes influenced by fashion, culture, and body image perception.

Regardless of this achievement, the degree of bra dissatisfaction is high [5]. Many women find it difficult to obtain a well-fitted, supportive and comfortable bra [6]; more so individuals with atypical breast shapes and sizes. These individuals may have voluminous breasts [7], significant breast asymmetry [8] and/or have undergone breast surgery. These breast surgeries include mastectomy [9], a surgical procedure for the treatment of breast cancer [9], and mammoplasty, a surgical procedure to reduce or increase the breast volume and/or correct the breast shape [8]. Since the primary purpose of a bra is to support the breast, it is important that the breast shape/size is taken into account at the time of production. However, mass-manufactured bras are highly unlikely to cater to individuals with atypical breast shape/size [10]. This is because they are based on a sample range that is not representative of the entire bra user population [10]. In addition, individuals with atypical breast shape/size are varied and unique; each person is different. The most limiting factors are body variations and experiences, which are not taken into account into the bra production. For example, studies by Vasquez show that no two lumpectomies (a partial form of mastectomy where only a part of the breast is removed) are the same [9].

Therefore, there is a need to re-think the bra industry such that it aligns with the changing needs of women [11]. According to the Canadian Cancer Society [12], 25,000 women will be diagnosed with a new breast cancer in 2021 and will require at least one lumpectomy procedure [11,12]. In addition, studies show that more women have larger breasts due to lifestyle changes over the last few decades [13–15]. Thus, more women are likely to end up with significant breast asymmetry making it more challenging to obtain a suitable bra in their lifetime. This is a challenge as it increases the degree and frequency of bra dissatisfaction. Additionally, increased awareness on body image both in terms of perception and appearance is leading to increased occurrences of aesthetic breast surgeries [13,15]. According to literature, predicted surgery outcomes are sometimes inaccurate, driving up the differences between a breast pair [16,17]. According to Sharland, individuals with atypical breast shape/size require special intervention [6]

Furthermore, the human body has variations from person to person; these variations include neck-to-collar bone, shoulder-to-shoulder, and nipple-to-nipple distance [18]. This contributes to the current bra dissatisfaction experienced by bra users. Individuals with atypical breast shape/size experience even greater dissatisfaction due to their experiences. For instance, studies show that as the breast size increases, breast asymmetry becomes significant [18,19].

In an attempt to fill the gap in terms of critical review of the current state of knowledge around the topic of bras, this paper provides a description of individuals with atypical breast shape/size, and details the challenges they face and current strategies employed in an attempt to solve them. It also proposes a roadmap and discusses fabric analysis, bra design and patterning, fitting procedure, wear trials of prototypes, and translational approaches to patients as possible strategies to solve the remaining problems.

2. Individuals with Atypical Breast Shape/Size

This section inventories some conditions that can lead to atypical breast shape/size, either naturally or as a result of a surgery.

2.1. Breast Asymmetry

Most women have some level of breast asymmetry. However, if the difference in size and shape between the two breasts is significant, it can result in a medical condition known as anisomastia [20]. According to Reiley, when one breast is 30% larger or more than the other one, it is considered asymmetric [21]. Significant breast asymmetry has been reported to be present in as much as 44% of women [22]. Breast asymmetry includes also differences in the degree of sagging, position on the chest wall, and nipples placement and shape [10,21,22]. For these individuals, finding the right bra is challenging as the breast pair could be of different cup sizes [20]. The frustration they encounter while shopping for bras often puts them in a position to consider surgery, with all the risks and issues associated with breast surgeries [23,24]. Hence, there is the need to develop a model for bra making on a case-by-case basis that allows for individuality. This could potentially derail individuals from undergoing surgeries due to bra dissatisfaction.

2.2. Macromastia

Macromastia is the medical term for women with bulky breast tissues (plus-sized) that make their breast abnormally large [25]. There are no clinically defined measurements of symptomatic macromastia due to the wide variation of body shapes and sizes [26]. However, women having excess breast tissue between 900 g and 2220 g are considered large breasted [21,26]. Within the industry, large bra sizes are typically cup D and above [21,25]. Breast asymmetry also becomes more obvious as the breast size increases [25]. This contributes to the challenges these individuals experience when shopping for bras. Pandarum et al. studied the dissatisfaction experienced by plus-sized women with poorly fitted bras [27]. They reported the inadequacies of the current sizing system used by bra manufacturers in catering to this group of women. A further complication arises when women have excess skin around the armpit and breast base [28].

2.3. Mastectomy

Mastectomy is a medical surgery used in the treatment of breast cancer [7]. The procedure involves the partial removal of cancerous lumps in the breast or the complete removal of one breast or both breasts in severe conditions [29,30]. After a mastectomy, many women find it difficult to obtain bras that fit their breasts, especially if they have become significantly asymmetric. In addition, the breast prosthesis will not behave the same way as the remaining breast [31,32]. This issue may affect their physical and psychological health [31]. Furthermore, mastectomy increases the natural body variation that exists from person-to-person. Studies have shown that custom-made bras provide better comfort to those using prostheses after a mastectomy [31–33].

2.4. Mammoplasty

The reduction or augmentation of the breast and/or correction of the breast shape is known as mammoplasty [8]. The reasons why women opt for mammoplasty include improving their sexuality, confidence and appearance [34]. Additionally, studies show that large breasted women go for reduction mammoplasty due to health issues such as back, neck and breast pain [34,35]. Furthermore, having large breasts has been associated with body shaming, reduction in self-esteem and negative bra shopping experiences [34–36]. These issues often prompt the need for mammoplasty. However, things can sometimes go wrong; the procedure may lead to unwanted outcomes, including breast asymmetry [37,38]. Indeed, mammoplasty, as any other surgery, has its own limitations such as poor projection of incisions and scars [38].

3. Challenges for Individuals with Atypical Breast Shape/Size

3.1. Bra Dissatisfaction

Bra users have often expressed dissatisfaction due to poor fitting, low support and overall discomfort [36,39]. Table 1 details some of the common issues reported with bras. Poor and ill fitted bras have been associated with reduction in quality of life [27], increased breast, back and chest pain [6,27], which is intensified during sporting activities [6,39], as well as bra displeasure [5]. In addition, research shows that wearing poor and ill-fitted bras negatively impacts sleep and sexual activity [27,36,39]. On the opposite, well-fitted bras have been linked with reduced instances of breast deformation and sagging that may arise from weakened support tissue after massive weight loss and surgeries [39,40]. They also lead to increased breast firmness [40] as well as reduced post-surgery complications [41]. In addition, well-fitted bras allow for improved mobility and reduce exercise-induced pain by as much as 85% [6,42]. In particular, custom-made bras have positive impacts on the psychological experience of women, thereby improving their mental well-being [34,42]. Furthermore, there is a huge focus on body image driven by the fashion and beauty industries [34]. Thus, more people are concerned about how they look and how others perceive them. In some cases, this leads to body shaming [34]. According to Swami & Furhan and Neto et al., having large breasts has been associated with body shaming and reduction in self-esteem, with impacts on mental, physical and psychological health [34,36].

Table 1. Description of some bra issues, and their common causes.

Issue	Description	Common Causes	Source
Bulging	Skin pushing over or underneath the cup, band or straps	Large breasts, bra size too small or too tight	[43,44]
Digging	Bra parts digging into the skin, especially in individuals with extra under arm and side skin	Bra too small, or too tight	[45]
Lifting away	Centre gore lifts creating space between the bra and the skin.	Under wires, stiff materials, bra too small or too tight	[46]
Gapping	Gaps in the center gore or cups	Large breasts, bra too big or too loose	[44,46]
Double breast	Occurs mostly in deep-V and push-up style bras used for enhancing cleavages. It occurs when a large part of the breast is outside of the cup.	Large breasts, improper fitting or sizing.	[46]

Individuals with atypical breast shape/size have also reported displeasure with the outcome of specialized bras, which would have been expected to solve their bra problems [31,32,47]. Fong conducted a study on the aftermath of mastectomy patients in terms of their adjustment to life using mastectomy bras and prosthesis [47]. The study reported that most post-mastectomy patients find their bras uncomfortable. Similar views on the displeasure of post-mastectomy patients were shared by Jetha et al. [31]. When describing their experiences and what they hoped to find, women who underwent mastectomy indicated a desire to obtain bras that are unique to them [29–31]. Participants of a study carried out by the Breast Cancer Organization expressed their realization of the fact that no two women are the same [30]. Additionally, an investigation of the use of existing commercial bras and prosthesis proved that custom-made bras provide more comfort and satisfaction [32,33]. These studies highlight the need for custom-made bras that are not only unique to each individual, but provide added functionality and overall value.

Complaints made by patients seeking mammoplasty, particularly in case of a reduction, include neck and back pain, embarrassment, harassment, and poor self-image [37]. In such instances, it is quite difficult to deter such persons from going for a mammoplasty. However, Wood et al. [48] suggested that a good support bra could potentially mitigate breast pain and provide adequate support needed by the breast. It will also aid in a stress-free physical activity for large breasted women and perhaps discourage them to go for breast reduction. Outcomes following reduction mammoplasty were studied by Neto et al. [34]. The study was done on two groups of 50 women each; Group A was composed of in-patient (post-operation) and Group B included outpatients (pre-operation) who formed the control sample. Group A women were administered questionnaires six months post-operation. The researchers reported a significant improvement in self-esteem and functional capacity, but noted that a special brassier would aid in the initial recovery process. Therefore, there is the need to have special post-recovery bras that can enhance mobility and keep the breasts in place during physical activities

3.2. Measurement, Fit and Mass-Manufacturing

Inaccurate measurement often results in ill fit [27]. This leaves bra users frustrated [33]. It is challenging to obtain a good fit when it comes to individuals with atypical breast shape/size. Fit is regarded as the accuracy of the measurement done on an individual [49]. Its efficiency is linked to the ability to customize garments [50]. It is done conventionally using a tape rule as the measuring tool. This means measuring an individual standing and transferring those numbers to the flat fabric on a cutting table. This mostly introduces errors; the degree of error often depends on the experience of the measurer [51]. Thus, fitting is highly dependent on the skill of the measurer/fitter and the posture of the individual during the measurement [52]. The process requires a unique skill set and experience, one that is not often found in the average sales person in the stores [50].

On the other hand, mass-manufacturing is the industrial process of making clothing for multiple users with sizing based on a select few [10]. Mass-manufacturing comes with its benefits such as increase in productivity and profitability to brands and manufacturers. For users, mass-manufacturing takes away individuality and the ability for users to contribute to their clothing [51]. With regard to bras, the customary process of selecting a few sizes is done through the use of models [10]. These models, selected using clusters of bra sizes, seem not to give a good representation of the entire population [51]. Because of that, bra dissatisfaction is not a thing of the past, and it clearly shows the limitations of mass-manufactured goods. In reality, not everyone fits in the mass-manufacturing models used for products [5]. Participants from a study carried out by the Breast Cancer Organization expressed how they realized through that experience that no two women are the same [40–42]. Two separate studies by Kubon et al. [32] and Shin et al. [33] compared the use of existing mass-manufactured bras and custom-made bras and their compatibility with breast prosthesis. They both reported that custom-made bras provided more comfort and satisfaction. Therefore, there is the need to produce bras that are not only unique to each individual, but takes their experiences into account.

3.3. Breast Pain

Pain of any sort reduces a person's quality of life, and breast pain is a common occurrence in women [40]. When assessing breast pain prevalence in women with small to medium breasts, Sharland reported that most women will experience breast pain in their lifetime [6]. Similarly, Scurr et al. evaluated the prevalence, severity and impact of breast pain in the average population [40]. They reported that at least 50% of women in the general population suffer from breast pain. This ratio increases to about 65% for women with breast cup size larger than DD. This occurs particularly as the breast volume increases, which suggests the need for well-fitted bras to reduce breast pain for large breasted individuals. Furthermore, Sharland's study evaluated the possibility of well-fitted bras to reduce breast pain [6]; they concluded that wearing a bra that fits well and provides adequate support will greatly reduce the risk of breast and chest pain.

Breast pain has also been linked to poor bra fit. Xiaomeng investigated the pressures and sensations caused by wearing a bra and the influence on bra fit on them [39]. The results showed that the pressure and sensations due to ill-fitted bras result in discomfort, and although they are not always the direct cause of breast pain, ill-fitted bras can aggravate the associated discomfort. As part of their results, the author reported that keeping the breast in position, especially during intense activities like exercising, is important. As echoed by White & Scurr [53] and Sharland [6], keeping the breast in position can significantly reduce breast pain intensity caused by motion. It becomes more critical for individuals with large breasts [54]. A firm breast support has been the most commonly reported solution to reduce breast pain [27,31,40]. It is estimated that an effective supporting bra can reduce instances of breast pain by 85% [54–56]. Wearing the right bra has even been suggested as potentially offsetting for many cases the need for reduction mammoplasty [56].

4. Current Strategies

The current strategies to improve bra satisfaction are mainly two folds: the use of sizing tools and software, and increasing bra styles and design.

4.1. 3D Scanners and Other Measuring Software

There are a number of sizing software/3D scanners that are used to automate the sizing/measuring process. Some of them such as VickingSlice, Physio easy, Posture print, SAPO, and Fisiometer originate from the medical field where they were designed for the purpose of posture and body measurements [57]. Most clothing manufacturers use conventional 3D body scanners, which are cheaper, more accessible and easier to use compared to the medical systems [50,58–61]. These sizing software/3D scanners help manufacturers by increasing the production speed. However, they do not appear to help bra users [61–64]. Other software such as mybraFit™ designed by the company Wacoal and Upbra designed for Upbra are widely used by bra manufacturers [65]. Although these sizing applications were designed for the company's specific products, a good number of shop owners and users have now adopted them. Reviews on these applications have indicated that they are not useful for everyone, and most customers make a return [43,57].

Researchers estimate that in most cases, 3D scanners have automated the measuring process, but because of their limitations, they have not improved bra fit [27,59]. The limitations of 3D scanners include the fact that people's breathing during scanning affects the accuracy of the measurement [27]. In addition, 3D scanners are not able to capture the full breast volume, especially above a cup F, as the breast is likely to be in "fall" position during scanning [60,62]. Furthermore, measurements by 3D scanners are impacted by postures and the eventual positioning of physical features such as the back curvature and collar bone [28,61]. Figure 1 shows the 3D scanning of individuals with different sizes/shapes of breasts [28]. They show differences in the nipple position on the chest, nipple to nipple distance, and nipple to collar bone distance. This ultimately leads to profiling and sampling [5,28,61]. Profiling groups individuals based on the assumption that the measurements of a few can be projected on larger populations without errors. However, research has shown that not only are there errors, but also that these errors are larger than is acceptable and result in bra dissatisfactions [10,28,51].

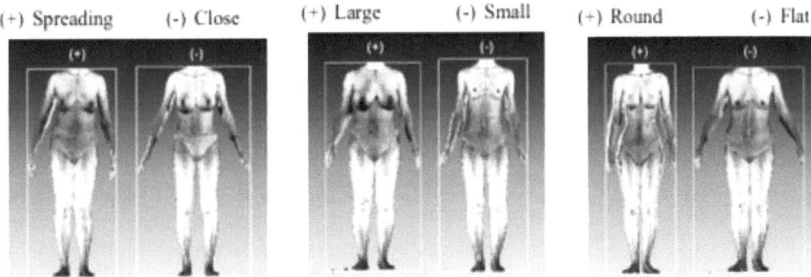

Figure 1. 3D scanning images showing differences in breast shape, size, and nipple position (reproduced from [28] with permission from Elsevier).

Bougourd et al. measured 22 young females using both 3D scanner and traditional tapes [49]. They found that in practice, issues when using 3D scanners increase rather than decrease when increasing the number of subjects. The issue appears to be related to the choice of posture and differences in sensitivity between instruments. In addition, some 3D scanners require the location of certain landmarks such as the collar bone to establish their focus [62]. This poses a major challenge because of the inherent variations in the location of these landmarks on the human body between different people [19]. In addition, greater variations in the position of these landmarks occur above cup F [60]. Studies by Catanuto et al. [63] and Bengston & Glicksman [5] also reported variations in 3D scanner measurements due to differences in breast shape even with breasts of identical volume. A similar study by Ancutiene targeting silhouette gowns concluded that due to different postures, it is difficult to achieve a measurement with 3D scanners that offers optimal fit [62]. In addition, different scanners may provide different results [63]. This can influence clothing sizes and contribute to the already existing confusion regarding chart sizing across brands and countries [64]. In general, statistical analysis suggested that most women are likely to wear a bra with an under-band that is too large or too small when measured with a 3D scanner [60].

4.2. Specialization and Production

There are a variety of specialized products commercially available that ought to cater to certain categories of the bra user population, including individuals with atypical breast shape/size. These include extra-large, sport and mastectomy bras. However, research has shown they do not mitigate bra problems to the level that brings consumer satisfaction.

"Extra-large" bras designated specifically for individuals with large breasts are increasingly common in stores. Unfortunately, these bras are only made bigger in terms of size but the design remains the same [66–68]. Additionally, they do not have the performance, for example in terms of strength, that is required for such "extra-large" bras to function properly [68]. Large breasted women need bras with added support to limit chest pain [48,54]. Articles by Eva's Intimate [66] and Forever Yours Lingerie [68] highlighted some of the issues associated with voluminous breasts:

- Back band rising: this is mostly due to the weight of the breast dragging it downwards with little or no support from the under band and/or strap;
- Falling straps: when the band is too narrow, too big or the shoulders of the individual are slopped, the straps are likely to fall;
- Displacement of the centerpiece (gore): the gore should rest on the sternum, otherwise the breast will put pressure on the gore causing it to bend, break or poke through the ends;
- Side boobs: this occurs when the side breast tissue is peeking out of the bra. In most cases, this adds an extra layer of excess skin to what may already exist under the armpit;
- Quad boobs: this happens when the cup is too small or the style does not fit. Unfortunately, women with big breasts often cannot afford to wear certain styles as their breasts require to be enclosed.

Sports bras have gained a large visibility and are one of the fastest selling bras. Today many top brands including lululemon, adidas and GymShark, who are key players in sports bras, still use mass-manufacturing. Their bras are marketed on the basis of fit and comfort; however, studies continue to show that bra users do not get the desired outcome [69]. Chung & Jang evaluated the comfort of sports bras with a focus on style and their corresponding compression levels after walking and running [70]. The study reported that "racker back" compression sports bras were uncomfortable due to high compression levels. The high compression led to an increase in skin pressure, thereby causing pain in the upper torso. On the other hand, they resulted in reduced breast displacement during exercise activities.

In another study, Lawson [71] assessed the comfort and support provided by eight selected sports bras of four different breast sizes (A–D). Lawson recorded that women with size D require extra support to significantly minimize breast displacement during intense activities. Thus, even for sports bras, women with large breasts have requirements that are different from small to medium breast sizes. Sports bra design, whether they include front zippers, back hooks, or ease of donning and doffing, has also a significant impact on the fit, comfort and wearability of the bra. Better fitting sports bras will also be critical to allow the proper operation of wearable vital sign monitoring. Navalta et al. [72] evaluated sports bras with integrated smart technology to monitor heart rate. They reported that proper fit are critical for the accuracy of the biometric data obtained.

Post-surgical bras are designed to cater to individuals who have undergone breast surgery [73]. Makers of post-surgical bras in the commercial space include Dale by Dale medical products Inc., CareFix by Tytex Inc., and Amoena by Amoena Ltd. However, reports show that their level of satisfaction is still low [33]. This research hypotheses that it is because the production process is still based on mass-manufacturing. Even though the benefits of these bras such as the reduction in infection rate and prevention of stitch raveling are well documented [73,74], the main challenge lies in finding one with the right fit for optimum functionality [74].

Hummel & Charlbois investigated the support level of post-surgical bras. The study had 20 post-cardiac surgery females who wore the bras between one to three weeks after surgery [74]. In the findings, they recorded that 77% of the participants indicated that the bra was helpful and would recommend it to a peer. However, about 65% of the participants reported displeasure regarding fit and wear-ability. In another study specific to post-mastectomy patients, Nicklaus et al. reported that bra fit is currently lacking in most post-surgical bras [75]. However, a well fitted bra can significantly improve the quality of life of those who have undergone breast surgeries. The authors also stated that available products do not meet the needs of breast cancer patients. Their findings also suggested that changes should be made, for example so that the bra does not irritate the scar.

Therefore, the existence of different bra styles and varieties without a change in the production process will make no significant difference to improve bra satisfaction. It is clear from the literature that the availability of specialized (custom-designed) bras is not enough. Custom-made bras are thus thought to be a better strategy towards improving the satisfaction of individuals with atypical breast shape/size.

5. Proposed Roadmap
5.1. Selecting the Right Type of Fabric Depending on the Bra Part

The primary raw material in making a bra is fabric. The fabric type and properties are key in determining the quality and performance of the bra delivered to users. Findings from the literature show that most, if not all, commercially available tight fitting garments including bras are made from knitted fabrics [51,76]. Because of their structure, knitted fabrics have certain characteristics that make them suited to this application. They have a very high stretch with the potential to recover fully after an extension provided that the elastic limit is not exceeded [76–78]. In addition, they are highly flexible and provide significant drape over the body [79]. This makes them well suited to applications where tight fitting is required. This includes socks, compression garments, bras and breast bands [79,80]. However, for individuals with atypical breast shape/size, knitted fabrics alone may not provide the properties required for optimum bra functionality. For instance, properties such as strength may not be fulfilled by knitted fabrics. This is particularly true in areas such as the back and side panels that provide support, form and structure to the bra.

Furthermore, a complete understanding of the fabric used is needed to improve the performance of the bra [10]. As already mentioned, it is important that the fabrics used in the production of bras provide the necessary support in terms of strength [81] and have adequate flexibility to drape over the breast tissue [81,82]. In addition, they should have adequate stretch to allow for good rotation of the arm [81], be breathable and well ventilated to allow for perspiration to be evacuated [81–83], and do not cause irritation or skin allergies [75].

Lawson and Lorentzen assessed seven commercial sports bras in terms of their performance for different breast sizes (small, medium and large) [84]. They reported that subjective measures of support and comfort were not correlated: bras that provided effective support were the least comfortable and vice versa. Having the proper balance in terms of support and comfort is thus quite complex. In another study, Norris et al. evaluated commercially available sports bras for their performance and function [85]. The study involved 77 females and 94 commercial bras. They reported that the bras did not provide enough support. In addition, if bras are supposed to provide support during exercise when the breast is in motion, support requirements are not the same for everyone [18]. These authors showed that support requirements increase as the breast volume increases.

Therefore, looking into other types of fabrics such as woven structures could be very useful. Woven fabrics have high strength due to their mode of construction [86]. However, since the bras are worn next to the skin, it is important that the fabrics have some degree of flexibility to allow for ease of movement while in motion [87]. Woven fabrics by themselves are rigid, stiff and have very little to no extensibility [88]. However, the addition of elastomeric yarns during the fabric production reduces the rigidity, increases its flexibility and improves hand feel [89]. These fabrics are known as stretch woven fabrics [85].

It is thus proposed to look at the bra and its parts from a functional lens in order to decide what fabric type is most suited to which part. Figure 2 shows the different parts of a typical bra. The different bra parts can be considered as individual modules in order to optimize their functions. This will ensure that the right fabric is employed depending on the intended function of each part. Table 2 shows the different parts and the fabric types that could be the most suited in order to optimize function. In this roadmap, we propose the use of stretch-woven fabrics in areas that provide support, form and structure such as the gore (4), back panel (6) and side panel (7) (Figure 2). On the other hand, knitted fabrics could be used for areas such as the cup (3), where drape, stretch and flexibility are key factors. We hypothesize that a bra made from a combination of these fabrics will deliver a more supportive, comfortable and durable bra that is well suited to individuals with atypical breast shape/size.

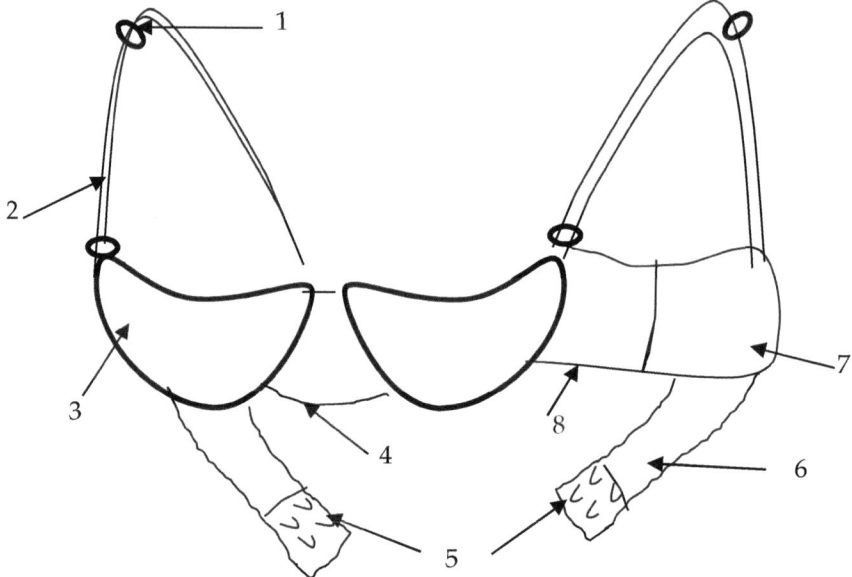

1. **Sliders**: to adjust the strap length.
2. **Straps**: link the cups to the bank panel.
3. **Cup**: hosts the breast.
4. **Gore**: connects the two cups.
5. **Hooks & eyes**: connect the two back panels & to adjust the panels
6. **Back panel**: provides support
7. **Back side panel**: enhances support & links the side & back panel.
8. **Front side panel**: inks the cup to the side panel &provides lift to the breast.

Figure 2. Front view of a typical bra.

Table 2. Different bra parts, their function, most suited fabric type, and rational for the fabric selection.

Bra Part	Function	Knitted	Stretch Woven	Rational for Fabric Selection
Strap (2)	Support the cups & back panels		✓	Good strength to ensure connections between the cup and the back panel
Cup (3)	Hold the breast	✓		Excellent drape to host the breast tissue, good breathability to prevent sweat & irritation between the breast and the upper stomach area
Gore (4)	Connect the two cups	✓		Good stretch to accommodate different breast sizes and breathability to promote ventilation
Back panel (6)	Support the side panels & straps		✓	Rigid enough to provide support
Side panel (7 & 8)	Link between the back panels & cups		✓	Good stretch and elasticity to provide good silhouette. This is especially important for those with excess skin around the armpit

Knitted and stretch woven fabrics come in varieties of styles and properties. A number of their characteristics such as weight, knit type, spandex content (in the case of stretch woven fabrics), and yarn density can impact the performance of the fabric [76,90]. Air permeability, stretch and recovery, tear strength, tensile strength, moisture management, and dimensional stability, for instance, are important in different measures depending on the bra part. The choice of the fabric type—knitted vs. stretch woven -is the first step. The next phase is to assess if these fabrics possess the required properties/performance required for the specific bra part they are intended for. Literature shows that selecting fabrics with adequate mechanical properties, comfort and durability will improve bra satisfaction [10]. The ASTM D7019 standard is a document that outlines the standard performance specifications for brassier, slip, lingerie and underwear fabrics [91]. It provides a guideline to manufacturers and other trade partners in terms of acceptable fabrics with respect to bras. Table 3 lists various mechanical performance (e.g., burst strength, tear strength and breaking strength) and durability properties (e.g., dimensional stability, pilling and color fastness) and the associated requirements found in ASTM D7019.

Table 3. Mechanical and durability requirements according to ASTM D7019 [91].

Property	Test	Requirement	
		Knitted	Woven
Mechanical	Burst strength	220N	-
	Tensile strength	-	111N
	Tear strength	-	6.7N
Durability	Pilling	Grade 4	Grade 4
	Dimensional stability	5%	5%
	Colorfastness: Laundering	Grade 4	Grade 4
	Colorfastness: Perspiration	Grade 4	Grade 4

Interestingly, the ASTM D7019 document makes no mention of comfort and stretch and recovery. However, these characteristics are critical to insure the performance of a bra and the satisfaction of the user. Stretch properties determine the clothing pressure on the skin [79] and the ability of the fabric to recover after stretch [92], and contributes highly to ease of motion [90]. It can also impact aesthetic properties and the production process in terms of cutting [82]. The stretch and recovery property of fabrics is influenced by weight, yarn type and structure amongst others [93,94]. Different standard test methods are available to assess the fabrics' stretch and recovery of textiles. This includes for example ASTM D4864, ASTM D3107, and ASTM D2594 [95–97]. The fabric comfort characteristics play a vital role as well [98,99]. Comfort properties include physical characteristics (e.g., air permeability, porosity, cover factor) and hand properties (e.g., smoothness, softness). In addition to being multifactorial, comfort is also high subjective, which makes it a relatively complex phenomenon [100]. However, researchers agree that clothing should maintain a good microclimate as a basis for providing comfort [74,101].

5.2. Bra Design

Bra design can vary greatly depending on the style and function [10]. Understanding that each part is designed to achieve a specific task is an integral part of the process. For example, the strap connects the cup and the back panel; while the back panel provides support (Table 2). As a result, the strap and back panels must have adequate strength to support the cup. Different requirements in terms of strength for the strap and back panels will exist for different cup sizes because of the weight of breast tissues. In addition, the bra should be designed in modules, such that the parts can be fitted to the individual body shape. Modularity is a concept that was imported from computer science and engineering into textiles to enhance textile design [102]. It is defined as the degree to which a product may be separated, combined and recombined [103]. The purpose is to enhance performance through versatility and accommodate differences [102].

Chen & Lapolla [104] explored the concept of modularity within Bye's "research through practice" design approach [105]. They described the benefits of modularity for textiles in terms of waste reduction in labor, time and material. They also mentioned that the infusion of modularity and design through practice optimizes garment design and fit. In this roadmap, we propose that modularity is used to optimize the overall performance of a bra beyond its design. An example of the implementation of modularity in bras design and beyond can be found in the patent by Krawchuck [106]. The bra parts are preassembled during production. They are adjusted during a one-fitting session before the final stitching is done. Designing bras as modules that are adjustable during the fitting process is a fundamental step in this roadmap. Each part of the bra is adjusted individually and relative to each other to fit different breast shapes/sizes and overall body shapes prior to final assembly.

Another important point of note in the design is interaction and compatibility. It is vital to consider how each part will interact with the others to enhance the compatibility between the different bra parts as well as the final fit, support, and comfort of the bra [102,107]. For instance, accessories such as the hook and eye (item 5 in Figure 2) and strap sliders (item 1 in Figure 2) must be compatible with the whole bra assembly. In particular, they should not cause any form of discomfort and/or limit the donning and doffing of the bra [108,109]. These accessories could potentially impact the wearability and useability of the bra. For example, in the case of an early post-surgery state, great care should be taken to favor the healing of the wound.

5.3. Customization and Fitting

Customization (also known as bespoke) is the process by which clothing is made for a specific individual [110,111]. This process takes individuality into account by tailoring a garment to a specific body type. This opposes to mass-production, where a few sizes and body types are selected to cater to all [10]. Customization is popular in suit making and less common in other forms of clothing [112]. This is due to certain limitations such as time, design and labor [53,112]. However, customized garments have been proven to offer better fit and clothing satisfaction [113].

In reality, not everyone fits into mass- manufactured goods that are based on models [111]. Custom-made bras as an alternative to mass-manufactured bras could mitigate the dissatisfaction of users [51]. Custom-made bras made with carefully selected fabrics for the individual bra parts are likely to provide maximum mechanical support and functionality. Furthermore, custom-made bras will factor in the experience of the user, such as breast surgery and weight loss. Based on the intended use of the bra, customization also allows users to determine the level of performance such as strength needed. This may be based on the type of activity, breast volume, and/or previous experience [54,55,114].

Mass-production of specialized bras such as sports bras and mastectomy bras limits their performance as it does not allow them to take into account each user's body type and/or breast shape/size. Thus, it reduces the potential to provide optimum fit, support, and comfort [113]. Therefore, we hypothesize that these bras could perform better if they were not mass-manufactured but custom-made.

A limitation to custom-made bras is higher cost and limited variety of styles [113]. However, in terms of cost, research has shown that consumers are willing to pay for value and comfort [114,115]. According to Tsorenko & Lo, custom-made bras are more comfortable [114]. Regarding styles, every woman has their "favorite bras "; these are bras that fit the most, and come in their desired style [56]). Getting one's favorite bra style may be difficult with currently available custom-made bras [69]. Thus, we hypothesis that increasing the choice of styles in custom-made bras will greatly improve the adoption of custom-made bras. In addition, the associated cost is generally not likely to be a limiting factor if there is a significant gain in fit, support, and comfort.

Finally, it is critical to differentiate between custom making and custom design as the two expressions are often used interchangeably. Custom making is bespoke, i.e., fitted to the wearer, while custom design is a unique concept, idea or innovation that can be applied to a variety of bra sizes during production.

5.4. Wear Trial

Wear trials can be conducted either by interviews or through online or paper-based surveys [116]. They have been successfully used to analyze the performance of finished goods. The purpose is to test functionality, compatibility and user-friendliness of the goods [117]. With regard to bras, very little information in terms of the results of wear trials has been documented or shared by manufacturers. However, some investigations have been performed for research purposes using mass-manufactured bras.

Bowles et al. administered a mail survey to quantify the breast support obtained from different bra styles in regular bras [118]. Their survey results showed that over 65% of the participants chose an encapsulating bra especially for intense activities, because it reduced breast motion significantly. However, in terms of fit, they went for other options such as a V-shaped bra. Furthermore, they reported that the difficulty to obtain a good fit increased with increasing breast volume. Similarly, Chan et al. conducted an interview survey with 80 women to gain women's perspective on bra design and common causes of discomfort [119]. The focus was on fit, support, cup shape, underwire, sizing, elasticity and fasteners. The survey results highlighted bra straps, cups and underwire as major areas of discomfort. Accordingly, Brown et al. carried out an online survey involving 30 marathon female athletes [55]. 75% of participants reported bra fit issues that included shoulder straps digging in, which was predominant in women with bigger breasts.

Naismith & Street compared the usefulness of a special post-surgical bra called *Cardibra* with that of a regular bra [120]. The method involved clinical trials where participants wore the bras and then answered interview questions. The results indicated that the Cardibra might have beneficial therapeutic effects on pain levels and wound healing up to day 14 after cardiac surgery due to the ability of the bra to keep the breast in position and provide the needed support. In another study, Bolling et al. used a paper-based investigator-developed survey to evaluate the satisfaction and compliance level of commercial mastectomy bras [121]. The study involved 60 post-cardiac surgery large-breasted women. They reported that the bras fell short of compliance and were uncomfortable, which made the clients dissatisfied. Furthermore, Hummel & Charlbois surveyed 20 post-cardiac surgery female patients wearing post-surgical bras for one to three weeks after surgery [73]. The participants answered survey questionnaires after each week. In their findings, over 75% of them acknowledged the usefulness of the post-surgical bra but indicated the need for fit. Additionally, Greenbaum et al. interviewed 103 women who were using a post-surgical bra and/or breast prosthesis [56]. They concluded that women who undergo surgery often experience difficulty in obtaining a well-fitted bra, which would significantly improve their quality of life after surgery. These studies show that different bra products can provide different levels of satisfaction. Thus, conducting a wear trial of products before their commercialization is important to ensure that they meet the users' needs.

Human trials thus appear as an effective way to test product performance and customer satisfaction. In addition, for individuals with atypical breast shape/size, it is critical to ensure that they obtain functional bras that respond to their needs.

5.5. Translational Approaches to Patients

The reduction in quality of life experienced by those with atypical breasts shape/size includes ill-fit, uncomfortable pressure sensations, and breast pain especially those with large breasts [39,40]. For those with large breasts, most studies have assessed the impact of bra fit and support on their overall comfort during intense activities, but rarely from a patient perspective. Individuals with voluminous breasts tend to have increased breast displacement as motion increases. McGhee et al. [122] assessed the bra-breast force and its implication on sports bra design. They found a significant correlation between breast mass and breast displacement. They concluded that developing bras that can significantly reduce breast displacement is critical to the comfort of these individuals during physical activities even in motion as low as walking [122]. Coltman et al. [19] conducted an analysis to assess which bra components contribute to incorrect bra fit in women across a range of breast sizes. Incorrect fit was predominantly due to the cups, front band and straps [19].

Regarding postmastectomy and/or mammoplasty patients, most studies have been from a clinical perspective [74]. Hummel & Charlebois [74] assessed the effectiveness of surgical support bras after a cardiac surgery. A total of 77% of the participants expressed that the bras contributed to a reduction in pain by providing support. However, some participants stated that the bras did not fit and was unwearable. They reported that appropriate training on bra use and making available a broader range of sizes would help improve fit and wear-ability [74]. Nicklaus et al. [75] assessed the need for undergarment for post-mastectomy patients as a key survivorship consideration. They concluded that ready-to-wear surgical bras do not adequately cater to the needs of postmastectomy patients.

This roadmap will have evident benefits on mastectomy patients and those undergoing various medical treatments affecting the breasts and the torso as it allows providing a more holistic perspective and improve their overall quality of life beyond the medical treatment. As illustrated in Figure 3, the process begins with determining the specific category of the intending bra user (T1), which defines the purpose of the bra. Next, the best fabric is selected for the production of the modular parts (T2), which are pre-assembled. The fitting (T3) is done within one session and allows that each client is catered for based on their specific needs. Finally, the modular pieces are stitched to their final position and the bra is delivered to the client who wears it to assess its fit, support, and comfort.

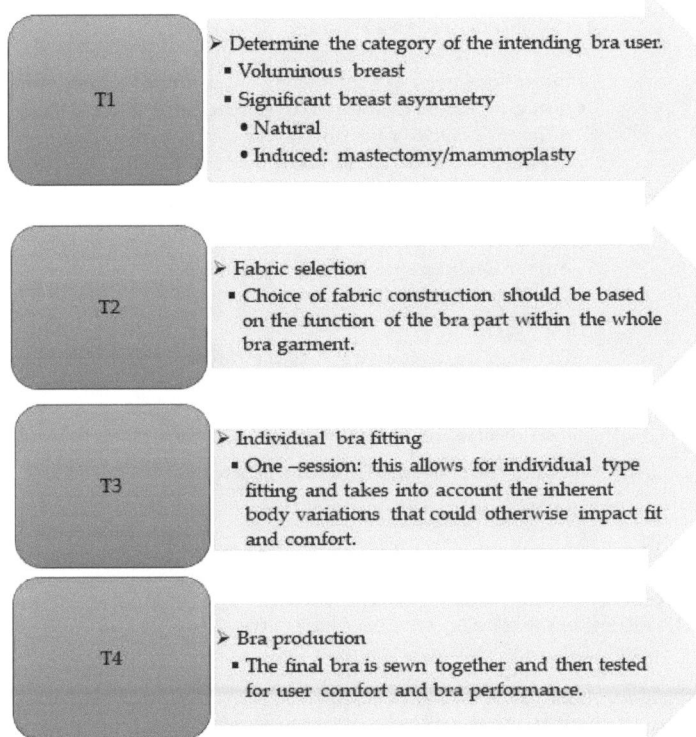

Figure 3. Translational pathway to bras well suited to those with atypical breast shape/size.

6. Conclusions

The objective of the paper is to contribute to filling the gap in terms of critical review of the current state of knowledge around the topic of bras and identify possible avenues of improvements in this area. The challenges experienced by bra users, especially those with atypical breast shape/size and body type are mainly related to issues of fit, support and comfort. As the current strategies to solve the challenges do not appear to have yield successful results yet, a need exists to re-evaluate bras from the user's perspective, especially within the context of individuals of the target audience. As a result, a road map is proposed to provide bras suited to individuals with atypical breast shape/size and body type, and improve the bra performance and functionality.

This roadmap takes the perspective of the bra user. As the problem is multidisciplinary, a human-centered interdisciplinary approach is key to ensure that all aspects are considered at all stages of the process: (1) fabric analysis, (2) bra design and patterning, (3) fitting procedure, and (4) wear trials of prototypes. The fabric should have the ability to provide the necessary mechanical support and provide breathability to increase comfort. In addition, the fabric choice should be based on the function and requirements of each bra part. Stretch woven fabrics offer a large potential in the production of bras to provide added support in areas such as the under band and back panels. Furthermore, custom-making the bra by fitting it directly on the user will increase the likelihood that optimum fit is achieved. Finally, it is important that the bra design, selected fabrics, fitting procedure, and performance of the bra as a whole assembly are tested by the intended user. This will ensure the useability, wearability, and durability of the bra.

The novelty of this review thus lies in the proposed scientific approach in the analysis of fabrics for the purpose of bra production from the lens of the bra parts and the function they perform, with the use of stretch woven fabrics to provide the needed strength and support required by bra users. This article also describes a new custom-fitting process for bespoke bras using pre-assembled modular pieces that is currently put to the trial. If current limitations exist in terms of cost and limited variety of styles, the change in paradigm brought by this new bespoke manufacturing process along with the implementation of the different aspects of the proposed roadmap, including in terms of translational approaches to patients, will hopefully allow progress to be made in improving the comfort, support and fit of bras and enhancing the quality of life of individuals with atypical breast shape/size and body type.

Author Contributions: Writing—original draft preparation, J.T.B.; writing—review and editing, P.I.D.; visualization, J.T.B.; funding acquisition, P.I.D. All authors have read and agreed to the published version of the manuscript.

Funding: This research was funded by Mitacs (Project IT17557) and Simply Best Underpinning Corp.

Institutional Review Board Statement: Not applicable.

Data Availability Statement: As this is a review paper, data is available in the articles cited.

Conflicts of Interest: The authors declare no conflict of interest.

References

1. Hawkins, A. The Evolution of the Bra. Good House Keeping Hearst Magazine Media. Available online: https://www.goodhousekeeping.com/beauty/fashion/g1291/bra-history,2015 (accessed on 28 May 2020).
2. Garber, M. The First Bra was Made of Handkerchiefs. November 2014. Available online: https://theatlantic.com/technology/archive/2014/11/the-first-bra-was-made-of-handkerchiefs/382283/ (accessed on 21 May 2020).
3. Wood, C. Bra Support Comes of Age: The History of the Bra, 1920–1930. Finery. Greater Bay Area Costumers Guild, 2013. Available online: http://gbacg.org/finery/2013/bra-support-comes-of-age-the-history-of-the-bra-1920-1930/ (accessed on 21 January 2020).
4. Farrel-Beck, J.; Gau, C. *Uplift: The Bra in America*, 1st ed.; University of Pennsylvania Press: Philadelphia, PA, USA, 2002.
5. Bengtson, B.P.; Glicksman, C.A. The standardization of bra cup measurements: Redefining bra sizing language. *Clin. Plast. Surg.* **2015**, *42*, 405–411. [CrossRef] [PubMed]
6. Sharland, E.L. The Development, Piloting, and Evaluation of a Bra Intervention for Women with larger Breasts Who Are Experiencing Breast Pain. Ph.D. Thesis, University of Portsmouth, Hampshire, UK, 2018.
7. Martin, L.J. Mastectomy. March 2019. Available online: https://www.webmd.com/breast-cancer/mastectomy (accessed on 29 May 2020).
8. Lalardrie, J.P.; Mouly, R. History of mammaplasty. *Aesthetic Plast. Surg.* **1978**, *2*, 167–176. [CrossRef]
9. Vasques, E.S.L. Auto-adjustable bra for women with a pronounced alteration in breast volume. In Proceedings of the TEI Conference, Tempez, AZ, USA, 17–20 March 2019.
10. Luk, N.; Yu, W. Bra fitting assessment and alteration. *Adv. Women's Intim. Appar. Technol.* **2016**, *53*, 110–133.
11. Vogue, N.D. The True History of the Ultimate Wardrobe Essentials–Your Bra. Available online: https://vogue.com/article/history-of-the-bra-vanity-fair-lingerie (accessed on 27 May 2020).
12. Canadian Cancer Society. Breast Cancer Statistics. Available online: https://cancer.ca/en/cancer-information/cancer-types/breast/statistics (accessed on 30 September 2021).
13. Tehrani, K. What's Trending in Breast Augmentation? Available online: https://www.plasticsurgery.org/news/blog/whats-trending-in-breast-augmentation (accessed on 3 June 2020).
14. Lindgren, J.; Dorgan, J.; Savage-Williams, J.; Coffman, D.; Hartman, T. Diet across the lifespan and the association with breast density in adulthood. *Int. J. Breast Cancer* **2013**, *2013*, 808317. [CrossRef]
15. Antonova, N.L.; Merenkov, A.T. Perceived personal attractiveness & self-improvement practices. *Chang. Soc. Personal.* **2020**, *4*, 91–106.
16. Waldman, A.; Maisel, A.; Weil, A.; Lyengar, S.; Sacotte, L.J.M.; Kurumety, S.; Shaunfield, S.L.; Reynolds, K.A.; Poon, E.; Robinson, J.K.; et al. Patients believe that cosmetic procedures affect their quality of life: An interview study of patients-reported motivations. *Am. Acad. Dermatol.* **2019**, *80*, 1671–1681. [CrossRef]
17. Spector, D.; Mayer, D.; Knafl, K.; Pusic, A. Not what I expected: Informational needs of women undergoing breast surgery. *J. Plast. Surg. Nurs.* **2010**, *30*, 70–74. [CrossRef]
18. Griffin, L.; Compton, C.; Dunne, L.E. An analysis of the variability of anatomical body references within ready-to-wear garment sizes. In Proceedings of the International Semantic Web Conference, Heidelberg, Germany, 12–16 September 2016.
19. Coltman, E.C.; Steele, J.R.; McGhee, D.E. Fit in women across a range of breast sizes. *Cloth. Text. Res. J.* **2017**, *36*, 78–90. [CrossRef]

20. Waltho, D.; McRae, M.; Thomas, A. Patient-reported measurement of breast asymmetry using Archimedes' principle in breast reduction mammaplasty: A retrospective study. *Cureus* **2020**, *12*, e6536. [CrossRef] [PubMed]
21. Reilley, A.F. Breast asymmetry: Classification and management. *Aesthetic Surg. J.* **2006**, *26*, 596–600. [CrossRef]
22. Rohrich, R.J.; Hartley, W.; Brown, S. Incidence of breast and chest wall asymmetry in breast augmentation: A retrospective analysis of 100 patients. *Cosmetics Asymmetry Breast Augment.* **2003**, *111*, 1513–1519. [CrossRef] [PubMed]
23. Myung, Y.; Choi, B.; Yim, S.J.; Yun, B.L.; Kwon, H.; Pak, C.S.; Heo, C.; Jeong, J.H. The originating pattern of deep inferior epigastric artery: Anatomical study and surgical considerations. *Surg. Radiol. Anat.* **2018**, *40*, 1187–1191. [CrossRef] [PubMed]
24. Greggianin, M.; Tonetto, L.M.; Brust-Renck, P. Aesthetic and functional bra attributes as emotional triggers. *Fash. Text.* **2018**, *5*, 1–12. [CrossRef]
25. Sridhar, G.R.; Sinha, M.J. Macromastia in adolescent girls. *Indian Pediatrics Brief Rep.* **1995**, *32*, 496–499.
26. Layon, S.A.; Pflibsen, L.R.; Maasarani, S.; Noland, S.S. Macromastia, a cause of chronic back pain. *J. Women's Health* **2021**, *30*, 1372–1374. [CrossRef]
27. Pandarum, R.; Yu, W.; Hunter, L. 3-D breast anthropometry of plus-sized women in South Africa. *Ergonomics* **2011**, *54*, 866–875. [CrossRef]
28. Zheng, R.; Yu, W.; Fan, J. Development of a new Chinese bra sizing system based on breast anthropometric measurements. *Int. J. Ind. Ergon.* **2007**, *37*, 697–705. [CrossRef]
29. Newman, T. What is a Mastectomy? MedicalNewsToday. July 2019. Available online: https://www.medicalnewstoday.com/articles/302035 (accessed on 29 May 2020).
30. Breast Cancer Organization. What to Expect? Available online: https://www.breastcancer.org/treatment/surgery/mastectomy/what-to-expect (accessed on 28 July 2022).
31. Jetha, Z.A.; Gul, R.B.; Lalani, S. Women's experience of using external breast prosthesis after mastectomy. *Asia Pac. J. Oncol. Nurs.* **2017**, *4*, 250–258. [CrossRef]
32. Kubon, T.M.; McClennen, J.; Fitch, M.I.; McAndrew, A.; Anderson, J. A mixed-method cohort study to determine perceived patient benefit in providing custom breast prosthesis. *Curr. Oncol.* **2012**, *19*, 43–52. [CrossRef]
33. Shin, K.; Leung, K.; Fred, H.; Jiao, J. Thermal & moisture control performance on different mastectomy bras & external breast prosthesis. *Text. Res. J.* **2020**, *90*, 824–837.
34. Neto, M.S.; Dematté, M.F.; Friere, M.; Garcia, E.B.; Quaresma, M.; Ferreira, L.M. Self-esteem and functional capacity outcomes following reduction mammaplasty. *Aesthetic Surg. J.* **2008**, *28*, 417–420. [CrossRef] [PubMed]
35. Schnur, P.; Schnur, D.; Hanson, T.; Weaver, A. Reduction mammaplasty: An outcome study. *Plast. Reconstr. Surg.* **1997**, *100*, 875–883. [CrossRef] [PubMed]
36. Swami, V.; Furnhan, A. Breast size dissatisfaction, but not body dissatisfaction, is associated with breast self-examination frequency and breast change detection in British women. *Body Image* **2017**, *24*, 76–81. [CrossRef] [PubMed]
37. Chopra, K.; Tadisina, K.K.; Singh, D.P. Breast reduction mammaplasty. *Interesting Cases.* **2013**, *19*, 1–4.
38. Graf, R.M.; Auersvald, A.; Bernardes, A.; Briggs, T.M. Reduction mammaplasty and mastopexy with shorter scar & better shape. *Aesthetic Surg. J.* **2000**, *20*, 99–106.
39. Xiaomeng, L. *An Investigation into the Pressure and Sensations Caused by Wearing a Bra and the Influence of These on Bra Fitting*; DeMonfort University: England, UK, 2008.
40. Scurr, J.; Hedgar, W.; Morris, P.; Brown, N. The prevalence, severity, and impact of breast pain in the general population. *Breast J.* **2014**, *20*, 508–513. [CrossRef]
41. Knackstedt, R.; Grobmyer, S.; Djohan, R. Collaboration between the breast and plastic surgeon in restoring sensation after mastectomy. *Breast J.* **2019**, *25*, 1187–1191. [CrossRef]
42. Breast Research Australia. Sports Bra Fitness. Available online: https://www.uow.edu.au/science-medicine-health/research/bra/ (accessed on 16 September 2021).
43. Rissa, A. What Causes Bra Bilge (and What Bras Fix It)? Available online: https://thebetterfit.com/what-causes-bra-bulge (accessed on 22 July 2022).
44. Vitenas, A. Let's Talk About Bra: What Causes Bra Buldge. June 2018. Available online: https://www.drvitenas.com/blog/lets-talk-about-bra-bulge#:~{}:text=Skin%20and%20fat%20can%20push,it%20or%20even%20eliminate%20it (accessed on 22 July 2022).
45. Bra Bands Dig Into Side or Back (The Bra Recyclers n.d.). Available online: https://www.amplebosom.com/blogs/bra-band-digs-into-side-or-back/ (accessed on 26 July 2022).
46. Gesauldi. Why does My Bra Cup Gap? Does it Mean it's the Wrong Size? 2022. Available online: https://glamorise.com/blogs/news/ (accessed on 25 July 2022).
47. Fong, L.T. *Mastectomy Bra and Prosthesis Design Innovation for Hong Kong Mastectomy Patients*; MPhil, The Hong Kong Polytechnic University: Hong Kong, China, 2017.
48. Wood, K.; Cameron, M.; Fitzgerald, K. Breast size, bra fit & thoracic pain in young women: A correlational study. *Chiropr. Osteopathy* **2008**, *16*, 1.
49. Bougourd, J.P.; Dekker, L.; Grant Ross, P.; Ward, J.P. A Comparison of Women's Sizing by 3D Electronic Scanning and Traditional Anthropometry. *J. Text. Inst.* **2000**, *91*, 163–173. [CrossRef]

50. Istook, C.; Hwang, S.-J. 3D body scanning system with application to the apparel industry. *J. Fash. Mark. Manag.* **2000**, *5*, 120–132. [CrossRef]
51. Shin, K. Intimate apparel: Designing intimate apparels to fit different body shapes. In *Designing Apparels for Consumers: The Impact of Body Shapes and Size*; Woodhead Publishing Limited: Cambridge, UK, 2014; p. 273.
52. Hung, P.C.-Y.; Witana, C.P.; Goonetilleke, R.S. Anthropometric measurements from photographic images. *Comput. Syst.* **2004**, 764–769.
53. White, J.; Scurr, J. Evaluation of professional bra fitting criteria for bra selection and fitting in the UK. *Ergonomics* **2012**, *55*, 704–711. [CrossRef] [PubMed]
54. Mundy, L.R.; Homa, K.; Klassen, A.F.; Pusic, A.; Kerrigan, C.L. Understanding the health burden of macromastia: Normative data for the breast-Q reduction module. *J. Plast. Reconstr. Surg.* **2017**, *139*, 846–853. [CrossRef] [PubMed]
55. Brown, N.; White, J.; Brasher, A.; Scurr, J. An investigation into breast support and sports bra use in female runners of the 2012 London Marathon. *J. Sports Sci.* **2014**, *32*, 801–809. [CrossRef] [PubMed]
56. Greenbaum, A.R.; Heslop, T.; Morris, J.; Dunn, K.W. An investigation of the suitability of bra fit in women referred for reduction mammaplasty. *Br. J. Plast. Surg.* **2003**, *56*, 230–236. [CrossRef]
57. Camelo, E.M.P.; Uchoa, D.M.; Santos-Junior, F.F.U.; Vaconcelos, T.B.; Macena, R.H.M. Use of softwares for posture assessment: Integrative review. *Coluna/Columna* **2015**, *14*, 230–235. [CrossRef]
58. Han, H.; Nam, Y.; Choi, K. Comparative analysis of 3D body scan measurements & manual measurement of size Korea adult females. *Int. J. Ind. Ergon.* **2010**, *40*, 530–540.
59. Wang, Z.; Suh, M. Bra underwire customization with 3-D printing. *Cloth. Text. Res. J.* **2019**, *37*, 281–296. [CrossRef]
60. Mitchel, N.A. Utilization of 3d Body Scanning Technology As a Research Tool When Establishing Adequate Bra Fit. Master's Thesis, Manchester Metropolitan University, Manchester, UK, 2013.
61. Chen, C.-M.; LaBat, K.; Bye, E. Physical characteristics related to bra fit. *Ergonomics* **2010**, *53*, 514–524. [CrossRef]
62. Ancutiene, K. Comparative analysis of real and virtual garment fit. *Text. Ind.* **2014**, *65*, 158–165.
63. Catanuto, G.; Spano, A.; Pennati, A.; Riggio, E.; Farinella, G.M.; Impoco, G.; Nava, M.B. Experimental methodology for digital breast shape analysis & objective surgical outcome evaluation. *J. Plast. Reconstr. Aesthetic Surg.* **2008**, *61*, 314–318.
64. Simons, P.K.; Istook, C.L. Body measurement techniques: Comparing 3D body scanning & anthropometric methods for apparel applications. *J. Fash. Mark. Manag. Int. J.* **2003**, *7*, 306–332.
65. Donelly. Why Does My Bra Gap in Front? October 2018. Available online: https://blog.parfaitlingerie.com/why-does-my-bra-gap-in-front/ (accessed on 21 July 2022).
66. Seven Big Bra Problems and How to Solve It, n.d. Eva's Intimate. Available online: https://www.evasintimates.com/blog/big-cup-bra-problems/ (accessed on 17 May 2021).
67. Lubitz, R. Here's How Hard It Is to Actually Design Beautiful Bras for People with Big Boobs. December 2015. Available online: https://www.mic.com/articles/127137/here-s-how-hard-it-is-toactually-design-beautiful-bras-for-people-with-big-boobs (accessed on 16 September 2021).
68. Bra Fit Problems, n.d. Forever Yours Lingerie. Available online: https://www.foreveryourslingerie.ca/pages/bra-fit-problems (accessed on 17 May 2021).
69. Tsarenko, Y.; Stizhakova, Y. What does a woman want? The moderating effect of age in female consumption. *J. Retail. Consum. Serv.* **2015**, *26*, 41–46. [CrossRef]
70. Chang, J.; Jang, Y. A study of sports bra by style & bra cup size. *Res. J. Costume Cult.* **2012**, *20*, 549–559.
71. Lawson, L.R. A comparison of Eight Selected Sports Bras: Biomedical Support, Overall Comfort Ratings and Overall Support Rating. Masters' Thesis, Utah State University, Logan, UT, USA, 1985.
72. Navalta, J.W.; Ramirez, G.G.; Maxwell, C.; Radzak, K.N.; McGinnis, G.R. Validity & reliability of three commercially available smart sports bras during treadmill walking and running. *Sci. Rev.* **2020**, *10*, 7397.
73. Sieradzki, A. Breast Dressing: A Critical Review of Post-Surgical Bras. Master's Thesis, City University of New York, New York, NY, USA, 2020.
74. Hummel, A.; Charlebois, A. Evaluation of surgical support bras in post-cardiac surgery female patients. *Can. J. Cardiol.* **2020**, *36*, S118. [CrossRef]
75. Nicklaus, K.M.; Bravo, K.; Liu, C.; Chopra, D.; Reece, G.P.; Hanson, S.E.; Markey, M.K. Undergarment needs after breast cancer surgery: A key survivorship consideration. *Supportive Care Cancer* **2020**, *28*, 3481–3484. [CrossRef] [PubMed]
76. Hunter, L.; Fab, J. Improving the comfort of clothing. In *Textiles and Fashion: Materials, Design, and Technology*; Woodhead Publishing Cambridge: Shaston, UK, 2015; pp. 739–761.
77. Ray, S.C. *Fundamentals & Advances in Knitting Technology*; Woodhead Publishing: New Delhi, India, 2012.
78. Ibrahim, N.A.; Khalifa, T.F.; El-hossamy, M.B.; Tawfik, T.M. Effect of knit structure and finishing treatments on functional and comfort properties of cotton knitted fabrics. *J. Ind. Text.* **2010**, *40*, 49–63. [CrossRef]
79. Amber, R.R.V.; Lowe, B.J.; Niven, B.E.; Laing, R.M.; Wilson, C.A. Sock fabrics: Relevance of fiber type, yarn, fabric structure and moisture on cyclic compression. *Text. Res. J.* **2015**, *85*, 26–35. [CrossRef]
80. Sang, J.S.; Lee, M.S.; Park, M.J. Structural effect of polyester SCY knitted fabric on fabric size, stretch properties, and clothing pressure. *Fash. Text.* **2015**, *2*, 22. [CrossRef]
81. Kaynak, H.K. Optimization of stretch and recovery properties of woven stretch fabrics. *Text. Res. J.* **2017**, *87*, 582–592. [CrossRef]

82. Tsai, I.D.; Cassidy, C.; Cassidy, T.; Shen, J. The influence of woven stretch fabric properties on garment design and pattern construction. *Trans. Inst. Meas. Control* **2002**, *24*, 3–14. [CrossRef]
83. Orgulata, R.T.; Mavruv, S. Taguchi approach for the optimization of the bursting strength of knitted fabrics. *Fibres Text. East. Eur.* **2010**, *18*, 78–83.
84. Lawson, L.; Lorentzen, D. Selected Sports Bras: Comparisons of Comfort and Support. *Cloth. Text. Res. J.* **1990**, *8*, 55–60. [CrossRef]
85. Norris, M.; Blackmore, T.; Horler, B.; Wakefield-Scurr, J. How the characteristics of sports bras affect their performance. *Ergonomics* **2020**, *64*, 410–425. [CrossRef]
86. Jahan, I. Effect of fabric structure on the mechanical properties of woven fabrics. *Adv. Res. Text. Eng.* **2017**, *2*, 1018. [CrossRef]
87. Jariyapunya, N.; Musilova, B. Analysis of stress and strain to determine pressure changes in tight fitting garments. *Autex Res. J.* **2020**, *20*, 49–55. [CrossRef]
88. Butterworth, G.A.M.; Platt, M.M.; Singleton, R.W.; Sprague, B.S. Prediction of stretch-woven fabric behavior from false-twist textured filament yarn properties. Pert I: The dependence of fabric stretch upon yarn shrinkage and stretch properties. In Proceedings of the 37th Annual Meeting of Textile Research Institute, New York City, NY, USA, 6 April 1986.
89. El-Messiry, M.; Mito, A.; Al-oufy, A.; El-Tanhan, E. Effect of fabric material & tightness on the mechanical properties of a fabric-cement composites. *Alex. Eng. J.* **2014**, *53*, 795–801.
90. Senthilkumar, M.; Anbumani, N.; Hayavadana, J. Elastane fabrics- A tool for stretch applications in sports. *Indian J. Fiber Text. Res.* **2011**, *36*, 300–307.
91. *ASTM D7019-14*; Standard Performance Specification for Brassiere, Slip, Lingerie and Underwear Fabrics. ASTM International: West Conshohocken, PA, USA, 2014.
92. Pannu, S.; Jamdigni, R.; Behera, B.K. Influence of weave design on shrinkage potential of stretch fabric. *Int. J. 360 Manag. Rev.* **2019**, *7*, 1025–1035.
93. Azim, A.Y.M.A.; Sowrov, K.; Ahmed, M.; Hassan; Fruque, A.A. Effect of elastane on single jersey knit fabric properties–physical & dimensional properties. *Int. J. Text. Sci.* **2014**, *3*, 12–16.
94. Kim, H. Mechanical properties and garment formability of PET/Spandex stretch fabric. *J. Korean Soc. Cloth. Text.* **2017**, *41*, 1098–1108. [CrossRef]
95. *ASTM D 4964*; Standard Test Method for Tension and Elongation of Elastic Fabrics (Constant-Rate-of-Extension Type Tensile Testing Machine). ASTM International: West Conshohocken, PA, USA, 2016.
96. *ASTM D 3107*; Standard Test Method for Stretch Properties of Fabrics Woven from Stretch Yarns. ASTM International: West Conshohocken, PA, USA, 2007.
97. *ASTM D 2594*; Standard Test Method for Stretch Properties of Knitted Fabrics Having Low Power. ASTM International: West Conshohocken, PA, USA, 2021.
98. Elnashar, E.A. Volume porosity and permeability in double layer woven fabrics. *AUTEX Res. J.* **2005**, *5*, 2017–2218.
99. Kumar, C.B.S.; Sivagnanam, J.S.; Kumar, B. Effect of in-lay yarn in moisture and thermal transmission properties of plaited double-knit fabric structures. *Int. J. Res. Appl. Sci. Eng.* **2020**, *8*, 770–774.
100. Wang, S.K.; Wang, L.; Li, Z.; Pan, R.; Gao, W. An objective fabric smoothness assessment method based on a multi-scale spatial masking model. *Inst. Electr. Electron. Eng.* **2019**, *7*, 73830–73840. [CrossRef]
101. Kamalha, E.; Zeng, Y.; Mwasiagi, J.I.; Kyatuheire, S. The comfort dimension: A review of perception in clothing. *J. Sens. Stud.* **2013**, *28*, 423–444. [CrossRef]
102. Baldwin, C.Y.; Clark, K.B. Modularity in the design of complex engineering systems. In *Complex Engineering Systems*; Braha, D., Minai, A., Bar-Yam, Y., Eds.; Springer: Berlin/Heidelberg, Germany, 2006; pp. 175–205.
103. Ribeiro, L.; Rui, P.; Pereira, M.; Trindade, I.; Lucas, J. Design of fashion accessories: Fabrics, modularity and technology. *Int. J. Manag. Cases* **2013**, 366–372.
104. Chen, C.; Lapolla, K. The Exploration of the Modular System in Textile and Apparel Design. *Cloth. Text. Res. J.* **2021**, *39*, 39–54. [CrossRef]
105. Bye, E. A direction for clothing and textile design research. *Cloth. Text. Res. J.* **2010**, *28*, 2015–2217. [CrossRef]
106. Krawchuck, T. A System and Method for Garment Fitting and Fabrication. U.S. Patent 8,549,763 B2, 8 October 2013.
107. Hardeker, C.H.M.; Fozzard, G.J.W. The bra design process–a study of professional practice. *Int. J. Cloth. Sci. Technol.* **2015**, *9*, 311–325. [CrossRef]
108. Chen, X.; Dong, D.; Wang, J.; Shi, H. Technology on Bra Pattern Structure Design. *Adv. Mater. Res.* **2012**, *569*, 256–259. [CrossRef]
109. Chen, C.-M.; LaBat, K.; Bye, E. Bust prominence related to bra fit problems. *International J. Consum. Stud.* **2011**, *35*, 695–701. [CrossRef]
110. Deceulaer, H. Entrepreneurs in the guilds: Ready-to-wear clothing and subcontracting in late sixteenth and early seventeenth-century antwerp. *Text. Hist.* **2000**, *31*, 133–149. [CrossRef]
111. Ghosh, S. Fashionably Intimate. June 2018. Available online: :https://www.fibre2fashion.com/industry-article/8089/fashionably-intimate (accessed on 13 July 2020).
112. Babin, B.K.; Darden, W.R.; Griffin, M. Work and/or fun: Measuring hedonic and utilitarian shopping value. *J. Consum. Resour.* **1994**, *20*, 644–656. [CrossRef]
113. Lanier, S. The Foundations of Lingerie Design and Manufacturing. December 2019. Available online: https://makersrow.com/blog/2015/05/the-foundations-of-lingerie-design-and-manufacturing/ (accessed on 13 July 2020).

114. Tsarenko, Y.; Lo, C.J. A portrait of intimate apparel female shoppers: A segmentation study. *Australas. Mark. J.* **2017**, *25*, 67–75. [CrossRef]
115. Busic, V.; Das-Gupta, R. Correctly fitted bra. *Br. Assoc. Plast. Surg.* **2004**, *57*, 588–594. [CrossRef]
116. Wang, G.; Zhang, W.; Postle, R.; Phillips, D. Evaluating wool shirt comfort with wear trials and the forearm test. *Text. Res. J.* **2003**, *73*, 113–119. [CrossRef]
117. Grujic, D.; Gersak, J. Examination of the relationship between subjective clothing comfort assessment and physiological parameters with wear trials. *Text. Res. J.* **2017**, *87*, 1522–1537. [CrossRef]
118. Bowles, K.-A.; Steele, J.R.; Munro, B. What are the breast support choices of Australian women during physical activity? *Br. J. Sports Med.* **2008**, *42*, 670–673. [CrossRef]
119. Chan, C.Y.C.; Yu, W.W.M.; Newton, E. Evaluation and analysis of bra design. *Des. J.* **2001**, *4*, 33–40. [CrossRef]
120. Naismitha, C.; Street, A. Introducing the Cardibra: A randomized pilot study of a purpose designed support bra for women having cardiac surgery. *Eur. J. Cardiovasc. Nurs.* **2005**, *4*, 220–226. [CrossRef] [PubMed]
121. Bolling, K.; Long, T.; Jennings, C.D.; Dane, F.C.; Cater, K.F. Bras for breast support after sternotomy: Patient satisfaction and wear compliance. *Am. J. Crit. Care* **2021**, *30*, 21–26. [CrossRef] [PubMed]
122. McGhee, D.E.; Steele, J.R.; Zealey, W.J.; Takacs, G.J. Bra-breast forces generated in women with large breasts while standing and during treadmill running: Implications for sports bra design. *Appl. Ergon.* **2013**, *144*, 112–118. [CrossRef] [PubMed]

Article

Turkey Red Oil as a Renewable Leveling and Dispersant Option for Polyester Dyeing with Dispersed Dyes

Jully Schmidt Pinto Filippi [1], Angelo Oliveira Silva [2], Cintia Marangoni [1,2], Jeferson Correia [2], José Alexandre Borges Valle [1] and Rita de Cassia Siqueira Curto Valle [1,*]

1. Textile Engineering Graduate Program, Federal University of Santa Catarina, Blumenau 89036-004, Brazil
2. Chemical Engineering Graduate Program, Federal University of Santa Catarina, Florianópolis 88040-900, Brazil
* Correspondence: rita.valle@ufsc.br

Abstract: The objective of this work was to evaluate Turkey red oil as a renewable dispersant and leveling option for dyeing polyester knitted fabric with disperse dyes. The dyeing results were evaluated by measuring the color at several positions of the dyed samples to verify the levelness. In addition, the amount of residual dye was evaluated. Migration tests were also carried out to evaluate the leveling effectiveness of Turkey red oil. Wet rubbing and washing fastness analysis, hydrophilicity, thermogravimetric analysis (TGA), surface analysis with scanning electron microscopy (SEM) and modification of functional groups by FTIR were also carried out. The results obtained in the analyses show that Turkey red oil is efficient as a dispersant and leveling agent when compared to the well-known sodium naphthalene sulfonate. It is concluded that Turkey red oil reduces the time of the dyeing process and consequently its energy consumption, and reduces the amount of effluent generated while improving hydrophilicity and fastness, thus being a renewable and sustainable option for current products based on petroleum.

Keywords: environmentally friendly dispersing agent; dye dispersion; equalization; textile sustainability; sustainable dyeing; turkey red oil

Citation: Filippi, J.S.P.; Silva, A.O.; Marangoni, C.; Correia, J.; Valle, J.A.B.; Valle, R.d.C.S.C. Turkey Red Oil as a Renewable Leveling and Dispersant Option for Polyester Dyeing with Dispersed Dyes. *Textiles* **2023**, *3*, 163–181. https://doi.org/10.3390/textiles3020012

Academic Editor: Young Il Park

Received: 31 March 2023
Revised: 14 April 2023
Accepted: 18 April 2023
Published: 22 April 2023

Copyright: © 2023 by the authors. Licensee MDPI, Basel, Switzerland. This article is an open access article distributed under the terms and conditions of the Creative Commons Attribution (CC BY) license (https://creativecommons.org/licenses/by/4.0/).

1. Introduction

The textile industry has faced strong competition from products imported at low prices from eastern countries. For this industry to remain competitive, textile companies have sought defensive strategies to differentiate their products [1].

As environmental pollution is a problem faced worldwide [2], there is a growing ecological concern with the textile dyeing and finishing industries, as they are responsible for high environmental impacts. This is due to high water consumption, high energy use and a large use of chemicals, many of which are dangerous to human health and the environment. For this reason, in recent years, efforts have been made to implement natural and safer molecules in these processes [3].

To guarantee the sustainable growth of the textile industry, the development of a dyeing process with ecological textiles is essential [4]. Therefore, it is important to establish environmental policies that encourage the development of new technologies that emphasize sustainability [5].

Polyester fibers, synthetic fibers formed from polymers that use resins derived from petroleum as a raw material [6,7], face a challenging dyeing process, with high energy consumption and a significant amount of water required. The most common dyes used to dye polyester fibers are disperse dyes, which are practically insoluble in water. Thus, it is necessary to use dispersing agents to keep them dispersed in water, keeping the dyes in fine dispersion and preventing the occurrence of dyeing particle coalescence. To obtain a dyeing with an adequate level of absorption, it is necessary to heat it up to temperatures close to 130 °C. Recent studies have looked for alternative processes for dyeing polyester with non-hazardous and more ecological compounds that also require less energy in the overall dyeing process [8–10].

There is a growing demand for renewable resources due to the depletion of oil reserves. Therefore, the importance of renewable resources is rapidly increasing, and one of the sources of renewable and sustainable fatty acids is seed oils, which can be used to prepare a series of oleochemicals, which replace petrochemicals for industrial applications [11].

Turkey red oil (TRO) is increasingly becoming an important bio-based feedstock for industrial applications. The presence of a hydroxyl group, a double bond, a carboxylic group and a long-chain hydrocarbon in ricinoleic acid (one of the main components of the oil) offers several possibilities to transform the compound into different materials. The oil is therefore a potential alternative to petroleum-based starting chemicals to produce materials with various properties [12].

According to Kosolia et al. [13] and Gharanjig et al. [14], there are several options under study for the replacement of derivatives of the compound sodium naphthalene sulfonate (SNS), which is a commonly used commercial dispersant derived from petroleum. Castor oil has the ability to moisten and disperse dyes, pigments and fillers [15]. Sulfonation, which is the reaction with sulfuric acid, is among the main chemical reactions of castor oil [16,17], where sulfuric acid esters are obtained, in which the hydroxyl group of ricinoleic acid is esterified, thus providing characteristics specific to the TRO, which is widely used in the textile industry [12].

The present work aims to evaluate TRO as a renewable dispersant and leveling option in polyester dyeing with disperse dye. Thus, the goal is to contribute to the community, seeking to study the variables that allow the development of a more sustainable process compared to the standard currently adopted by the industries.

2. Materials and Methods

This research used Interlock knitted fabric, trade name Dry Sport, with 100% PES (polyester) composition, average weight of 130 g/m^2, yarn count of 75 DEN/72 filaments and purged at 60 °C with 1.0 g·L^{-1} of WK Fiberclean LC 8 (anionic surfactant supplied by the company Werken Química, Indaial, Brazil), supplied by the company Texneo, Indaial, Brazil. The dye used was Large Molecule-Disperse Orange 29 Brown Colorpes D-FRL 200%. Turkey red oil (TRO) and sodium naphthalene sulfonate (SNS), supplied by Werken Química, Indaial, Brazil, were used for comparative tests as dispersants and leveling agents.

A statistical design of rotational central composite was carried out, with a factorial experiment base 2^2, with $\alpha = \sqrt{2}$, plus four points of combinations of levels −1 and +1 and the central point (Table 1) to achieve greater model robustness. The tests were carried out at a standard temperature of 130 °C and pH of 4.5–5.0, and the values obtained in the color analysis were considered as the response. All treatments were performed in triplicate. Subsequently, the study of temperature variation was carried out. Statistical analysis of results was performed using analysis of variance (ANOVA) with 5% significance, with the aid of Statistica 13.0 software.

Table 1. Levels and variables studied in the factorial design.

Variables	Level (−α)	Level (−1)	Level (0)	Level (+1)	Level (+α)
(X_1) Process Time (min)	8.8	15.0	30.0	45.0	51.2
(X_2) TRO Concentration (g·L^{-1})	0.6	1.0	2.0	3.0	3.4

The dyeing tests were performed in a Laboratory Dyeing Machine, model IR Dyer, brand Texcontrol. The dyeing process was started at room temperature, with 5 g of polyester, 1.0% owf (on weight of fiber) of dye. The material-to-liquor ratio was 1:10. As the standard sample, 2.0 g·L^{-1} of SNS was used, while for other tests, the concentration of TRO was varied, with heating at 35 °C/min until it reached 130 °C, then remaining for various process times, as shown in Table 1. After the process time, the dyeing solution was cooled and removed (which was reserved for color measurements by absorbance). A new solution was used for reductive washing, containing 2.0 g·L^{-1} of 100% sodium hydroxide and 4.0 g·L^{-1} WK Redux ECO 120, heated to 70 °C, with a gradient of 5 °C/min, remaining for 20 min at 70 °C. Then, the reductive solution was released and a new bath was added containing 0.6 g·L^{-1} acetic acid, remaining at room temperature for 10 min. The solution was discharged, and the polyester fabric was dried at 120 °C and subsequently thermofixed with a stenter at 160 °C for 30 s.

To verify the influence of temperature on the dyeing process, tests were carried out with dyeing at 120 °C and 30 min. Furthermore, to understand the exhaustion process, dyeing tests were carried out at temperatures of 80 °C, 100 °C and 115 °C. Table 2 shows the chemicals and processes.

Table 2. Products used in the dyeing process at different temperatures and washes.

Process/Products	Un	Control	Standard	TRO Tests
Dyeing				
SNS	g·L^{-1}	-	2.0	-
TRO	g·L^{-1}	-	-	2.0 *
Acetic Acid 100%	g·L^{-1}	0.3	0.3	0.3
Brown Colorpes D-FRL 200%	% owf	1.0	1.0	1.0
Process Time	min	30	30	30 *
Reductive Washing—70 °C—20 min				
WK Redux ECO 120	g·L^{-1}	4.0	4.0	4.0
Sodium Hydroxide	g·L^{-1}	2.0	2.0	2.0
Final Washing				
Room Temperature—10 min				
Acetic Acid 100%	g·L^{-1}	0.6	0.6	0.6

* Except for the dyeing at 130 °C, which varied the TRO concentration and time according to Table 1.

After the dyeing, the baths' absorbance was measured to verify the depletion of the dyeing solution and to determine the residual dye concentration.

The migration test was conducted starting from a material weight of 10 g. Initially, 50% of the sample (5 g) was placed in the dyeing equipment, with dye and at a material-to-liquor ratio of 1:10 and 100% acetic acid to adjust pH at 4.5–5.0. Then, the sample was heated with a gradient of 3.5 °C/min up to 130 °C for 30 min and cooled to room temperature. Later, without removing the solution or the dyed material, 50% of the missing material (another 5 g) was added, as well as the leveling agent under test, plus water to keep the material-to-liquor ratio of 1:10. Heating was performed again, with a gradient of 3.5 °C/min, remaining at a level of 30 min, cooled to 60 °C and the solution was released, samples were removed, dried at 120 °C and subsequently thermofixed.

A hydrophilicity test by capillarity was carried out based on the JIS L 1907:2004 standard. To determine the degree of dispersion under specific conditions in aqueous medium, the AATCC Test Method-146-2001 was used.

For the wetting test, three beakers were prepared, one with water only, one with 2.0 g·L^{-1} of SNS and the other with 2.0 g·L^{-1} of TRO (Turkey red oil). The fabric was 100% PES, with an average weight of 130 g/m^2, yarn count of 75 DEN/72 filaments, purged and cut in the size of 2 × 2 cm. The time for submersion in each of the solutions was recorded and the tests were performed in triplicate.

An alternative to measure wetting efficiency is to check how long it takes for a drop of solution placed on the textile material to be absorbed. High-efficiency wetting is when the contact angle of the droplet and textile material is zero. To facilitate the visualization, a solution with 0.1 g·L^{-1} of disperse dye (Turquoise Colorpes GL 200%) was prepared to be the blank test (control). Additionally, two more solutions were prepared, one with 2.0 g·L^{-1} of SNS and another with 2.0 g·L^{-1} of TRO. The solutions were dropped onto the

PES fabric and the time was recorded until total absorption of each drop of the solutions. The test was performed in triplicate.

The measurements were performed in a UV absorbance spectrophotometer VIG–M51 from BEL Photonics. A scan was performed to check the maximum absorption wavelength of the dye solution. Readings were taken at 5 points of different dye concentrations (10 mg·L^{-1} to 50 mg·L^{-1}) to determine the calibration curve, which was used to obtain the correlation between the absorbance reading and the dye concentration. To investigate the ionic demand of the products used in the solution (SNS and TRO), a particle charge detector equipment, Mutek PCD 05, was used.

Readings of the coloristic properties of the fabrics after dyeing were performed with the DATACOLOR 500 spectrophotometer. Datacolor Tools software version 2.3.1 was used to analyze the CIE L*a*b* coordinates, ΔE and K/S. The rubbing fastness test was carried out under the ABNT NBR ISO 105-X12 standard. Washing fastness tests were conducted according to the ABNT NBR ISO 105-C06 standard.

The morphological analysis of the samples was performed using a conventional scanning microscope with a tungsten filament, with acceleration voltage from 0.5 to 30 kV and magnification between 25× and 300,000×. For the chemical characterization of functional groups, the samples were analyzed by an ATR module, making 20 scans from 400 to 4000 cm^{-1} with a resolution of 4 cm^{-1}. To evaluate the thermal stability of polyester through a thermogravimetric analysis, Netzsch equipment, model STA 449 F3 Jupiter, was used, using an alumina crucible, gas flow of 60 mL/min, a heating rate of 10 °C/min, a nitrogen atmosphere and a temperature range from 30 to 700 °C.

In addition to the analysis of dyed substrates and dyeing solutions, comparison analyses were also carried out between SNS and TRO products through filtration dispersion tests, wetting tests and ionic charge tests. The dispersion test was based on the test method AATCC-146-2001 (dispersibility of dispersed dyes: Filter Test) and evaluated the residues present on the filters using the residue scale (Figure 1)

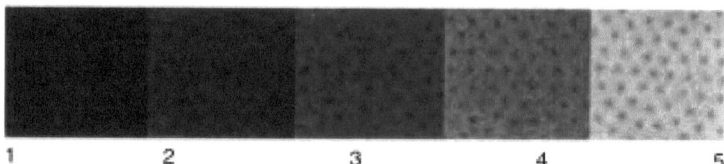

Figure 1. Filter residue scale (scale: 5: excellent, 3: intermediate values are interpolated, 1: poor).

3. Results
3.1. Dyeing

Dyeing tests were carried out at 130 °C, with 1% Orange 29 disperse dye, with standard (2.0 g·L^{-1} SNS), control (without dispersant) and different concentrations of TRO at different process times; all tests were performed in triplicate. The color measurements of the dyed polyester mesh samples were carried out with the sample folded, taking readings at four different points of the sample, from which an average value was determined. Table 3 shows the CIE coordinates L*a*b* and K/S (color strength) reading values of the dyed samples.

The ΔE values (color deviation), in relation to the standard, indicate that the color of polyester dyed at 130 °C did not show major changes for the control, for the different TRO concentrations or for different process times. Knowing that the color difference is only perceptible to the human eye with ΔE above 1, only in samples from treatments 5 and 6 would it be possible to perceive the color difference. In Figure S1 (Supplementary Material), it is possible to visualize the photos of the samples from the 13 treatments, as well as the standard and the control sample. An important observation is that the samples treated with 2.0 g·L^{-1} TRO and 15 min showed K/S values slightly higher than the standard treated with 30 min. With the K/S data, a statistical analysis was carried out to effectively verify the response of the concentration of TRO. It was observed that both the TRO concentration factor and the process time factor are significant. Tables 4 and 5 show the effects of the parameters studied and the results of the analysis of variance (ANOVA).

Table 3. CIE L*a*b* coordinates, ΔE and K/S of the dyed samples.

Treatment	Process Time (min)	TRO (g·L^{-1})	L*	a*	b*	ΔE in Relation to the Treatment with SNS	K/S
2.0 g·L^{-1} SNS	30.0	0.0	59.75 ± 0.13	38.59 ± 0.15	53.83 ± 0.09	-	123.7 ± 0.5
Control	30.0	0.0	59.73 ± 0.34	37.95 ± 0.15	53.54 ± 0.21	0.72 ± 0.29	124.6 ± 0.3
1	8.8	2.0	60.07 ± 0.21	38.88 ± 0.29	54.21 ± 0.28	0.49 ± 0.18	122.0 ± 2.5
2	15.0	1.0	60.08 ± 0.23	39.18 ± 0.50	54.46 ± 0.24	0.74 ± 0.01	122.4 ± 0.9
3	15.0	2.0	59.98 ± 0.19	39.16 ± 0.31	54.57 ± 0.23	0.86 ± 0.26	124.2 ± 1.5
4	15.0	3.0	59.84 ± 0.29	39.02 ± 0.12	54.29 ± 0.30	0.61 ± 0.13	124.3 ± 0.2
5	30.0	0.6	59.92 ± 0.28	39.20 ± 0.18	54.62 ± 0.02	1.07 ± 0.02	126.6 ± 0.1
6	30.0	1.0	59.64 ± 0.10	39.51 ± 0.09	54.90 ± 0.04	1.39 ± 0.02	127.1 ± 0.9
7	30.0	2.0	59.83 ± 0.06	39.11 ± 0.15	54.26 ± 0.16	0.80 ± 0.12	126.0 ± 0.5
8	30.0	3.0	60.10 ± 0.07	38.95 ± 0.08	54.45 ± 0.21	0.81 ± 0.27	124.4 ± 1.8
9	30.0	3.4	59.91 ± 0.15	38.99 ± 0.23	54.29 ± 0.13	0.54 ± 0.06	123.6 ± 0.9
10	45.0	1.0	60.33 ± 0.31	38.64 ± 0.30	54.35 ± 0.21	0.98 ± 0.14	124.0 ± 1.3
11	45.0	2.0	60.29 ± 0.20	38.78 ± 0.15	54.29 ± 0.05	0.78 ± 0.18	124.3 ± 0.7
12	45.0	3.0	60.53 ± 0.08	38.46 ± 0.11	54.16 ± 0.12	0.79 ± 0.05	121.7 ± 0.6
13	51.2	2.0	59.92 ± 0.27	39.03 ± 0.23	54.33 ± 0.21	0.63 ± 0.16	123.5 ± 1.3

Table 4. Effects of studied parameters and analysis of variance of TRO concentration and time.

Factor	Effect	p-Value	Error
Average Values	126.3258	0.000000	0.625462
(1) TRO Concentration (L)	−1.4944	0.016298	0.541904
TRO Concentration (Q)	−1.5359	0.093588	0.848942
(2) Time (L)	0.1353	0.808066	0.545578
Time (Q)	−4.1654	0.000344	0.866837
1 L by 2 L	−1.9975	0.035844	0.853391

Table 5. Results of the analysis of variance (ANOVA).

Factor	SS	df	mS	F	p-Value
(1) TRO Concentration (L)	11.07658	1	11.07658	7.60466	0.016298
TRO Concentration (Q)	4.76776	1	4.76776	3.27332	0.093588
(2) Time (L)	0.08953	1	0.08953	0.06147	0.808066
Time (Q)	33.63296	1	33.63296	23.09082	0.000344
1 L by 2 L	7.98001	1	7.98001	5.47870	0.035844
Lack-of-Fit	9.91707	7	1.41672	0.97266	0.489424
Pure Error	18.93517	13	1.45655		
SS Total	84.04264	25			

The ANOVA analysis demonstrates that the TRO concentration (L), time (Q) and the interaction between the studied factors are significant. It is observed that they appear as negative values, which indicates that when varying the factors from their lowest values (−1) to values of higher levels (+1), there is a reduction in the value of K/S. This result demonstrates that the values studied were properly chosen and that intermediate values of both factors are sufficient to maximize the color values of polyester dyeing using Turkey red oil (TRO) as a dispersing agent, as can be seen in Figure 2.

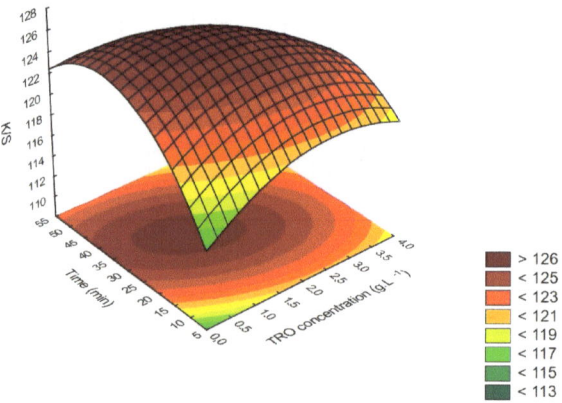

Figure 2. K/S response surface as a function of TRO concentration and treatment time at 130 °C.

Values in red in Tables 4 and 5 indicate the relevant factors at a significance level of 5%. If the p-value is greater than the significance level, the test detects no fit (lack-of-fit test). Observing the K/S (color strength) response surface as a function of TRO concentration and treatment time, it is verified that there are central points with better K/S results.

Analyzing the K/S values in Figure 2, it can be seen that the highest K/S values are found in the central conditions, between 1.0 and 2.0 g·L^{-1} TRO in the process times between 30 and 40 min, and the critical values are 1.41 g·L^{-1} of TRO and 32.3 min.

From the statistical analysis of the data, it was possible to present the statistical model that generated the response surface curve, according to Equation (1).

The R^2 value of the response surface curve was 0.6567, a value that is justified because there are deviations related to the dyeing process itself, especially in terms of washing steps. The R^2 value accounts for 65% of the variation due to the proposed factors (TRO concentration and time) and 45% can be by virtue of the fluctuation of the reductive washing. However, the efficiency of surface dye removal and the degree of influence on the properties varies significantly with the particular dye [18].

In the case of polyester, reductive washing in the laboratory has limitations in terms of reproducibility. Analyzing the residual variance, the lack-of-fit test was not significant (p-value > 0.05) for the predicted model, implying that the experimental data were well fitted.

$$Y = 126.3 - 0.747\, x_1 + 0.0676\, x_2 - 0.768\, x_1^2 - 2.0827\, x_2^2 - 0.998\, x_1 x_2 \qquad (1)$$

3.1.1. Dyeing Levelness Analysis

Dyeing levelness indicates the efficiency of dye absorption in the process. The smaller the presence of undesirable stains, the higher the quality of the product.

To evaluate the uniformity of the dyeing, color measurements were performed on the dyed samples at four different points (four quadrants): upper front, lower front, upper back and lower back; then, the average was determined.

All measurements performed on the dyed samples showed ΔE below 1, in relation to the average of the sample itself, indicating that the dyeing is even. According to the results obtained for ΔE (color deviation) and K/S of the dyed samples, it was decided to select the dyeing samples at 130 °C for 30 min of treatment with 2.0 g·L^{-1} of TRO (Treatment 7), from the standard (SNS) and the control (without dispersant), for the comparison of results using the same amount of product. Table 6 shows the readings by quadrants of the samples as well as the ΔE values and mean standard deviation of each sample, and Figure 3 shows the residual dyeing solutions of these treatments.

Table 6. Color readings by quadrants. Note: [a] ΔE was calculated concerning the mean of its own sample.

SNS—Sample 01	L*	a*	b*	ΔE [a]
1st Quadrant Reading	59.73	38.11	53.07	0.47
2nd Quadrant Reading	60.27	38.04	53.42	0.30
3rd Quadrant Reading	60.04	38.26	53.75	0.31
4th Quadrant Reading	59.93	38.19	53.61	0.17
Average of 4 Quadrants	59.99	38.15	53.46	0.31
Mean Standard Deviation	0.23	0.10	0.29	0.13
SNS—Sample 02	**L***	**a***	**b***	**ΔE [a]**
1st Quadrant Reading	59.75	38.30	53.60	0.22
2nd Quadrant Reading	59.93	38.52	53.89	0.23
3rd Quadrant Reading	59.99	38.25	53.58	0.20
4th Quadrant Reading	60.00	38.34	53.87	0.16
Average of 4 Quadrants	59.92	38.35	53.74	0.20
Mean Standard Deviation	0.12	0.12	0.17	0.03
SNS—Sample 03	**L***	**a***	**b***	**ΔE [a]**
1st Quadrant Reading	59.90	38.13	53.65	0.22
2nd Quadrant Reading	59.75	38.12	53.65	0.08
3rd Quadrant Reading	59.46	38.24	53.88	0.35
4th Quadrant Reading	59.60	38.03	53.32	0.33
Average of 4 Quadrants	59.68	38.13	53.63	0.25
Mean Standard Deviation	0.19	0.09	0.23	0.13
Control—Sample 01	**L***	**a***	**b***	**ΔE [a]**
1st Quadrant Reading	59.07	37.96	53.15	0.27
2nd Quadrant Reading	59.10	38.05	53.35	0.24
3rd Quadrant Reading	59.15	37.68	53.28	0.19
4th Quadrant Reading	59.58	37.69	53.55	0.44
Average of 4 Quadrants	59.23	37.85	53.33	0.28
Mean Standard Deviation	0.24	0.19	0.17	0.11
Control—Sample 02	**L***	**a***	**b***	**ΔE [a]**
1st Quadrant Reading	59.66	37.41	53.10	0.08
2nd Quadrant Reading	59.47	37.40	53.07	0.20
3rd Quadrant Reading	59.20	37.85	53.36	0.64
4th Quadrant Reading	60.21	37.01	53.16	0.70
Average of 4 Quadrants	59.64	37.42	53.17	0.40
Mean Standard Deviation	0.43	0.34	0.13	0.31
Control—Sample 03	**L***	**a***	**b***	**ΔE [a]**
1st Quadrant Reading	59.22	37.58	53.20	0.27
2nd Quadrant Reading	59.02	38.00	53.47	0.30
3rd Quadrant Reading	59.67	37.83	53.63	0.52
4th Quadrant Reading	58.94	37.72	53.24	0.31
Average of 4 Quadrants	59.21	37.78	53.39	0.35
Mean Standard Deviation	0.33	0.18	0.20	0.11
TRO—Sample 01	**L***	**a***	**b***	**ΔE [a]**
1st Quadrant Reading	59.41	39.46	54.53	0.52
2nd Quadrant Reading	59.82	39.20	54.39	0.06
3rd Quadrant Reading	60.00	38.99	54.46	0.25
4th Quadrant Reading	60.04	38.97	54.35	0.30
Average of 4 Quadrants	59.82	39.16	54.43	0.28
Mean Standard Deviation	0.29	0.23	0.08	0.19

Table 6. *Cont.*

TRO—Sample 02	L*	a*	b*	ΔE [a]
1st Quadrant Reading	59.90	38.63	54.13	0.28
2nd Quadrant Reading	59.75	39.06	54.51	0.41
3rd Quadrant Reading	60.14	38.66	54.27	0.19
4th Quadrant Reading	60.23	38.76	54.47	0.26
Average of 4 Quadrants	60.01	38.78	54.35	0.29
Mean Standard Deviation	0.22	0.20	0.18	0.09
TRO—Sample 03	**L***	**a***	**b***	**ΔE [a]**
1st Quadrant Reading	59.81	39.35	54.72	0.44
2nd Quadrant Reading	59.96	39.35	54.58	0.30
3rd Quadrant Reading	60.21	38.71	54.21	0.48
4th Quadrant Reading	60.13	38.95	54.26	0.25
Average of 4 Quadrants	60.03	39.09	54.44	0.37
Mean Standard Deviation	0.18	0.32	0.25	0.11

Figure 3. Residual dyeing solution: control, SNS and treatment 7—130 °C.

The concentrations of the dyeing effluents were determined from the measurement of the absorbance at a wavelength of λ = 310 for all treatments, were is possible to compare visually the dye concentration (Figure 4). The absorbances of the dispersants/leveling agents in water and without dye, and of the solutions with 2.0 g·L^{-1} of SNS and 0.6 g·L^{-1}, 1.0 g·L^{-1}, 2.0 g·L^{-1} and 3.4 g·L^{-1} of TRO were also measured.

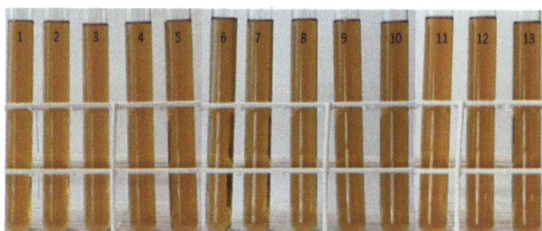

Figure 4. Residual dyeing solution from treatments 1 to 13—130 °C.

With the absorbance data of the residual dyeing solutions, deducted from the value obtained from the solutions containing only the dispersants, the actual residual concentration of dye in each solution was calculated, as shown in Table 7.

Table 7. Residual dye concentration obtained from absorbance reading.

Treatment	Plateau Time (min)	TRO $(g \cdot L^{-1})$	$g \cdot L^{-1}$ of Residual Dye According to Absorbance Reading	Mean Standard Deviation
2.0 $g \cdot L^{-1}$ SNS	30.0	0.0	0.41	0.01
Control	30.0	0.0	0.41	0.00
1	8.8	2.0	0.40	0.00
2	15.0	1.0	0.43	0.00
3	15.0	2.0	0.41	0.01
4	15.0	3.0	0.42	0.01
5	30.0	0.6	0.42	0.01
6	30.0	1.0	0.41	0.01
7	30.0	2.0	0.40	0.00
8	30.0	3.0	0.43	0.01
9	30.0	3.4	0.44	0.01
10	45.0	1.0	0.40	0.00
11	45.0	2.0	0.40	0.01
12	45.0	3.0	0.40	0.01
13	51.2	2.0	0.40	0.00

As seen in Figure 3, it is observed that the residual dyeing solution with SNS presents greater color intensity. This is due to the color depth of the SNS. The comparison between the solutions with dispersants (without dye) with the concentration of the dye is shown in Table 8.

Table 8. Comparison of the dispersant with the residual dye concentration.

Solution	Residual Dye $(g \cdot L^{-1})$	Mean Standard Deviation
2.0 $g \cdot L^{-1}$ SNS	1.50	0.01
0.6 $g \cdot L^{-1}$ TRO	0.01	0.00
1.0 $g \cdot L^{-1}$ TRO	0.01	0.00
2.0 $g \cdot L^{-1}$ TRO	0.01	0.00
3.0 $g \cdot L^{-1}$ TRO	0.02	0.00
3.4 $g \cdot L^{-1}$ TRO	0.02	0.00

It is observed that when adding SNS, there is a significant increase in the color of the effluent, while with the addition of TRO, this increase is extremely small; considering the mean standard deviations of the solutions, it can be considered insignificant.

3.1.2. Dyeing at a Temperature of 120 °C—30 min

Dyeing tests were performed at 120 °C for 30 min, with 1% Orange 29 Disperse Dye, with standard (2.0 $g \cdot L^{-1}$ SNS), control (without dispersant) and 2.0 $g \cdot L^{-1}$ of TRO. Color strength (K/S) measurements of the dyed polyester mesh samples were taken at four different points with the folded sample, and then an average value was determined. In Figure 5, it is possible to visualize photos of the dyed fabric samples dyed at 120 °C and the residual dyeing solutions.

The results obtained for K/S and dye concentration in the residual dyeing solutions at 120 and 130 °C for 30 min are shown in Figures 6 and 7, respectively.

Analyzing the results, it is possible to verify that with the decrease in temperature, there is a loss of K/S, both in the standard treatment with SNS (sodium naphthalene sulfonate—control) and in the treatment with TRO (Turkey red oil). This demonstrates that there is lower dye incorporation with decreasing temperature, which is corroborated by the dye concentration in the residual solution. At 120 °C, there was an increase in the dye concentration in the residual solution.

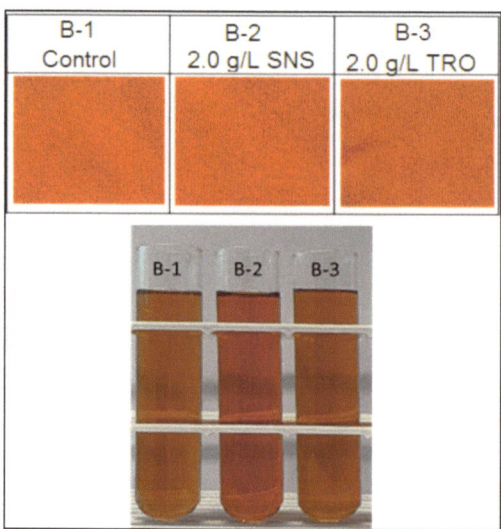

Figure 5. Photos of the knitted fabrics and the residual dyeing solutions at 120 °C—30 min.

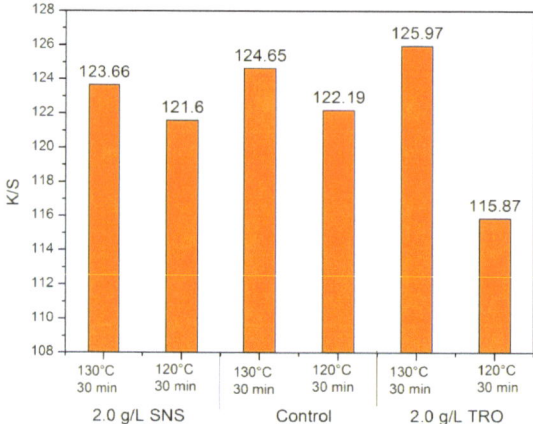

Figure 6. K/S values of dyeing samples at 130 and 120 °C.

3.1.3. Dyeing at Temperatures of 80, 100 and 115 °C

In Figure S2 (Supplementary Material), it is possible to visualize the fabrics dyed at 80, 100 and 115 °C, as well as the residual solutions of dyeing at 80, 100 and 115 °C. Table 9 presents the K/S data and dye concentration in the residual solutions.

Figures 8 and 9 show the dye adsorbed and K/S values for treatments with 2 g·L^{-1} of SNS and TRO, along with the control. It is observed that with the increase in temperature, there was greater adsorption of dye and an increase in K/S for the control test and SNS. However, it is also observed that for the control sample at lower temperatures, the adsorption of the dye was greater than the standard.

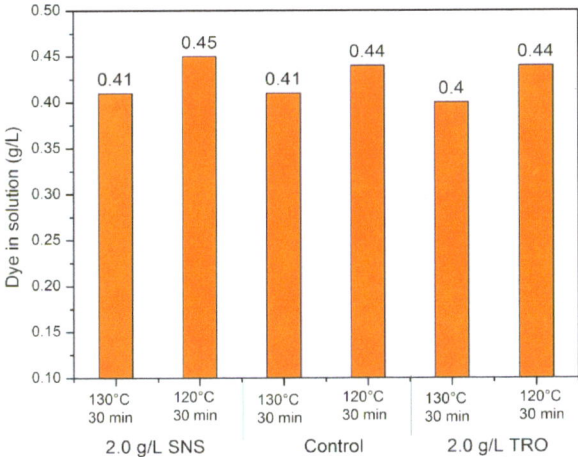

Figure 7. Dye concentration in the residual dyeing solutions at 130 and 120 °C.

Table 9. K/S results and dye concentration in residual solutions for dyeing at 130 °C for 30 min and 80, 100 and 115 °C.

Process	Treatment	K/S	Mean Standard Deviation	$g \cdot L^{-1}$ of Residual Dye	Mean Standard Deviation
130 °C—30 min	2.0 $g \cdot L^{-1}$ SNS	123.7	0.48	0.41	0.01
	Control	124.6	0.26	0.41	0.00
	2.0 $g \cdot L^{-1}$ TRO	126.0	0.46	0.40	0.00
115 °C	2.0 $g \cdot L^{-1}$ SNS	67.7	0.10	0.71	0.01
	Control	73.9	0.39	0.61	0.01
	2.0 $g \cdot L^{-1}$ TRO	71.7	0.48	0.64	0.01
100 °C	2.0 $g \cdot L^{-1}$ SNS	21.5	0.37	0.86	0.01
	Control	28.3	0.63	0.76	0.01
	2.0 $g \cdot L^{-1}$ TRO	24.0	0.49	0.81	0.01
80 °C	2.0 $g \cdot L^{-1}$ SNS	15.1	0.47	0.91	0.01
	Control	17.0	0.55	0.89	0.01
	2.0 $g \cdot L^{-1}$ TRO	13.0	0.36	0.92	0.01

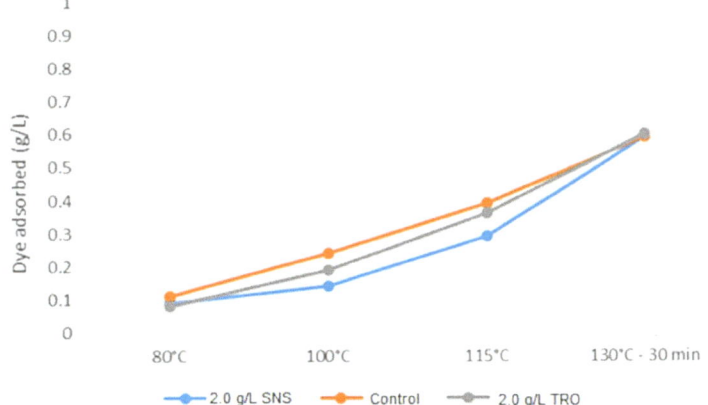

Figure 8. Adsorbed dye concentration for dyeing at 130 °C for 30 min and 80, 100 and 115 °C.

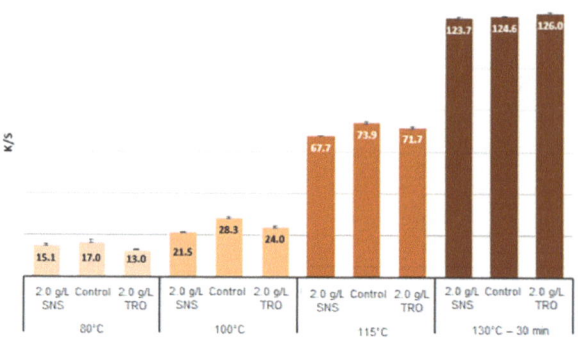

Figure 9. K/S values for dyeing at 80 °C, 100 °C, 115 °C and 130 °C for 30 min.

The association between dye and the leveling agent is established through a balance that gradually dissolves. This happens with the release of dye molecules as they are fixed on the fiber while the equilibrium is changed, considering a free behavior in the dyeing solution. The leveling agent that has an affinity with the fiber is, in general, a molecule smaller than the dye, capable of diffusing more easily inside the fiber. Therefore, at the beginning of the dyeing process, the leveling agent occupies the freest access points in the fiber. The presence of leveling agents is normally transitory, and when the dye reaches its position, it moves and becomes fixed due to the greater affinity it has with the fiber. Dye fixation is consequently delayed, promoting a bigger migratory effect.

3.2. Wet Rubbing and Washing Fastness

The analyses were conducted for wet rubbing fastness of the SNS (sodium naphthalene sulfonate) standard samples, control and samples treated with 2.0 g·L^{-1} of TRO (Turkey red oil). Figure S3 (Supplementary Material) presents photos of the staining on cotton and polyester fabrics. According to the results of Table 10, the TRO did not negatively interfere with the rubbing fastness—it even showed a slight improvement in the CO fabric.

Table 10. Assigned grades to the samples for wet rubbing and washing fastness using a grayscale.

Treatment	Rubbing		Washing		
	Witness CO	Witness PES	Witness CO	Witness PES	Color Alteration
Control	4	4/5	4	5	5
2.0 g·L^{-1} SNS	4	4/5	4	5	5
2.0 g·L^{-1} TRO	4/5	4/5	4/5	5	5

Washing fastness analyses of standard SNS samples, control and samples treated with 2.0 g·L^{-1} TRO were also conducted, as shown in Figure S4. Again, it is evidenced that TRO does not negatively interfere with washing fastness; instead, it shows a slight improvement in the CO.

3.3. Scanning Electron Microscopy (SEM)

Through the SEM technique, four samples were analyzed: without dyeing, control, SNS standard (2.0 g·L^{-1} SNS dyed at 130°C—30 min) and TRO test (2.0 g·L^{-1} TRO dyed at 130 °C—30 min). The images in Figure 10 were obtained with 1000× magnification, making it possible to see that the morphology of the samples is very similar. They present small particles deposited on the fibers, which is probably due to the presence of oligomers. As these particles are evidenced both in the undyed and control samples, as well as in the SNS standard and TRO tested samples, it is evident that both the SNS and the TRO did not show efficiency for the removal of oligomers. In Figure S5, the micrographs are shown at 3000× magnification.

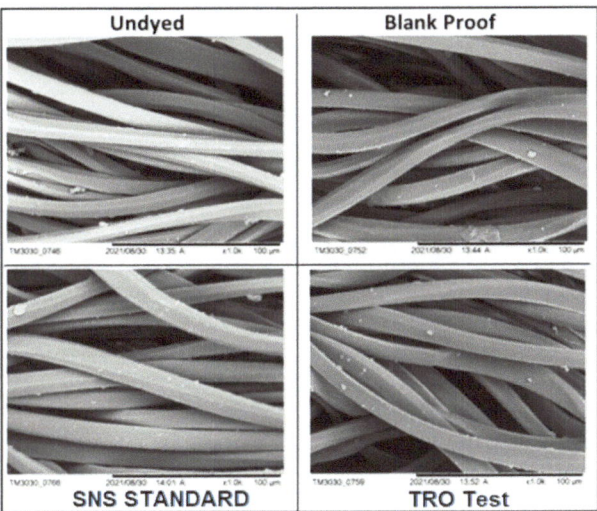

Figure 10. SEM images of fabrics at 1000× magnification.

3.4. Fourier Transform Infrared Spectroscopy (FTIR)

FTIR tests were performed to identify structural characteristics, mainly regarding functional groups and bonds presented in the fabrics. There were four samples tested: without dyeing, control (dyeing without dispersant), standard SNS (dyeing with 2.0 g·L^{-1} SNS) and TRO test (dyeing with 2.0 g·L^{-1} TRO), with dyeing performed at 130 °C for 30 min.

The spectra presented in Figure 11 have the characteristic peaks associated with the polyester fiber, according to the evaluations already presented in the literature by Assis [19]. No significant structural change is observed through the FTIR analysis in the treated samples when compared to the sample without dyeing.

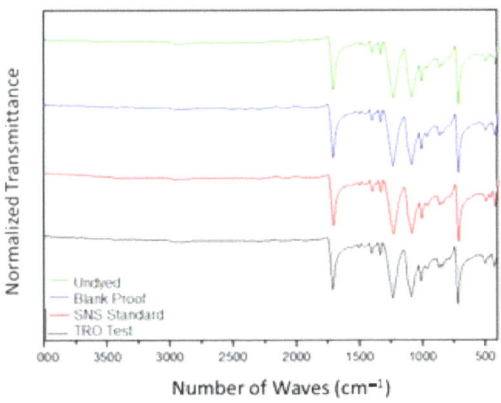

Figure 11. FTIR analysis—comparative chart between the samples.

3.5. Thermogravimetric Analysis (TGA)

Thermogravimetric analysis (TGA) is widely used to investigate the temperature degradation of polymers such as polyester and to determine the kinetic parameters of thermal degradation [18,20,21]. It is used to gain a better understanding of the thermal stability of polymers. Figure 12 shows the graphs with the results obtained in the TGA analysis of the samples: without dyeing, control, SNS standard and TRO test (sample treated with 2.0 g·L^{-1} of TRO at 130 °C for 30 min). The final mass loss percentage results, as well as the onset (extrapolated start of the thermal event) and endset (extrapolated end of the thermal event) values for these samples, are summarized in Table 11.

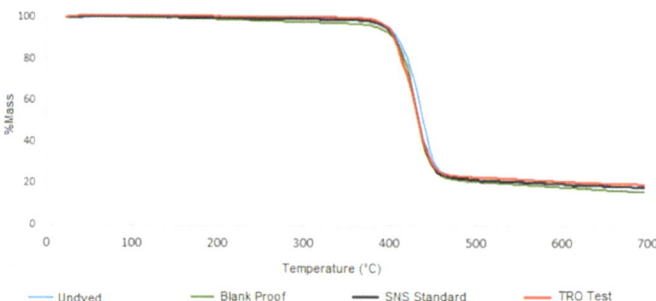

Figure 12. TGA analysis—comparative chart between samples.

Table 11. TGA analysis results—onset, endset and mass loss (%).

Analysis/Samples	Undyed	Control	SNS Standard	TRO Test
Onset Temperature	413.8 °C	410.4 °C	403.9 °C	404.3 °C
Endset Temperature	452.9 °C	447.7 °C	449.6 °C	449.5 °C
% Mass Loss (695 °C)	81.54%	83.94%	81.75%	80.15%
% Mass Residual (695 °C)	18.53%	16.13%	18.22%	19.85%

It can be observed that the thermal stability is very similar among the samples. The highest proportion of mass loss occurs for all samples from 400 °C. The TRO test sample was the one that presented the lowest onset temperature, but it also presented the lowest percentage of final total mass loss.

In addition to the direct thermogravimetric relationship (% mass loss x temperature), the DTG graph was also made to visualize the data more precisely [22]. Figure 13 shows the DTG curves (%/min) of the undyed, blank proof, SNS standard and TRO test samples.

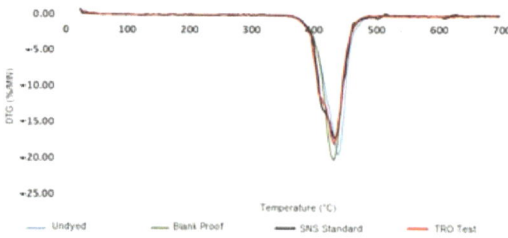

Figure 13. DTG of the fabrics.

With the TG/DTG data, it is possible to verify that the treatment with TRO did not interfere negatively with the thermal stability of the polyester fabric, and even presented a lower percentage of mass loss when compared to other samples.

3.6. Migration and Levelness

Faced with dyeing in ideal situations in the laboratory, with excellent levelness results, such as those presented in the previous tests, the migration test has the objective of verifying the real potential of a leveling agent in a situation of little equalization, such as cases of reprocessing. Figure 14 shows photos of reprocessing treated with TRO and the standard with SNS, making it possible to visually verify the difference between the test without a leveling agent and the test with the SNS standard.

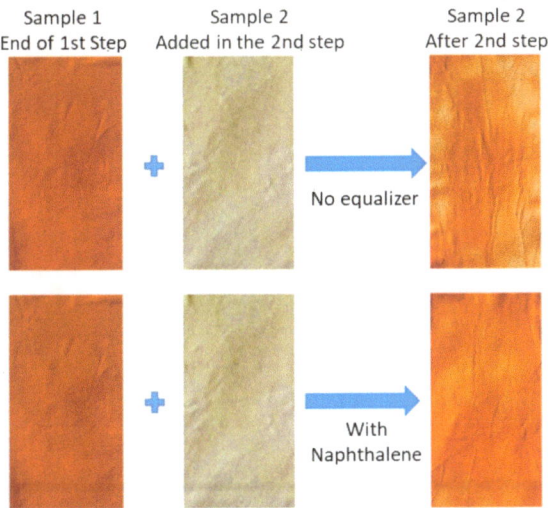

Figure 14. Migration test without equalizer x SNS standard.

The tests were performed with 2.0 g·L^{-1} of sodium naphthalene sulfonate and different concentrations of TRO (0.6 g·L^{-1}, 1.0 g·L^{-1}, 2.0 g·L^{-1}, 3.0 g·L^{-1} and 3.4 g·L^{-1}), plus a control sample (second step without the addition of leveling agent). In Figure 15, it is possible to visualize the photos of the reprocessed fabrics (after the second step) of the control sample with an uneven aspect. The standard (SNS) has a more even aspect than the samples with TRO, where a significant improvement in levelness is observed for the samples with 2.0 g·L^{-1}, 3.0 g·L^{-1} and 3.4 g·L^{-1}.

Color measurements of the dyed polyester fabric samples were taken in four different areas of the samples (that is, four quadrants: upper front, lower front, upper back and lower back). The readings were performed at four different points in each area, from which an average value is shown in Table 12. It is possible to verify that the TRO presents better performance than the standard leveling agent SNS at the same concentration. The average of ΔE specific to the SNS standard is equal to 2.22, while the test with 2.0 g·L^{-1} of TRO presented an average ΔE of 0.86.

Table 12. Average values of ΔE and K/S of the samples for the migration test.

Samples	Overall Average Own ΔE	Overall Average K/S
Control	4.41	67.4
SNS	2.22	77.0
0.6 g·L^{-1} TRO	2.18	77.9
1.0 g·L^{-1} TRO	1.29	91.7
2.0 g·L^{-1} TRO	0.86	94.4
3.0 g·L^{-1} TRO	1.01	103.7
3.4 g·L^{-1} TRO	1.31	107.3

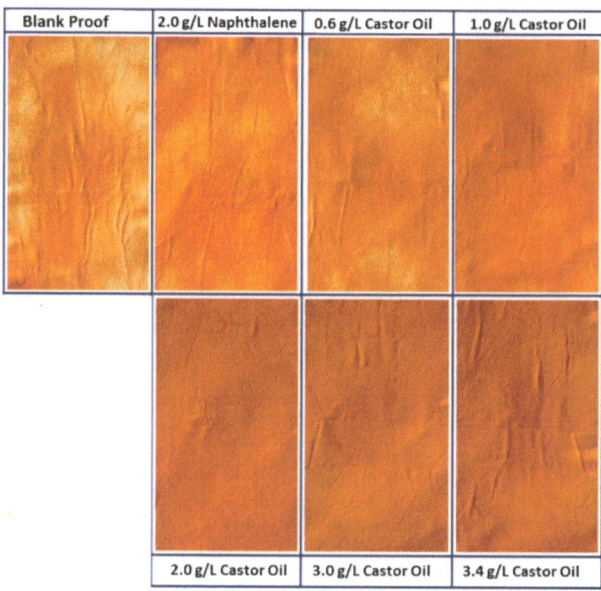

Figure 15. Migration test with reprocessed fabric samples.

3.7. Dispersion, Hydrophilicity, Wetting and Ionic Demand Tests

To evaluate the efficiency of the dispersants, dispersibility tests were conducted by measuring the filtering time of dye solutions, with and without dispersant. The residues present in the filters were evaluated under standard filtering conditions. The results obtained are shown in Table 13 and Figure 16.

Table 13. Dispersion, wetting and ionic demand tests.

Sample	Dispersion Test Time (s)	Wetting Test		Ionic Demand (eq/g)
		Drop Absorption	Submersion	
Dye Only	152 ± 20	15 s	10 s	-
Standard—10 g·L^{-1} SNS	107 ± 13	15 s	10 s	−2465 ± 24
10 g·L^{-1} TRO	88 ± 10	1 s	2 s	−1206 ± 21

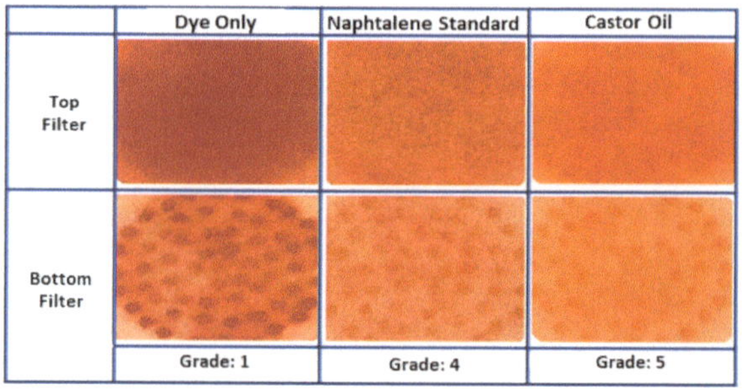

Figure 16. Filters with filtration residues.

According to the scale, the control sample presented a score of 1, where a large presence of dye particles on the filter is verified. On the other hand, the filter with SNS presented a score of 4, and the filter with TRO presented a score of 5, related to a lesser presence of dye particles on the filter.

With these results, it is possible to state that the TRO has a very satisfactory dispersing power, even presenting better results than the SNS standard.

Hydrophilicity analysis by capillarity of the SNS standard, control and samples treated with 2.0 g·L^{-1} TRO was carried out. It is possible to see, through Figure 17, a greater and more uniform capillarity in the sample with TRO. The results demonstrate the efficiency of TRO as a surfactant due to the evenness and better capillarity.

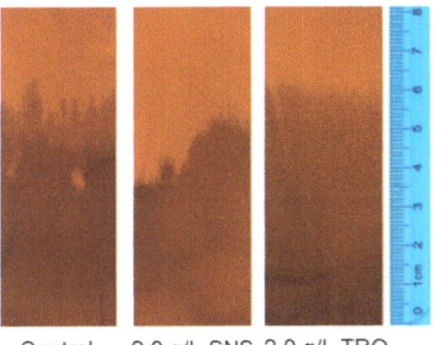

Control   2.0 g/L SNS   2.0 g/L TRO

Figure 17. Capillarity of control samples, SNS standard and 2.0 g·L^{-1} TRO.

Wetting tests were also carried out on standard SNS samples, control and samples treated with 2.0 g·L^{-1} TRO. Each test verified how long it takes for a drop of the solution placed on the textile material to be absorbed. As shown in Figure 18 and Table 13, the TRO presented the best result.

Solution	Contact Time 1 second	Contact Time 5 seconds	Contact Time 10 seconds	Contact Time 15 seconds
Blank Proof Dye only		●	●	●
STANDARD DYE with 2.0 g/L SNS		●	●	●
DYE TEST with 2.0 g/L TRO		●	●	●

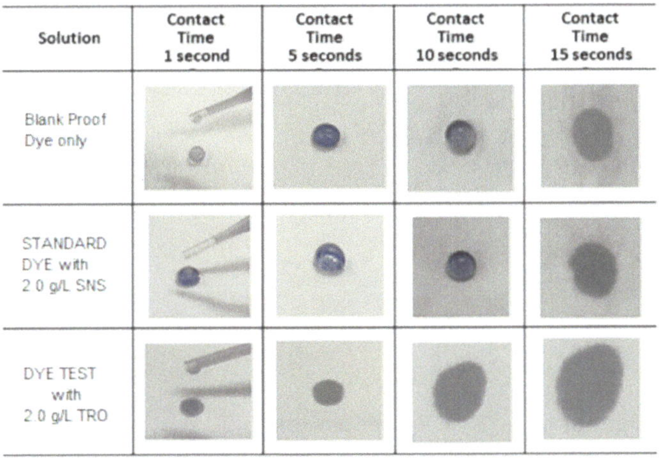

Figure 18. Wetting power: drop test.

Another wetting test was performed by submersion; as shown, the TRO was also the one that presented the best result. With these results presented, it is possible to demonstrate and confirm the efficiency of TRO as a surfactant, with good wetting power for polyester fibers.

The results obtained in the ionic demand tests carried out using the Mutek equipment are shown in Table 13, where it is observed that both the SNS and the TRO are presented as strongly anionic products. The SNS standard presents a higher ionic demand value than TRO. However, TRO also has a significant ionic demand value, thus confirming that it is an anionic dispersant. This confirms that anionic surfactants have the necessary ability to disperse dyes.

4. Conclusions

The analyses conducted demonstrate the possibility of replacing a petroleum-based product with a renewable alternative in the process of dyeing polyester with disperse dye. TRO was also better than SNS in terms of residual coloration of the dyeing solution, generating an effluent with a lower dye concentration. Additionally, the TRO performed better in the filtration tests.

For the dyeing tests at 130 °C, all measurements performed on dyed samples showed ΔE below 1, indicating good evenness. Moreover, higher K/S values were obtained with TRO compared to TRO under a given condition.

In this sense, with the results of dyeing, with variations in the concentration of TRO, time and temperatures, the levelness and dispersion tests show that the TRO presents efficient power in both dispersing and equalizing when compared to the standard sodium naphthalene sulfonate. It even presented the potential to use a smaller amount of product or decrease the processing time, which leads to lower energy consumption. In the evaluation of the residual dyeing solutions, the results show that the solutions with TRO present less color residue in the effluent than the solutions with SNS.

As for the ionic demand, the TRO showed a significant value, thus confirming that it is an anionic dispersant.

According to the analysis of washing and wet rubbing fastness of the dyed fabrics, the TRO even showed an increase in these properties in comparison to the SNS standard or the control. With the TGA technique, it was verified that the treatment with TRO presented a lower percentage of mass loss when compared to other samples. According to the hydrophilicity analysis by capillarity and wetting, the surfactant power of TRO stood out, with an increase in the hydrophilicity power of the material after dyeing and the wetting potential in solution; this was not evidenced with the SNS.

Therefore, it is concluded that TRO presents promising results for application as a dispersant and leveling agent for polyester dyeing, thus being a renewable and sustainable option for petroleum-based products.

Supplementary Materials: The following supporting information can be downloaded at: https://www.mdpi.com/article/10.3390/textiles3020012/s1, Figure S1: Photos of the fabric samples from treatments at 130 °C; Figure S2: Fabrics and residual dyeing solutions at 80 °C, 100 °C and 115 °C; Figure S3: Samples of cotton and polyester after wet rubbing fastness test; Figure S4: Fabric samples after the washing fastness; Figure S5: SEM images of samples at 3000× magnification.

Author Contributions: Conceptualization, J.S.P.F., J.C., J.A.B.V. and R.d.C.S.C.V.; methodology, J.S.P.F., J.C., J.A.B.V. and R.d.C.S.C.V.; validation, J.S.P.F., A.O.S., C.M., J.C., J.A.B.V. and R.d.C.S.C.V.; formal analysis, J.S.P.F., A.O.S., C.M., J.C., J.A.B.V. and R.d.C.S.C.V.; investigation, J.S.P.F., J.C., J.A.B.V. and R.d.C.S.C.V.; writing—original draft preparation, J.S.P.F., A.O.S., C.M., J.C., J.A.B.V. and R.d.C.S.C.V.; visualization, J.S.P.F., A.O.S., C.M., J.C., J.A.B.V. and R.d.C.S.C.V.; supervision, C.M., J.C., J.A.B.V. and R.d.C.S.C.V.; project administration, C.M., J.C., J.A.B.V. and R.d.C.S.C.V. All authors have read and agreed to the published version of the manuscript.

Funding: This research was funded by Coordenação de Aperfeiçoamento de Pessoal de Nível Superior (CAPES, Brazil), grant number 01.

Data Availability Statement: The data used to support the findings of this study are included within the article and its Supplementary Material.

Acknowledgments: The authors are grateful to the Coordenação de Aperfeiçoamento de Pessoal de Nível Superior (CAPES, Brazil, grant number 01) and to the Laboratório Central de Microscopia Eletrônica (LCME-UFSC) for the SEM images.

Conflicts of Interest: The authors declare no conflict of interest.

Abbreviations

PES	polyester
FTIR	fourier transform infrared spectroscopy
TGA	thermogravimetric analysis
SEM	scanning electron microscopy
TRO	turkey red oil
SNS	sodium Naphthalene Sulfonate
K/S	color strength
ΔE	Color deviation

References

1. Lucato, W.; Vieira Junior, M.; Vanalle, R.M.; Silva, R.C. Gerenciamento da transferência internacional de tecnologia: Estudo de caso na indústria têxtil brasileira. *Gestão Produção* **2015**, *22*, 213–228. [CrossRef]
2. Dilarri, G.; de Almeida, É.J.R.; Pecora, H.B.; Corso, C.R. Removal of dye toxicity from an aqueous solution using an industrial strain of saccharomyces cerevisiae (Meyen). *Water Air Soil Pollut.* **2016**, *227*, 1–11. [CrossRef]
3. Pasquet, V.; Perwuelz, A.; Behary, N.; Isaad, J. Vanillin, a potential carrier for low temperature dyeing of polyester fabrics. *J. Clean. Prod.* **2013**, *43*, 20–26. [CrossRef]
4. Varadarajan, G.; Venkatachalam, P. Sustainable textile dyeing processes. *Environ. Chem. Lett.* **2016**, *14*, 113–122. [CrossRef]
5. Assis AH, C.; Munaro, M. Melhoria no processo de tingimento de fibras de poliéster após hidrólise por enzima lipase. *Rev. Evidência* **2016**, *15*, 113–128. [CrossRef]
6. Barbosa, M.C.; Rosa, S.E.S.; Correa, A.R.; Dvorsak, P.; Gomes, G.L. Setor de fibras sintéticas e suprimento de intermediários petroquímicos. *Complexo Têxtil BNDESSet. Rio DeJan.* **2004**, *20*, 77–126.
7. Miúra, M.; Munoz, S.P.V. *Manual Técnico Têxtil e Vestuário: Fibras Têxteis*; SENAI: São Paulo, Brazil, 2015.
8. Radei, S.; Carrión-Fité, F.J.; Ardanuy, M.; Canal, J.M. Kinetics of low temperature polyester dyeing with high molecular weight disperse dyes by solvent microemulsion and agrosourced auxiliaries. *Polymers* **2018**, *10*, 1–11. [CrossRef] [PubMed]
9. Chakraborty, J.N. *Fundamentals and Pratices in Colouration of Textiles*, 2nd ed.; Woodhead Publishing India: New Delhi, India, 2014.
10. Gomes, J.N.R. *Corantes Dispersos, Química Qualidade Materiais Têxteis*; Universidade do Minho: Minho, Portugal, 2008.
11. Kamalakar, K.; Satyavani, T.; Mohini, Y.; Prasad, R.B.N. Synthesis of thumba, castor e sal fatty ethanolamide-based anionic surfactants. *J. Surfactants Deterg.* **2013**, *17*, 637–645. [CrossRef]
12. Mubofu, E.B. Castor oil as a potential renewable resource for the production of functional materials. *Sustain. Chem. Process.* **2016**, *4*, 1–12. [CrossRef]
13. Kosolia, C.; Varka, E.M.; Tsatsaroni, E. Effect of surfactants as dispersing agents on the properties of microemulsified inkjet inks for polyester fibers. *J. Surfactants Deterg.* **2011**, *14*, 3–7. [CrossRef]
14. Gharanjig, H.; Gharanjig, K.; Khosravi, A. Effects of the side chain density of polycarboxylate dispersants on dye dispersion properties. *Color. Technol.* **2019**, *135*, 160–168. [CrossRef]
15. Dehankar, P.B.; Bhosale, V.A.; Patil, S.U.; Dehankar, S.P.; Deshpande, D.P. Turkey red oil from castor oil using sulphonation process. *Int. J. Eng. Res. Technol.* **2017**, *10*, 293–296.
16. Jadhav, Y.; Deshpande, S.D.; Akash, M.; Sindhikar, A. Manufacturing of sulphated castor oil (turkey red oil) by sulphonation process. *Int. J. Adv. Res.Sci. Eng.Technol.* **2018**, *5*, 6384–6389.
17. de Paula Queiroga, V.; da Silva, O.R.R.F.; da Cunha Medeiros, J. *Tecnologias Utilizadas no Cultivo da Mamona (Ricinus communis) Mecanizada*; AREPB: Béziers, France, 2021; 228p.
18. ul Aleem, A.; Christie, R.M. Christie, The clearing of dyed polyester. Part 1. A comparison of traditional reduction clearing with treatments using organic reducing agents. *Color. Technol.* **2016**, *132*, 280–296. [CrossRef]
19. Assis, A.H.C. Avaliação Das Mudanças Ocorridas Em Fibras de Poliéster Submetidas a Tratamento Alcalino e Enzimático. Master's Thesis, Universidade Federal do Paraná, Curitiba, Brazil, 2012; 116p.
20. Xu, S.; Chen, J.; Wang, B.; Yang, Y. An environmentally responsible polyester dyeing technology using liquid paraffin. *J. Clean. Prod.* **2016**, *112*, 987–994. [CrossRef]
21. Ferreira, B.T.M.; Espinoza-quiñones, F.R.; Borba, C.E.; Módenes, A.N.; Santos, W.L.F.; Bezerra, F.M. Use of the β-cyclodextrin additive as a good alternative for the substitution of environmentally harmful additives in industrial dyeing processes. *Fibers Polym.* **2020**, *21*, 1266–1274. [CrossRef]
22. Zohdy, M.H. Cationization and gamma irradiation effects on the dyeability of polyester fabric towards disperse dyes. *Radiat. Phys. Chem.* **2005**, *73*, 101–110. [CrossRef]

Disclaimer/Publisher's Note: The statements, opinions and data contained in all publications are solely those of the individual author(s) and contributor(s) and not of MDPI and/or the editor(s). MDPI and/or the editor(s) disclaim responsibility for any injury to people or property resulting from any ideas, methods, instructions or products referred to in the content.

Article

Use of Rotary Ultrasonic Plastic Welding as a Continuous Interconnection Technology for Large-Area e-Textiles

Christian Dils [1,*], Sebastian Hohner [1] and Martin Schneider-Ramelow [2]

1. Fraunhofer IZM (Institute for Reliability and Microintegration), 13355 Berlin, Germany
2. Microperipheric Center, Technical University Berlin, 10623 Berlin, Germany
* Correspondence: christian.dils@izm.fraunhofer.de; Tel.: +49-30-46403-208

Abstract: For textile-based electronic systems with multiple contacts distributed over a large area, it is very complex to create reliable electrical and mechanical interconnections. In this work, we report for the first time on the use of rotating ultrasonic polymer welding for the continuous integration and interconnection of highly conductive ribbons with textile-integrated conductive tracks. For this purpose, the conductive ribbons are prelaminated on the bottom side with a thermoplastic film, which serves as an adhesion agent to the textile carrier, and another thermoplastic film is laminated on the top side, which serves as an electrical insulation layer. Experimental tests are used to investigate the optimum welding process parameters for each material combination. The interconnects are initially electrically measured and then tested by thermal cycling, moisture aging, buckling and washing tests, followed by electrical and optical analyses. The interconnects obtained are very low ohmic across the materials tested, with resulting contact resistances between 1 and 5 mOhm. Material-dependent results were observed in the reliability tests, with climatic and mechanical tests performing better than the wash tests for all materials. In addition, the development of a heated functional prototype demonstrates a first industrial application.

Keywords: electronic textiles (e-textiles); interconnection technology; ultrasonic welding

Citation: Dils, C.; Hohner, S.; Schneider-Ramelow, M. Use of Rotary Ultrasonic Plastic Welding as a Continuous Interconnection Technology for Large-Area e-Textiles. *Textiles* **2023**, *3*, 66–87. https://doi.org/10.3390/textiles3010006

Academic Editor: Laurent Dufossé

Received: 22 December 2022
Revised: 12 January 2023
Accepted: 25 January 2023
Published: 28 January 2023

Copyright: © 2023 by the authors. Licensee MDPI, Basel, Switzerland. This article is an open access article distributed under the terms and conditions of the Creative Commons Attribution (CC BY) license (https://creativecommons.org/licenses/by/4.0/).

1. Introduction

Textiles have gained ubiquitous importance due to their scalability in manufacturing and numerous beneficial mechanical and protective properties. The first electrically heatable textiles and patents have been known since 1910 [1]. However, intensive research into the integration of electronics in textiles has only been carried out since the late 1990s [2] fueled by the progressive miniaturization of electronics and new material developments in combination with the megatrends of pervasive computing and the internet of things. Jumping to today, a large business growth for electronic textiles (e-textiles) is predicted within the next decade [3], whereby the potential application areas are constantly expanding [4] and are not limited by geometric sizes. Wicaksono et al. predict that the scale of future e-textiles will range from microns to kilometers [5]. Today, e-textiles are already used for flexible heaters [6], wearable antennas [7] or large-area illumination [8], and also for monitoring bio-signals [9], vital data [10] and in situ structural health monitoring of composite materials [11]. Yet, there are still challenges in the production of large-area e-textiles that stand in the way of further market penetration. These include, in particular, textile-compatible, scalable integration and interconnection technologies [12], process automation for cost reduction [13] and also improved washability [14] and sustainability [15].

Since textiles and electronics differ greatly in their material properties and dimensions, standard processes from electronics manufacturing cannot be readily transferred to textiles. For example, although some metallic textile conductors can be soldered even with low contact resistances, these brittle connections do not provide reliable mechanical contact. In recent years, numerous more textile-compatible interconnection processes have been developed, but most of them focus on contacting rigid electronic components or assemblies by means of interposers on a flexible textile substrate [16]. New e-textile production developments include, for example, the ZSK RACER 1W embroidery machine with an integrated circuit board laying unit, which can be used to realize automated embroidered contacts for a rigid interposer with conductive yarns [17]. Another solution was presented by Fraunhofer IZM, where a large-format bonder was developed [18], enabling the electrical and mechanical integration of multi-I/O modules on a textile substrate of up to 1 m × 1 m by non-conductive adhesive bonding [19].

While science and industry have shown that it is possible to realize scalable and reliable textile integration of various types of hard components through novel joining technologies, purely fiber- or yarn-based interconnects still represent a major challenge for the development of textile-based electronic systems. In textile-heavy e-textiles, the electronic function is essentially provided by the textile, which provides more comfort and reliability than integrated systems made of hard and soft materials. Applications include the realization of conductive tracks and sensors made of conductive yarns, the integration of RFID chips directly into the yarn [20], or the development of fiber-based electronic, optoelectronic, energy-collecting, energy-storing and sensory devices that can be integrated into multifunctional e-textile systems [21].

Dhawan et al. presented the first results on studies of fiber-to-fiber contacts [22]. They integrated copper yarns orthogonally into a plain weave and modeled and experimentally determined the contact resistances at the crossing points of the conductive yarns. They found a strong influence of an acting force, which can also occur due to the movement of the textile, on the value of the contact resistance and recommended a further joining process to realize a mechanically stable interconnection. Using resistance welding (RW), they were able to achieve a crossover point interconnect resistance of 31 mOhm. Zhang, for example, uses the pressure-dependent contact resistance of multiple yarn-to-yarn contacts to develop a knitted strain sensor [23]. Suchý et al. describe that embroidered hybrid conductive yarns contacted by RW achieve a contact resistance of 10 mOhm [24]. The contacts showed stable values after dry heat, temperature shock and bending tests, but failed after washing tests. To optimize the washability, they propose additional encapsulation of the brittle contacts. For this purpose, thermoplastic films were subsequently applied to the contacts by ultrasonic welding or by lamination. While the ultrasonic welding damaged the brittle contacts to the point of failure, the laminated-over contacts were found to reliably withstand the washing tests [25]. Locher presents a different approach, in which a woven fabric was developed with intersecting copper wires insulated from each other by polyester threads [26]. Individual junctions, called textile vias, are first exposed by a laser ablation process. A drop of conductive adhesive is applied to the ablated area connecting the two crossing wires, and then epoxy resin is added as mechanical and electrical protection for the joint. The DC contact resistance obtained is 14.1 mOhm.

To reduce the very labor-intensive, multi-step process required to achieve a mechanically and electrically stable yarn-to-yarn contact, researchers have investigated ultrasonic plastic welding (UPW). UPW is a very efficient and ultra-fast joining process that is already established in the textile and packaging industries. Since many textiles have thermoplastic components, or thermoplastics can be easily added to an e-textile system in numerous forms, for example, as an electrical insulation layer or film-like structure, the otherwise necessary stripping, interconnection and encapsulation steps can be reduced to a single ultrasonic welding step. Slade and Winterhalter first described this approach for developing a selectively enabled wiring in textiles toolkit as follows: "This welding process melts the insulation around the wires and the polymer material (e.g., nylon or polyester) in the yarn, allowing the conductive cores of the wires to come into contact with each other. When the flow of ultrasonic energy into the weld site is stopped the polymer material surrounding the wires rapidly cools and hardens. As a result, the conductive cores of the wires remain locked in contact with each other, insulated from the environment. The result is a durable, easily achieved interconnection amongst selected conductors in an E-textile fabric, garment, or textile article. By forming an ultrasonic weld at a defined point, conductive elements within the fabric can be made to form permanent connection with one another, for instance across a seam or between warp and weft yarns" [27]. The first experimental studies on ultrasonically welded contacts for e-textiles were only published in recent years. Thurner describes the UPW of electrically conductive adhesive nonwovens with silver-coated yarns, where low contact resistances of 10 mOhm were obtained [28]. It is also reported that the welded contacts can withstand high ampacity loads of 10 A and mechanical tests such as bending, flexing or torsion for 10,000 cycles without major resistance changes. Micus et al. investigated the UPW for contacting a knitted conductive textile with copper stranded wires, measuring a contact resistance of 1.95 Ohm [29]. After the washing tests, the contact resistances increased threefold, which the authors attributed mainly to the damage of the selected silver-coated yarns. Dils et al. report on electrical and mechanical contacting of rigid interposers with silver-metallized nonwovens using a thermoplastic film and UPW [30]. By adapting the contact structures and process parameters, they realized electrical contacts in the range < 20 mOhm, providing material-dependent results in reliability. Another study by Dils et al. investigated the interconnection of two embroidered hybrid conductive yarns at their crossing point [31]. In their study, by adapting the embroidered contact pad design and using an experimental setup, process parameters for different material combinations could be investigated. Simultaneous encapsulation of the contacts was achieved by using a TPU cover layer pre-placed over the crossing point. The average contact resistance of the samples was below 2.5 mOhm and showed no changes even after mechanical and environmental reliability tests.

In addition to the presented investigations using ultrasonic spot welding for yarn-to-yarn interconnects, there are also publications on the continuous ultrasonic plastic welding (CUPW) process for e-textiles but they only investigate the bonding of conductive materials on textile substrates. Atalay et al. use the CUPW process to integrate various conductive yarns into a waterproof polyester fabric as signal lines for e-textiles applications [32]. They examined the influence of the welding parameters on the conductivity and seam strength of the selected materials. Leśnikowski investigated the textile integration of nickel-coated polyester strips as a single or double layer using CUPW to fabricate transmission signal lines [33]. He reported that direct welding caused damage in the conductive tape or short circuits that did not occur when a double-sided textile outer layer was added. Furthermore, the samples made in this way allowed the construction of signal lines capable of transmitting DC and AC signals at frequencies up to several hundred MHz.

The aim of this work is to investigate and analyze the continuous interconnection of textile conductor materials by CUPW. To the best of our knowledge, this is the first study on the subject. Based on the results from the presented studies on spot interconnects of crossing embroidered hybrid conductive yarns as well as the literature review, we target the use of metal strand-based textile conductors, since damage to metal-coated yarns and fabrics has often been reported. Several additional materials are investigated as adhesion agents and simultaneous electrical and mechanical insulation of the contacts, for example thermoplastic films without conductive particles, in order to consider requirements for resource efficiency and reparability. For large-area textile-based electronic systems, we focus on integrating embroidered conductors into the textile substrate and welding a conductive ribbon orthogonal to them, so that the multiple conductors can make electrical and mechanical contact with each other.

2. Materials and Methods
2.1. Welding Tools

The welding process was carried out on an 8312 flatbed ultrasonic machine from PFAFF Industriesysteme und Maschinen GmbH, which is equipped with a 10 mm wide steel sonotrode wheel. The ultrasonic generator has a working frequency of 35 kHz and a maximum power of 800 W. On the machine, a gap dimension of 0–2 mm with a fine adjustment of 1/50 mm can be set. The parameters welding speed (0.5–20.0 m/min), welding force (0–400 N) and amplitude (from 18 μm to 34 μm for the sonotrode used (or 50–100%)) can also be adjusted. A knurled steel wheel with a width of 10 mm (PFAFF article number 95-256 126-05) was used as the anvil, which achieved the best welding results in preliminary tests. A pyramid wheel also showed suitability for welding but achieved inferior electrical results compared with the knurled wheel and produced wavy structures in the welded specimens. The smooth, zigzag or circular shaped anvil wheels did not produce repeatable mechanical and electrical connections in different test materials. Figure 1a shows a detailed view of the topography of the knurled wheel, and Figure 1b shows the welding of a conductive ribbon onto a conductive knit with the ultrasonic machine.

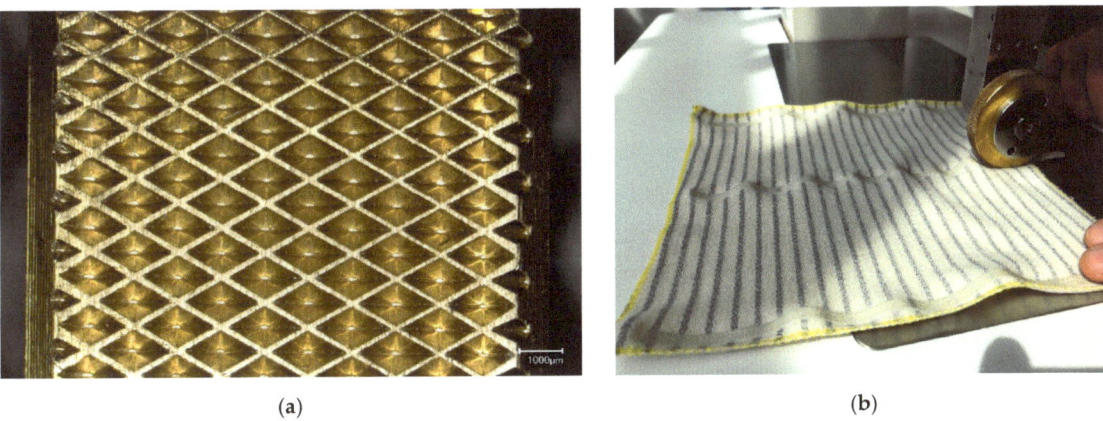

(a) (b)

Figure 1. Ultrasonic welding tools: (**a**) detailed image of the surface structure of the knurled wheel; (**b**) anvil wheel of the Pfaff 8312 welding machine and welded specimen.

2.2. Substrate Materials

Elastic knitted fabrics were more difficult to process than non-elastic woven textiles and shrank during welding. Even the use of embroidery hoops to mechanically fix the elastic textiles did not lead to any improvements. Therefore, woven substrates were used for the trials in this study. Cotton is particularly suitable as a textile substrate material, as textiles made of polymers can melt during ultrasonic welding, which could lead to an undesired stiffening of the welding area. Therefore, a 100% plain weave cotton fabric with a weight per unit area of 190 g/m^2 was selected as the substrate.

Electrically conductive yarns were integrated into the textile substrate, with heating, conductor tracks, or sensors as possible functions. In total, four commercially available conductive yarns from the supplier VÚB a.s. and its Clevertex brand were selected for testing. All the yarn variants are characterized by the fact that they consist of a twisted yarn of textile polymer fibers and metal strands and are, therefore, referred to as hybrid conductive yarns. The hybrid conductive yarns are marketed as suitable for embroidery, knitting and weaving. Table 1 gives an overview of the four selected yarns, their composition and their properties according to the information provided by the manufacturer Clevertex.

One yarn type was available in two versions, without (Y08) and with thermoplastic insulation (Y09). Another yarn (Y12) was made of silver-metallized polymer threads in addition to silver-plated copper strands, which makes it more ductile overall. The last yarn selected (Y13) consisted of high-impedance stainless steel filaments.

2.3. Conductive Ribbon

While there are numerous weldable adhesive tapes, no electrically conductive ones are known. One exception was the conductive hotmelt adhesive non-woven e-Web 140 from the manufacturer imbut GmbH, which is available in rolls. However, the material cannot be processed with an ultrasonic welding device as it was observed in our tests that using it damaged or destroyed both the conductive silver coating on the polymer fibers and the nonwoven material. We suspect that this is due to the impact of ultrasonic vibrations on the material. Due to this lack of suitable welding tape, we therefore, looked for conductive tapes or ribbons that are made weldable by means of a subsequent lamination with thermoplastic film.

Table 1. Properties of the selected conductive yarns as provided by the manufacturer Clevertex.

ID	Material Composition	Coating/Insulation	Linear Resistance (Ω/m)	Fineness (tex)	Diameter (mm)	Dry Strength (cN/tex)	Dry Elongation (%)
Y08	4× PES threads, 8 × 0.03 mm Cu/Ag filaments	None	2.85	78	0.24	21.85	13.3
Y09	4× PES threads, 8 × 0.03 mm Cu/Ag filaments	TPA	3.3	76	0.23	22.07	13.4
Y12	4 × 110 dtex/f24 SilverStat threads, 4 × 0.03 mm Cu/Ag filaments	None	6.6	84.9	0.29	16.11	53
Y13	4× PES threads, 4 × 0.02 mm stainless steel filaments	None	560.5	37.6	0.22	46.91	15

PES: polyester; TPA: thermoplastic polyamide; Cu: copper; Ag: silver.

Four different types of conductive ribbons with a linear resistance significantly lower than 1 Ohm/m were selected for the experiments, which are presented in Table 2.

Table 2. Properties of the selected conductive ribbons.

ID	Manufacturer	Material Composition	Linear Resistance (Ω/m)	Width (mm)	Thickness (mm)
Elasta Vestil	ELASTA VESTIL a.s.	weft: 12/88 PET threads; warp: 34× Y08 conductive yarns	0.14 (0.11 when embedded into TPU film)	8	0.5
High Flex 3981	Karl Grimm GmbH & Co. KG	33 × 1 PET threads (74 dtex) stranded with thin Cu/Ag foil	0.41 (0.42 when embedded into TPU film)	4	0.2
Amotape 3587	AMOHR Technische Textilien GmbH	PET threads, hotmelt yarns and 6× Cu/Ag strands	0.07 (0.09 when embedded into TPU film)	9	0.54
Amotape 46050	AMOHR Technische Textilien GmbH	weft: 6× Cu/Ag strands, hotmelt yarns and PET threads (84 dtex); warp: PET threads (140 dtex)	0.03	9	0.6

PET: polyethylene terephthalate; TPU: thermoplastic polyurethane.

Amotape 3587 was only used in the first trial, and Amotape 46050 is an optimized version and was only used in the following main trial. The main difference between the two versions is that in the second variant, the six conductive strands cross each other regularly to improve electrical characteristics and that it incorporates further hotmelt yarns which make an additional lamination of the TPU film unnecessary. Besides the electrical, material and size differences, the ribbons partially differ in the textile production. Except for the High Flex ribbon, which is braided, the other three ribbons are all woven. X-ray and cross-sectional images of the conductive ribbons are shown in Figures 2–6.

For sufficient bonding between textile-integrated conductor materials and the conductive ribbon, an additional adhesive was needed. In our study, we laminated an adhesive film to the conductive ribbon prior welding. Covestro Platilon 4201 AU polyurethane (ether) film in 100 μm thickness was used as an adhesive [34]. The material has proven itself in recent years as a suitable substrate material for soft and stretchable circuit boards [35–37] as well as an adhesion agent for e-textile non-conductive adhesive bonding technology [38]. Before the welding process, the TPU film was cut to a width of 10 mm and laminated to the conductive ribbon on both sides using a Sefa HP45 2PS thermal press at 180 °C and 2 bar for 45 s. The laminate was afterwards cooled under pressure for a few minutes.

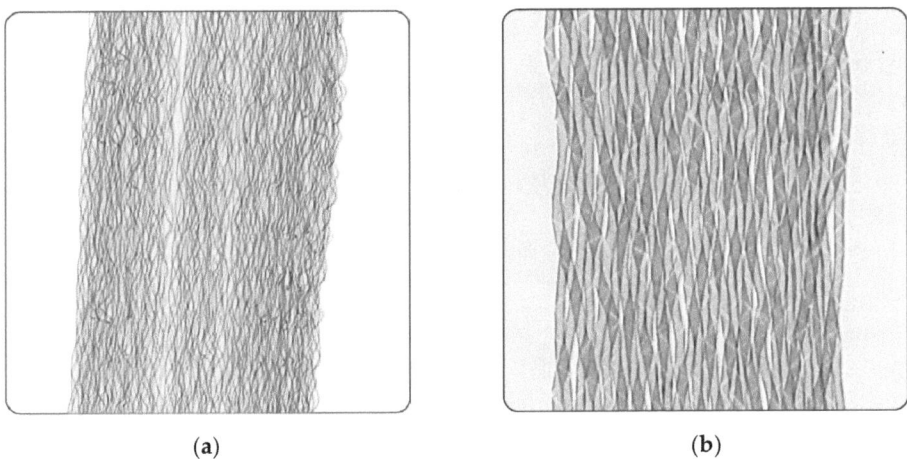

Figure 2. X-ray images of the selected conductive ribbons: (**a**) Elasta Vestil; and (**b**) High Flex 3981.

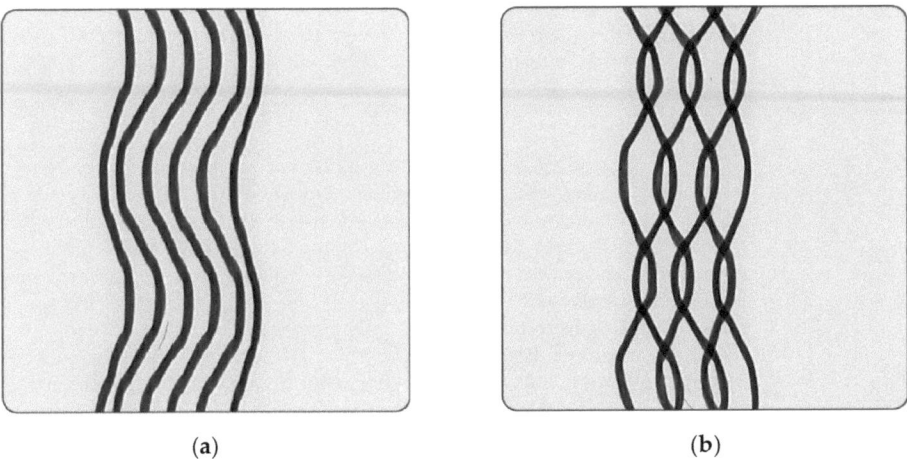

Figure 3. X-ray images of the selected conductive ribbons: (**a**) Amotape 3587; and (**b**) Amotape 46050.

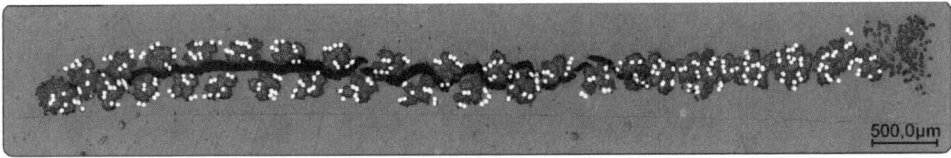

Figure 4. Cross-section image of the Elasta Vestil woven ribbon with 34 hybrid conductive yarns distributed over the entire area.

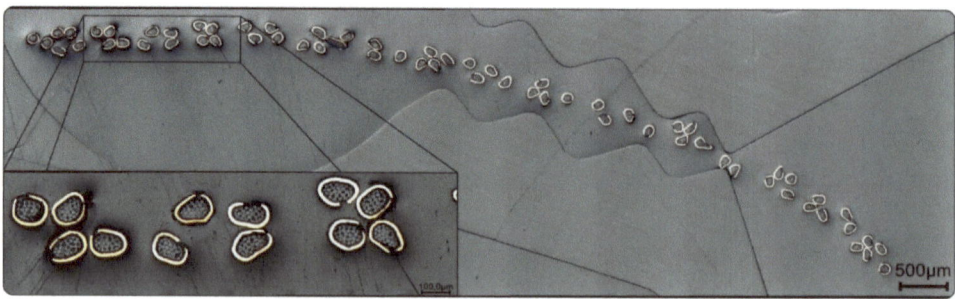

Figure 5. Cross-section image of the High Flex 3981 braided ribbon. In the zoom window, the individual yarns covered with thin silver-plated copper foil are shown for better visualization.

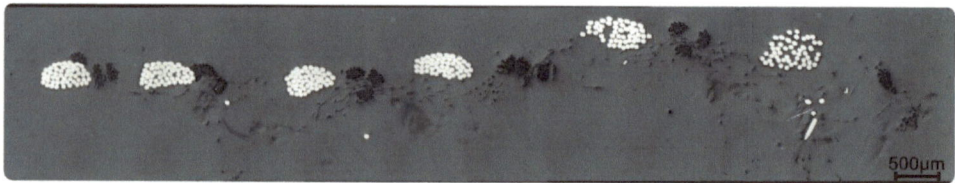

Figure 6. Cross-section image of the Amotape 3587 woven ribbon with six parallel-running bundles of silver-plated copper strands. The Amotape 46050 woven ribbon with six crossed bundles of silver-plated copper strands and additional hotmelt yarn is not displayed as the cross-sections of both Amotape versions are comparable.

2.4. Peel Strength of Welded Materials

To determine the adhesion strength, the optimum welding parameters were determined by the adhesion force between the conductive ribbons and the textile substrate. The conductive ribbons, partially pre-laminated with TPU, were ultrasonically welded onto the cotton substrate in a wide process window by modifying the welding parameters gap size, power, speed and force. Subsequently, a T-peel test was carried out according to IPC TM 650 standard [39], the peel forces were recorded, and the fracture pattern evaluated. Differing from the standard, peeling was not actuated at an 90° angle, but at an 180° angle due to the limp textile properties. The test parameters were a peeling speed of 50.8 mm/min and 3 repetitions per parameter and material combination, with a sample size of 10 mm × 80 mm. The peeling force F_p (N) was measured and standardized over the width a (mm) for better comparability. The resulting peeling force F_a (N/mm) was calculated according to Equation (1).

$$F_a = \frac{F_p}{a} \qquad (1)$$

2.5. Shrinkage of Conductive Ribbons

To determine shrinkage effects due to thermal and mechanical impact during the welding process, the conductive ribbons were welded onto the cotton substrate using the best welding parameters determined from the peel tests. The samples were measured before and after welding and the shrinkage was calculated according to Equation (2). The results were then used in the adjustment of the sample lengths for the main tests.

$$S = \frac{L_0 - L}{L_0} \cdot 100 \qquad (2)$$

Here, S is shrinkage in %, L_0 is initial length in mm and L is the resulting length in mm after welding.

2.6. Electrical Characterization of Materials and Interconnections

All the electrical measurements, both for the initial values of the conductive yarns and ribbons, as well as for the ultrasonically welded contact resistance R_c, were measured with the four-point method using a Keithley 2010 multimeter. This measurement method eliminates the influence of the resistance of embroidered conductor tracks, the resistance of the contact pins and the resistance of the connecting cables, so that only the linear resistance of the materials or the contact resistance of the welded interconnects were measured. For the initial material resistance measurements, the yarn and fabric procedures from the standard DIN EN 16812:2016-11 were applied [40].

In electrical characterization, the interconnections must be part of an electronic circuit and are considered as electrical resistances. The permissible value of the contact resistance depends on the electronic system and the desired application and is, therefore, not generally defined [41]. However, the lower the contact resistance, the better the signal quality and the lower the heat generation. A gradual increase in contact resistance, as determined by aging and durability tests, is also a good indicator of contact fatigue and can be caused by conductor dissolution, delamination of coatings, and cracks in coatings or conductors that can lead to electronic system failure over time [42,43].

In our study, a sudden loss of electrical connection, e.g., due to interruption of a previously stable conductor or a gradual increase in resistance above a threshold contact resistance R_{th} of 50 mOhm, was defined as a contact fault. This threshold value was based on the large contact area between the contact materials, which allows for low contact resistance, and on the initial measurements from the preliminary tests, where R_c values between 1 and 20 mOhm could be achieved, depending on the material. If the R_c value rises (steadily) to 50 mOhm, one can already assume a gradual degradation of the contact resistance.

2.7. Design of Test Pattern

A uniform test pattern was designed for easy evaluation and comparison of the effects of the welding process parameters and materials on the contact resistance of the welded joints. The test pattern was developed for the electrical four-point measurement and is shown in Figure 7a. The hybrid conductive yarns were first embroidered onto the cotton substrate to obtain the required 2D pattern, with the ends being designed as a measurement pad for easy electrical measurements. Then, the contact ribbons were linearly and continuously welded over the embroidered yarns, with the electrical interconnection being made at the point of intersection between the two contact partners (see Figure 7b).

2.8. Proposed Interconnection Process

The novel interconnection process for e-textiles is based on CUPW and works analogously to the welding of an adhesive seam tape onto textiles. Here, instead of joining two textile parts together, textile-integrated conductor materials are mechanically and electrically joined with a conductive ribbon or tape. The proposed interconnection process is shown schematically in Figure 8a,b. Friction welding generated by ultrasonic vibrations leads to strong molecular movements in the material interface layers, thereby generating heat in the joining zone and causing a reduction in the viscosity of the thermoplastic materials, resulting in local softening of the polymers. Due to the simultaneous application of force F, the conductor materials are pressed through the softened polymers and touch, creating an electrical contact. The material is then transported out of the welding zone by the wheel at a defined speed v and solidified again. Since the ultrasonic welding process is very fast, runs at room temperature, and does not require additional additives, such as solder flux or conductive particles in the adhesive, it is very efficient and resource saving and, therefore, well suited for industrial production.

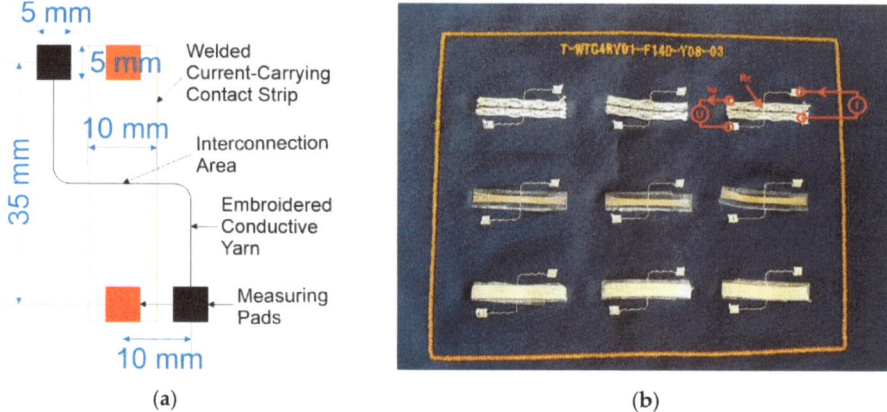

Figure 7. Test structure design: (**a**) pattern design; and (**b**) welded pattern sheet with one type of conductive yarn and three different types of conductive ribbons.

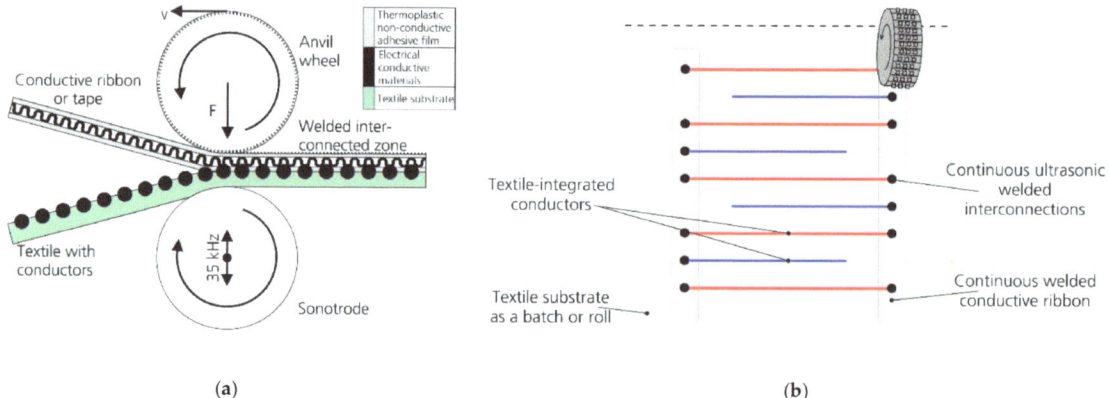

Figure 8. Schematic representation of the proposed continuous interconnection process using rotary ultrasonic welding: (**a**) the conductive ribbon is continuously welded to the textile substrate with the integrated conductors by means of an adhesive with contacts made at the crossing points; and (**b**) view of typical structures to be interconnected, here with conductors contacted on both sides (colored red) or interdigital structures contacted on one side (colored blue).

2.9. Reliability Tests of the Welded Interconnections

Reliability tests were applied to determine the lifetime and failure mechanisms of the welded interconnections. Due to the novelty of e-textiles, testing standards are not yet available or only for very limited specifications [44,45]. The challenge in the selection of reliability tests, therefore, also consisted of researching alternative existing standards and evaluating whether they are suitable for the materials, technologies and applications to be investigated. In addition, self-developed test methods were used if no existing standards were available.

Climatic tests were used to investigate thermal and moisture influence on the materials. The focus lay on both expansion and swelling of the polymers as well as the oxidation behavior of the metals, which can lead to delamination or impurity layers and thus to degradation of the welded interconnection. For reliability tests, the IEC 60068-2-14 and IEC 60068-2-1 (thermal cycling) as well as IEC 60068-2-78 (damp heat test) standards were applied, which are specified as general interconnection tests for electronic systems [46–48].

In the temperature-cycling test, the test specimens were held at 65 °C and −5 °C for 10 min each in ESPEC Corp's TSA-102ES climatic shock chamber and cycled 500 times. In the humidity-heat test, the samples were maintained at a constant 40 °C and 93% RH for 240 h. This was done in the climate test chamber VC^3 7034 from Voetsch.

Due to the lack of standardized e-textile tests, a new method for mechanical reliability tests was necessary since conventional bending tests with limp textile materials lead to very large bending radii and can, therefore, only provide very limited information about mechanical reliability under more realistic buckling and folding loads. In our self-developed method for buckling load tests, the test specimen was fixed with rigid clamps, as shown in Figure 9a. Fixed clamps made of FR-4 PCB substrate material, were attached to both sides of the textile, and the distance between the upper and lower clamping elements determined the resulting bending angle. In the following experiments, a bending angle of 70° could be realized, as shown in Figure 9b. The tests were carried out on an INSTRON 5000 tensile testing machine at a buckling frequency of 1 Hz for 10,000 cycles per sample.

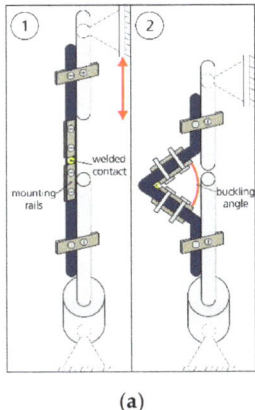

Figure 9. Test rig for mechanical buckling load of welded interconnections: (**a**) schematic representation of the buckling test; and (**b**) test sample in the buckling test with 70° buckling angle.

To date, there is no specific wash testing standard for e-textiles according to which standardized tests can be carried out. As a result, the test methods currently used by academia and industry are very different and thus not comparable. In addition, the existing textile washing standards are only suitable for e-textiles to a limited extent. In current research results, experts agree that gentle washing is necessary as a basis for a future standard for e-textiles [49,50]. A new test protocol based on gentle household washing and the ISO 6330 standard was published, tested for cleanability, and used for the wash tests in this study. The following washing conditions were taken from [51] (pp. 8–9):

- Washing program: main wash, intermediate spinning (500 rpm), two rinsing cycles, spinning (800 rpm); duration: 40 min; temperature: 30 °C; on-time: 40%; water volume: 12 l;
- load: additional PES base load items for a total weight of 2 kg including test samples;
- detergent: 30 g of ECE-2 standard powder detergent (adjusted amount according to used hard water);
- drying: air drying;
- number of cycles: 20.

3. Results

3.1. Determination of the Welding Process Parameters

For the peel tests, the conductive ribbons were welded onto the cotton fabric according to a test matrix with varying welding parameters for power P of 100–400 W, speed v of

0.2–1 m/min, force F of 100–400 N, and a material-adapted gap size s of 0.1–0.75 mm. The amplitude could not be changed in the manual power welding method with which method the tests were performed. The material-dependent highest measured peel forces are listed in Table 3 and compared with the required adhesive strengths of flexible electronic substrates.

Table 3. Highest determined peel forces from the welding test matrix for each conductive ribbon compared with required adhesion strength of conventional, flexible electronic substrates.

Material Combination	Standard	Peel Force F_H [N/mm]
Cotton/Elasta Vestil	None	0.7
Cotton/High Flex 3981	None	1.7
Cotton/Amotape 3587	None	0.4
Cotton/Amotape 46050	None	0.6
FR-4/Cu (35 µm)	DIN EN 60249-2-4	>1.4
PET/Cu (35 µm)	DIN EN 60249-2-8	>0.7
PI/Cu (35 µm)	DIN EN 60249-2-13	>0.8

PI: Polyimide.

Without additional TPU adhesives, the peel forces F_H for Elasta Vestil and High Flex 3981 would be only 0.0–0.1 N/mm and thus insufficient for integration and interconnection. The adhesion forces of the Amotape ribbons could be significantly higher with an additional adhesive, but this was not applied to save the additional material and lamination process step as this would be the most economically preferred option. Testing of the peel strength of standardized values for electronic substrates (where the adhesion strength between laminated copper foils and carrier material is determined) with the ultrasonically welded contact ribbons provided comparable results. In some cases, the specimens already met the required adhesion strength of conventional, flexible electronic substrates.

Adhesion fractures dominated the fracture pattern evaluation, except for the Amotape tapes not prelaminated with TPU, where mixed fractures were observed (See also Figure 10a,b).

(a)

(b)

Figure 10. Visual fracture determination of the samples after the peel tests: (**a**) mixed fracture for Amotape 3587; and (**b**) adhesion fracture for High Flex 3981 prelaminated with TPU.

From the results of the peel tests, it was confirmed that some optimum welding parameters are the same for all different samples: P = 100 W, F = 100 N, and v = 0.2–0.4 m/min. At higher powers, carbonization of the materials could be observed, which led to failures both visually and functionally. In addition, excessively high gap sizes and speeds do not allow melting and bonding of the contact partners. If the gap distances are too small, the materials are pressed too hard, which is noticeable as audible squeaking and can lead to damage to the materials and welding tools. With the determined welding parameter windows, the next tests were carried out to measure the shrinkage of the conductive ribbons, with the results listed in Table 4.

Table 4. Determination of the shrinkage behavior of the welded conductive ribbons.

Value	Elasta Vestil	High Flex 3981	Amotape 3587
Initial length L_0 (mm)	39.5	39.4	38.7
Length after welding L (mm)	36.8	36.4	38.7
Shrinkage S (mm)	2.7	3	0
Shrinkage S (%)	6.8	7.6	0

The TPU-laminated conductive ribbons shrank between 6.8 and 7.6% due to the ultrasonic welding process. Therefore, in the following interconnection welding tests, the lengths of the Elasta Vestil and High Flex 3981 ribbons were adjusted.

3.2. Interconnection Welding Tests and Contact Resistances

Cotton substrates were embroidered with the four selected conductive hybrid yarns by the external partners VÚB a.s. and University of West Bohemia and provided for this study. The pre-determined welding parameters for each yarn are listed in Table 5. While power (P = 100 W) and force (F = 100 N) remained the same in the welding interconnection tests, the speed v and gap size s were varied depending on the different diameters of the yarns as well as the thickness of the ribbons.

Table 5. Process parameters for determining speed and gap size values for optimization of welded contact resistances.

Ribbon	Yarn	Speed v (m/min) for Parameter 1–3	Speed v (m/min) for Parameter 4–5	Gap Size s (mm) for Parameter 1	Gap Size s (mm) for Parameter 2	Gap Size s (mm) for Parameter 3	Gap Size s (mm) for Parameter 4	Gap Size s (mm) for Parameter 5
Elasta Vestil	Y08	0.4	0.4	0.2	0.3	0.4	0.6	0.8
	Y09	0.4	0.4	0.3	0.4	0.5	0.7	0.9
	Y12	0.4	0.4	0.35	0.45	0.55	0.75	0.95
	Y13	0.4	0.4	0.15	0.25	0.35	0.55	0.75
High Flex 3981	Y08	0.2	0.4	0.1	0.2	0.3	0.1	0.2
	Y09	0.2	0.4	0.2	0.3	0.4	0.2	0.3
	Y12	0.2	0.4	0.25	0.35	0.45	0.25	0.35
	Y13	0.2	0.4	0.05	0.15	0.25	0.05	0.15
Amotape 3587	Y08	0.2	0.4	0.1	0.2	0.3	0.1	0.2
	Y09	0.2	0.4	0.2	0.3	0.4	0.2	0.3
	Y12	0.2	0.4	0.25	0.35	0.45	0.25	0.35
	Y13	0.2	0.4	0.05	0.15	0.25	0.05	0.15

Figures 11–14 show the obtained and measured results on the welded contact resistances of the corresponding conductive ribbon/yarn combinations. The results are presented as box plots since with 10 or 11 individual tests per combination, sufficient values were available for a statistical evaluation. An auxiliary line at 50 mOhm is drawn to clearly indicate the target value for the contact resistance formulated in Section 2.6. The Y13 yarn variant developed for heating applications consists of only a very few stainless-steel filaments, and therefore, has a very high resistance, which is why the measured contact resistances here clearly exceed the 50 mOhm threshold value of R_c. The combination of Y13 and High Flex 3981 did not achieve an electrical contact in any test; therefore, no diagram is provided. For a better overview of the results obtained, the Y-axis in several diagrams was scaled differently and drawn in two colors. A blue ordinate axis indicates a contact resistance in the lower ohm range, a red one in the higher ohm range and the sign "X" stands for an electrical fault in all diagrams.

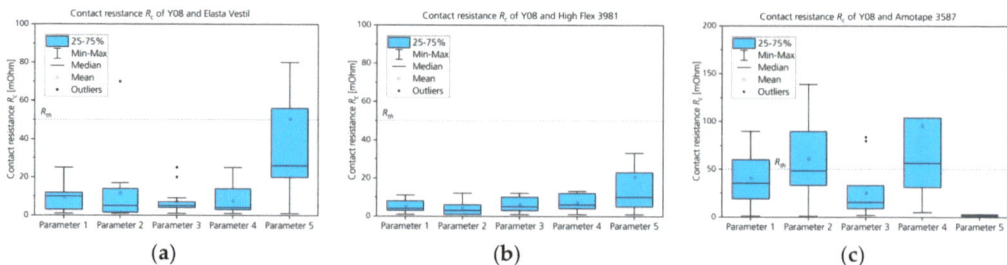

Figure 11. Welded interconnection of hybrid conductive yarn Y08 with three types of conductive ribbons: (**a**) Elasta Vestil; (**b**) High Flex 3981; and (**c**) Amotape 3587.

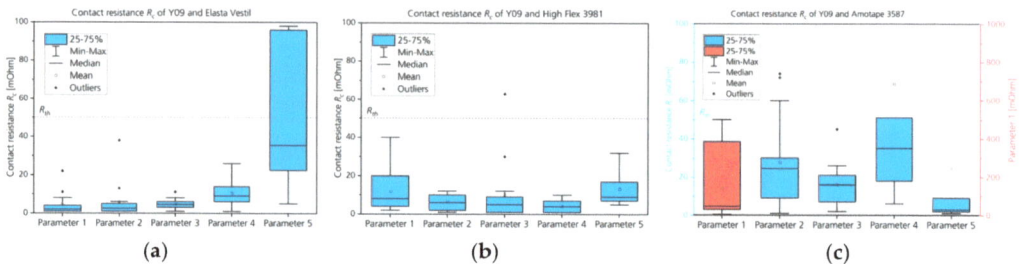

Figure 12. Welded interconnection of hybrid conductive yarn Y09 with three types of conductive ribbons: (**a**) Elasta Vestil ribbon; (**b**) High Flex 3981; and (**c**) Amotape 3587.

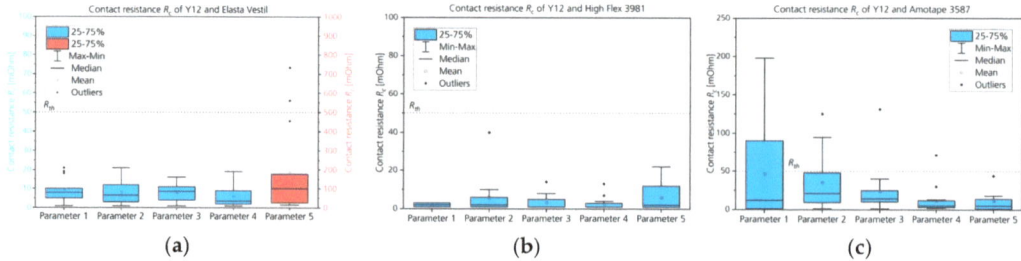

Figure 13. Welded interconnection of hybrid conductive yarn Y12 with three types of conductive ribbons: (**a**) Elasta Vestil; (**b**) High Flex 3981; and (**c**) Amotape 3587.

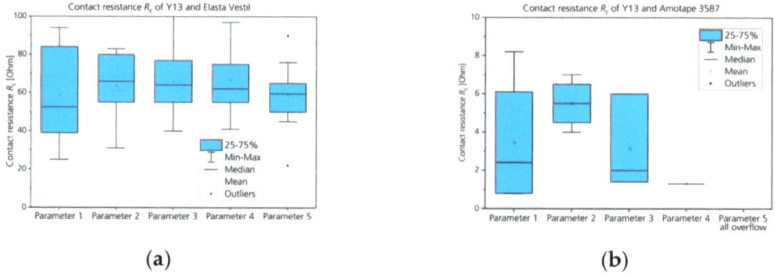

Figure 14. Welded interconnection of hybrid conductive yarn Y13 with two types of conductive ribbons: (**a**) Elasta Vestil; and (**b**) Amotape 3587.

Table 6 shows the percentage of contact failures for all results, which indicates the quality of the ultrasonic welding process for each material and parameter, except the high impedance Y13 contact resistances. A failure indicates a contact resistance over 50 mOhm or no contact measured. Note that, as listed in Table 5, each material combination has different parameters for speed and gap size for numbers 1–5.

Table 6. Failure rate of welded ribbons interconnected with conductive yarns for all parameters.

Conductive Ribbon	Interconnected with	Failure Rate (%) for Parameter 1	Failure Rate (%) for Parameter 2	Failure Rate (%) for Parameter 3	Failure Rate (%) for Parameter 4	Failure Rate (%) for Parameter 5
Elasta Vestil	Y08	14.3	14.3	0	0	42.9
	Y09	0	0	0	7.1	50
	Y12	0	0	0	0	57.1
High Flex 3981	Y08	0	0	0	0	6.7
	Y09	0	0	6.7	0	0
	Y12	0	0	0	0	0
Amotape 3587	Y08	26.7	46.7	13	60	100
	Y09	33.3	20	0	26.7	33.3
	Y12	40	20	6.7	6.7	13.3

In the welding interconnection tests, the Elasta Vestil ribbon performed well as low contact resistances could be determined over a wide window of welding parameters with low failure rates. After TPU embedding, High Flex 3981 not only offered the best adhesion to the textile substrate, but of the three conductive ribbons tested, it achieved the best electrical results, which could be due to the material structure having a large metallic surface contact area and a ductile polymer core. In addition, not only are the contact resistances low, but the process yield is also high over a large parameter window. With Amotape 3587, however, the defect rate is higher with a relatively high contact resistance compared with the other two ribbon variants.

A few optimizations were carried out for the second final batch of test vehicles. First, AMOHR Technische Textilien GmbH developed and provided the new Amotape 46050, which consists of crossing, silver-plated copper strands and thus, like the other ribbons, has an advantageous parallel connection of the individual metal strands resulting in lower conductor resistances. Furthermore, additional hotmelt adhesive yarns were incorporated into the conductive ribbon, which improved the mechanical adhesion to the textile substrate.

The best values from the evaluations of the first batch were selected as final welding parameters and are listed in Table 7. Due to the similar material thicknesses, the welding parameters determined for Amotape 3587 were also used for Amotape 46050. However, to reduce the test matrix and the testing workload, only two yarn variants were used in the final batch, Y08 and Y12.

Table 7. Overview of the determined welding parameters for the final batch.

Conductive Yarn	Interconnected with	Power P (W)	Force F (N)	Speed v (m/min)	Gap Size s (mm)
Y08	Elasta Vestil	100	100	0.4	0.4
	High Flex 3981	100	100	0.2	0.2
	Amotape 46050	100	100	0.2	0.2
Y12	Elasta Vestil	100	100	0.4	0.55
	High Flex 3981	100	100	0.2	0.25
	Amotape 46050	100	100	0.2	0.35

3.3. Results of the Reliability Tests

After ultrasonic welding, the initial contact resistances were measured and then the specimens were tested using the test methods described in Section 2.9. For this purpose, the batch of samples was evenly divided into four parts. After completion of the reliability tests, all contact resistances were again determined by means of a four-point measurement.

The results for the final batch are shown in the following Figures 15–17. For a simplified overview, all the initial contact resistances are listed as mean values. For welding of Y08 with Elasta Vestil, the mean value is 3.75 mOhm, for High Flex 3981 it is 1.0 mOhm and for Amotape 46050 it is 1.1 mOhm. For contacting Y12, the mean values are 5.0 mOhm for Elasta Vestil, 4.25 mOhm for High Flex 3981 and 1.0 mOhm for Amotape 46050.

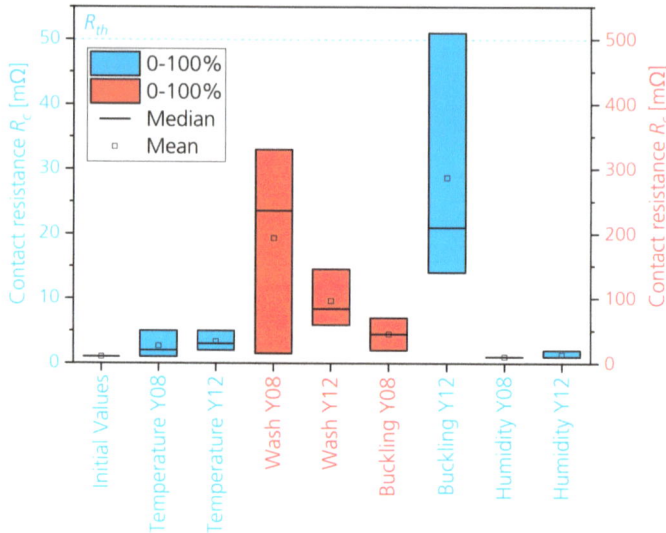

Figure 15. Influence of reliability tests on contact resistance for welding with Elasta Vestil.

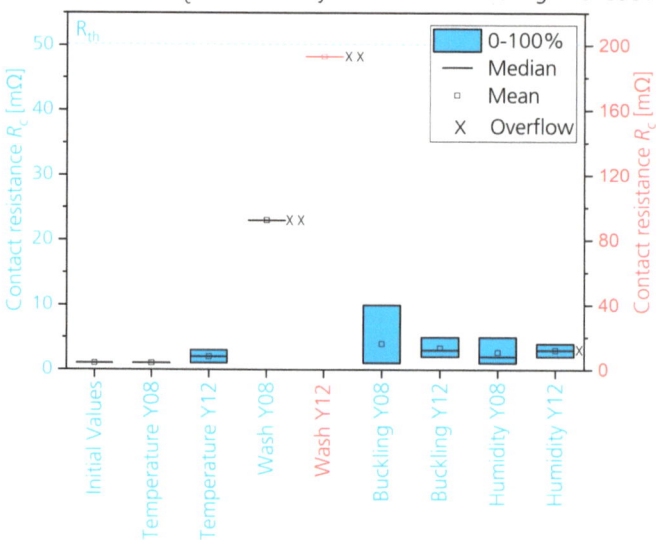

Figure 16. Influence of reliability tests on contact resistance for welding with High Flex 3981.

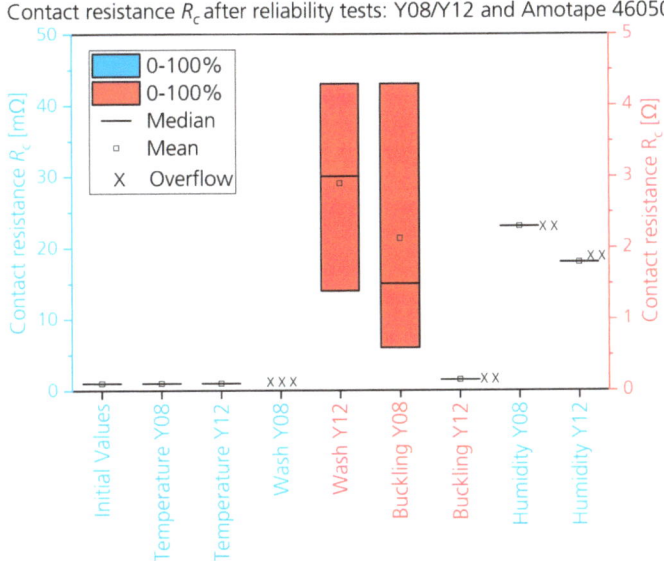

Figure 17. Influence of reliability tests on contact resistance for welding with Amotape 46050.

Based on a five-point rating scale, the individual test results are summarized in Table 8.

Table 8. Evaluation of the influence of the reliability tests on the welded contact resistances.

Reliability Test	Conductive Yarn	Elasta Vestil	High Flex 3981	Amotape 46050
Temperature Cycle	Y08	excellent	excellent	excellent
	Y12	excellent	excellent	excellent
Damp/Heat	Y08	excellent	excellent	good
	Y12	excellent	good	good
Buckling	Y08	fair	excellent	poor
	Y12	good	excellent	bad
Washability	Y08	poor	poor	bad
	Y12	poor	bad	bad

3.4. Optical Analysis of the Electrical Interconnection

To gain an initial insight into the ultrasonically welded contact structures, cross-sections of unstressed specimens were prepared and examined on a Keyence VHX-6000 digital light microscope. Figure 18 shows a cross-section of a contact between the individual conductors of the High Flex 3981 ribbon and Y08 yarn. Material compression and deformation of the metal strands in the contact area and the stamp imprint of the knurled welding wheel on the material surface are clearly visible.

All the samples were further inspected with a Phoenix nanome x 180 X-ray microscope before and after the reliability tests. No conductor breakages that could occur due to the ultrasonic vibrations were detected in any of the samples. An exemplary image of a contacted sample is shown in Figure 19a. Using X-ray microcomputed tomography (µ-CT) with a Phoenix nanotom m 180, selected samples were scanned three-dimensionally to gain a more detailed insight into the contact structure. A µ-CT image is shown in Figure 19b.

3.5. Functional Samples with Continuously Ultrasonic Welded Contacts

Various electrically heatable functional samples were prepared to demonstrate the developed interconnection technology as heatable textiles also account for the largest share of the e-textile market [3] (p. 11). First, textile samples with embroidered or knitted

Y08 heating yarns were produced, and then the integrated yarns were connected with conductive ribbons using US welding. Figures 20 and 21 show two such demonstrators. The flat, flexible structures of the contact tapes, which bond seamlessly to the textile substrate, as well as the low contact resistance, can be seen in the attached (thermal) images. Figure 22 shows the final demonstrator, which consists of a 190 cm × 90 cm large knitted bed sheet with knitted heating conductors made of Y08 yarn, onto which Elasta Vestil ribbons were welded on the upper and lower sides so that all parallel-running heating conductors were electrically and mechanically interconnected.

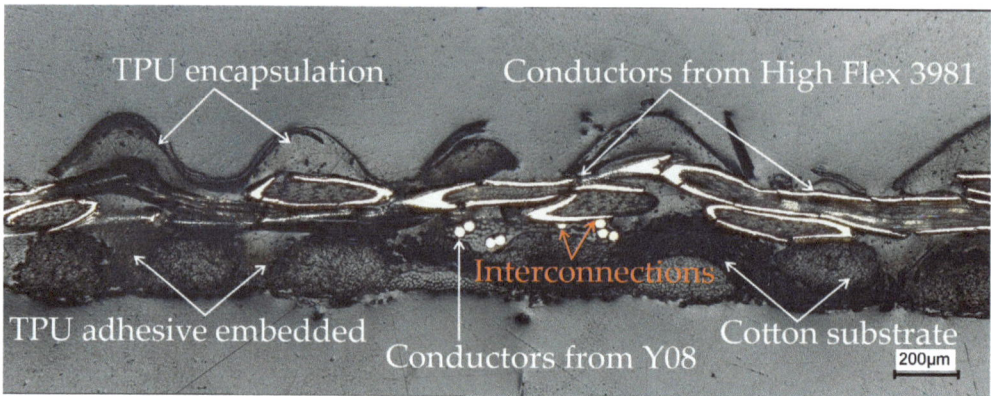

Figure 18. Cross-section of an ultrasonically welded interconnection between High Flex 3981 ribbon and Y08 yarn.

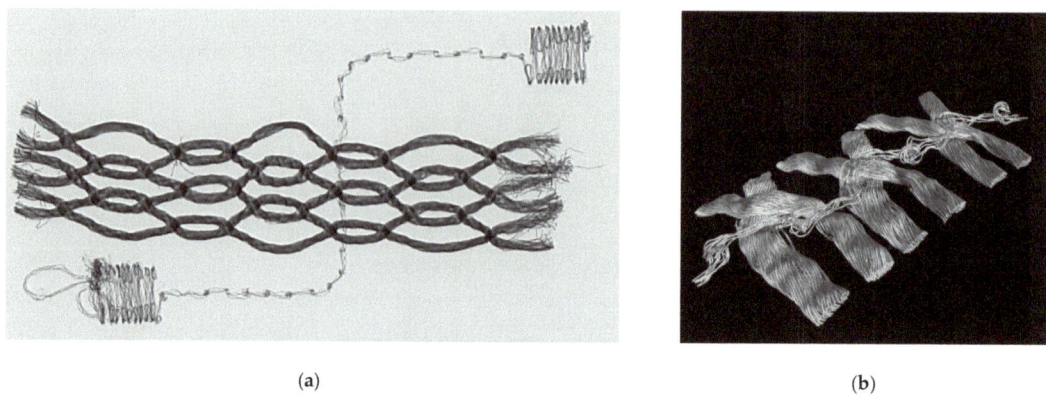

Figure 19. Non-destructive optical analysis of the welded interconnections: (**a**) X-ray image of a welded contact between Amotape 46050 and Y08; and (**b**) µ-CT scan of a welded contact between Amotape 46050 and Y08.

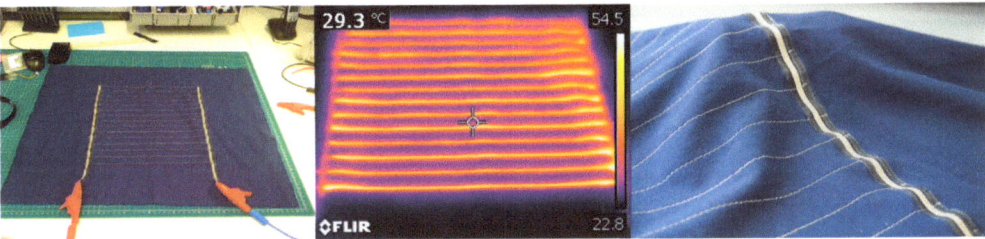

Figure 20. Heatable functional demonstrator made of embroidered Y08 yarn and welded contact ribbon High Flex 3981.

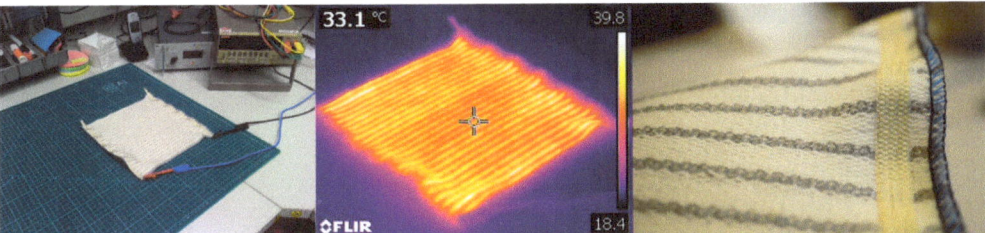

Figure 21. Heatable functional demonstrator made of knitted Y08 yarn and welded contact ribbon Elasta Vestil.

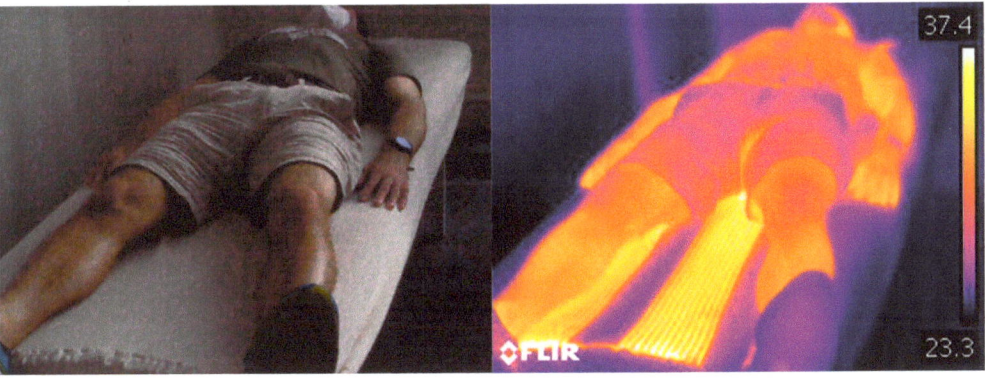

Figure 22. A 90 cm × 190 cm knitted bed sheet with integrated Y08 yarns, ultrasonically welded with conductive ribbon Elasta Vestil as a continuous contact element for the interconnection of the heating yarns.

4. Discussion

A new joining method for continuous electrical interconnections of textile-integrated conductors with conductive ribbons attached by means of rotary ultrasonic plastic welding has been proposed, tested and investigated. For this study, four selected hybrid conductive yarns were embroidered onto a cotton substrate and then woven or braided commercial conductive ribbons were welded onto the yarns by a flatbed ultrasonic welding machine equipped with a knurled steel anvil wheel and sonotrode as welding tools. Thermoplastic films laminated onto the ribbons or hot melt yarns already woven into the ribbons were tested as adhesives for a stable mechanical joint. As the TPU-coated ribbons shrank by 7–8% during ultrasonic welding, the ribbon length had to be individually adjusted for the experiments.

Empirical experiments were used to determine the optimum welding parameters. By measuring the highest peel forces of the welded conductive ribbons on cotton fabric, the welding power P of 100 W, welding force F of 100 N and welding speed v between 0.2 and 0.4 m/min were determined as optimal, and no influence of the conductive ribbon or adhesive on these results was observed. In contrast, the different material thicknesses strongly influenced the welding parameter gap size s. In extensive tests, the optimal gap sizes could be determined for each material combination by means of a four-point resistance measurement of the contact resistances achieved. Thus, a continuous ultrasonic welding process setup could be determined for each material combination, with the best results having a 100% yield and low contact resistances in the range of 1 - 10 mOhm.

In evaluating the reliability tests, washability remains the biggest challenge for e-textiles, whereas the ultrasonically welded contacts showed no or only minor ageing or failure after the climatic and mechanical tests. Some samples showed delamination between the textile substrate and conductive ribbon during the buckling or washing test, mainly with Elasta Vestil and Amotape 46050, which also had significantly lower peel forces than High Flex 3981. In the case of High Flex 3981, the TPU top layer was damaged during washing, causing the then no longer encapsulated conductive ribbon to twist and so partially loosen from the textile structure during the washing cycles.

The X-ray analysis did not show any broken conductors, but in the μ-CT images, several conductor fractures could be identified where the metal wires from the ribbons came into contact with those from the hybrid conductive yarns. This indicates that the brittle metal wires could break in some cases during ultrasonic welding. However, due to the redundant design with numerous conductors running in parallel for the conductive ribbons and yarns, this did not result in contact failure. The broken conductors as well as all the conductive materials are embedded so well in the textile and thermoplastic matrix that even extreme loads such as the buckling test did not lead to contact failure.

5. Conclusions

This study has proven that the continuous ultrasonic welding technique is suitable for the manufacturing of multiple interconnects for large-area e-textiles. Although washability has not yet been achieved, we see the greatest advantage in realizing a flat, flexible as well as thermally and mechanically stable interconnection and thus realizing a reliable contact between hybrid conductive yarns and potentially endless conductive ribbons. The reparability option and possible separation of the conductive ribbons in a recycling process at the end of the life cycle—due to thermoplastic adhesives, which can be softened again and thus detached from the textile substrate—also adds to the sustainability of the interconnection method.

The use of commercial materials as well as an established textile manufacturing technique enables a fast technology transfer into industry. The first areas of application for continuous ultrasonic welding could be in the production of large-area heating textiles, for example as seat heating in cars, as indoor wall heaters, and also for cut-detection sensor fabrics in truck tarpaulins or house roofs.

In the future, further development of the conductive ribbons could achieve higher reliability and thus open up new fields of application, for example, in smart clothing. For this purpose, investigations into suitable thermoplastics are planned, in particular, the use of different and thicker TPU films for better encapsulation of the metallic conductors, tests with low-melting films to increase the process speed and tests to increase adhesion through the integration of suitable TPU-based yarns into the welding area of the textile carrier.

Author Contributions: Conceptualization, methodology, writing—original draft preparation, project administration, funding acquisition, C.D.; investigation, visualization, S.H. and C.D.; writing—review and editing, S.H. and M.S.-R.; supervision, M.S.-R. All authors have read and agreed to the published version of the manuscript.

Funding: The IGF research project (278 EN) of the research association Forschungskuratorium Textil e.V., Reinhardtstraße 14-16, 10117 Berlin was funded via the AiF within the program for supporting the "Industrial Collective Research" (IGF) from funds of the Federal Ministry of Economic Affairs and Climate Actions (BMWK) on the basis of a decision by the German Bundestag.

Institutional Review Board Statement: Not applicable.

Informed Consent Statement: Not applicable.

Data Availability Statement: The data presented in this study are available on request from the corresponding author.

Acknowledgments: The authors would like to thank VÚB a.s. and the RICE research center of the Faculty of Electrical Engineering at the University of West Bohemia for providing test samples with embroidered or knitted hybrid conductive yarns. We would also like to thank the company AMOHR Technische Textilien GmbH for the development and provision of the Amotape 46050, which was carried out after joint discussions within the framework of this study. Further thanks go to our colleagues Lukas Werft for performing the buckling tests and Sigrid Rotzler for proofreading this manuscript.

Conflicts of Interest: The authors declared no potential conflicts of interest with respect to the research, authorship and/or publication of this article. The funders had no role in the design of the study; in the collection, analyses, or interpretation of data; in the writing of the manuscript; or in the decision to publish the results.

References

1. Hughes-Riley, T.; Dias, T.; Cork, C. A Historical Review of the Development of Electronic Textiles. *Fibers* **2018**, *6*, 34. [CrossRef]
2. Fernández-Caramés, T.M.; Fraga-Lamas, P. Towards The Internet of Smart Clothing: A Review on IoT Wearables and Garments for Creating Intelligent Connected E-Textiles. *Electronics* **2018**, *7*, 405. [CrossRef]
3. Hayward, J. *E-Textiles & Smart Clothing 2021–2031: Technologies, Markets and Players*; IDTechEx Reports: Cambridge, UK, 2021.
4. Koncar, V. (Ed.) *Smart Textiles and Their Applications*; Woodhead Publishing: Sawston, UK; Cambridge, UK, 2016.
5. Wicaksono, I.; Cherston, J.; Paradiso, J.A. Electronic Textile Gaia: Ubiquitous Computational Substrates Across Geometric Scales. *IEEE Pervasive Comput.* **2021**, *20*, 18–29. [CrossRef]
6. Repon, M.R.; Mikučionienė, D. Progress in Flexible Electronic Textile for Heating Application: A Critical Review. *Materials* **2021**, *14*, 6540. [CrossRef]
7. Almohammed, B.; Ismail, A.; Sali, A. Electro-textile wearable antennas in wireless body area networks: Materials, antenna design, manufacturing techniques, and human body consideration—A review. *Text. Res. J.* **2021**, *91*, 646–663. [CrossRef]
8. Shi, X.; Zuo, Y.; Zhai, P.; Shen, J.; Yang, Y.; Gao, Z.; Liao, M.; Wu, J.; Wang, J.; Xu, X.; et al. Large-area display textiles integrated with functional systems. *Nature* **2021**, *591*, 240–245. [CrossRef]
9. Chen, G.; Xiao, X.; Zhao, X.; Tat, T.; Bick, M.; Chen, J. Electronic Textiles for Wearable Point-of-Care Systems. *Chem. Rev.* **2022**, *122*, 3259–3291. [CrossRef] [PubMed]
10. Vieira, D.; Carvalho, H.; Providência, B. E-Textiles for Sports: A Systematic Review. *JBBBE* **2022**, *57*, 37–46. [CrossRef]
11. Koncar, V. Smart Textiles for Monitoring and Measurement Applications. In *Smart Textiles for In Situ Monitoring of Composites*; Koncar, V., Ed.; Woodhead Publishing: Duxford, UK, 2019; pp. 1–151.
12. Stanley, J.; Hunt, J.A.; Kunovski, P.; Wei, Y. A review of connectors and joining technologies for electronic textiles. *Eng. Rep.* **2022**, *4*, e12491. [CrossRef]
13. Gehrke, I.; Tenner, V.; Lutz, V.; Schmelzeisen, D.; Gries, T. *Smart Textiles Production: Overview of Materials, Sensor and Production Technologies for Industrial Smart Textiles*; MDPI: Basel, Switzerland, 2019.
14. Rotzler, S.; Kallmayer, C.; Dils, C.; von Krshiwoblozki, M.; Bauer, U.; Schneider-Ramelow, M. Improving the washability of smart textiles: Influence of different washing conditions on textile integrated conductor tracks. *J. Text. Inst.* **2020**, *111*, 1766–1777. [CrossRef]
15. Eppinger, E.; Slomkowski, A.; Behrendt, T.; Rotzler, S.; Marwede, M. Design for Recycling of E-Textiles: Current Issues of Recycling of Products Combining Electronics and Textiles and Implications for a Circular Design Approach. In *Recycling—Recent Advances*; Saleh, H.M., Hassan, A.I., Eds.; IntechOpen: London, UK, 2022; Available online: https://www.intechopen.com/online-first/84034 (accessed on 22 December 2022). [CrossRef]
16. Simegnaw, A.A.; Malengier, B.; Rotich, G.; Tadesse, M.G.; Van Langenhove, L. Review on the Integration of Microelectronics for E-Textile. *Materials* **2021**, *14*, 5113. [CrossRef]
17. Hoerr, M. Reliable Mass Production of E-Textiles by using Embroidery Technology. Presented at IPC E-Textiles, San Diego, CA, USA, 23 January 2023.
18. Garbacz, K.; Stagun, L.; Rotzler, S.; Semenec, M.; von Krshiwoblozki, M. Modular E-Textile Toolkit for Prototyping and Manufacturing. *Proceedings* **2021**, *68*, 5. [CrossRef]

19. Linz, T.; von Krshiwoblozki, M.; Walter, H.; Foerster, P. Contacting electronics to fabric circuits with nonconductive adhesive bonding. *J. Text. Inst.* **2012**, *103*, 1139–1150. [CrossRef]
20. Benouakta, S.; Hutu, F.D.; Duroc, Y. Stretchable Textile Yarn Based on UHF RFID Helical Tag. *Textiles* **2021**, *1*, 547–557. [CrossRef]
21. Seyedin, S.; Carey, T.; Arbab, A.; Eskandarian, L.; Bohm, S.; Kim, J.M.; Torrisi, F. Fibre electronics: Towards scaled-up manufacturing of integrated e-textile systems. *Nanoscale* **2021**, *13*, 12818–12847. [CrossRef]
22. Dhawan, A.; Seyam, A.M.; Ghosh, T.K.; Muth, J.F. Woven fabric-based electrical circuits: Part I: Evaluating interconnect methods. *Text. Res. J.* **2004**, *74*, 913–919. [CrossRef]
23. Zhang, H. Flexible textile-based strain sensor induced by contacts. *Meas. Sci. Technol.* **2015**, *26*, 105102. [CrossRef]
24. Suchý, S.; Kalaš, D.; Kalčík, J.; Soukup, R. A comparison of resistance spot and ultrasonic welding of hybrid conductive threads. In Proceedings of the 43rd International Spring Seminar on Electronics Technology (ISSE), Demanovska Valley, Slovakia, 14–15 May 2020; pp. 1–5. [CrossRef]
25. Suchý, S.; Rostás, K.; Soukup, R. Encapsulation Methods for Resistance-Welded Contacts in Smart Textiles. In Proceedings of the 45th International Spring Seminar on Electronics Technology (ISSE), Vienna, Austria, 11–15 May 2022; pp. 1–5. [CrossRef]
26. Locher, I. Technologies for System-on-Textile Integration. Doctoral Thesis, ETH, Zürich, Switzerland, 2006. [CrossRef]
27. Slade, J.R.; Winterhalter, C. Electro-textile garments for power and data distribution. In Proceedings of the Display Technologies and Applications for Defense, Security, and Avionics IX; and Head- and Helmet-Mounted Displays XX (SPIE Defense + Security), Baltimore, MD, USA, 21 May 2015. [CrossRef]
28. Thurner, F. New methods for reliable contacts of conductive textile substrates. *Tech. Text.* **2019**, *62*, 296–298.
29. Micus, S.; Rostami, S.G.; Haupt, M.; Gresser, G.T.; Meghrazi, M.A.; Eskandarian, L. Integrating Electronics to Textiles by Ultrasonic Welding for Cable-Driven Applications for Smart Textiles. *Materials* **2021**, *14*, 5735. [CrossRef]
30. Dils, C.; Kallmayer, C.; Gerhold, L.; Schneider-Ramelow, M. Investigations into ultrasonic plastic welding as an innovative contacting technology for the integration of electronics into textiles. *Join. Plast.* **2020**, *14*, 104–110.
31. Dils, C.; Kalas, D.; Reboun, J.; Suchy, S.; Soukup, R.; Moravcova, D.; von Krshiwoblozki, M.; Schneider-Ramelow, M. Interconnecting embroidered hybrid conductive yarns by ultrasonic plastic welding for e-textiles. *Text. Res. J.* **2022**, *92*, 4501–4520. [CrossRef]
32. Atalay, O.; Kalaoglu, F.; Bahadir, S.K. Development of textile-based transmission lines using conductive yarns and ultrasonic welding technology for e-textile applications. *J. Eng. Fibers Fabr.* **2019**, *14*, 1–8. [CrossRef]
33. Leśnikowski, J. Research into the Textile-Based Signal Lines Made Using Ultrasonic Welding Technology. *Autex Res. J.* **2022**, *22*, 11–17. [CrossRef]
34. Platilon TPU Films. Available online: https://solutions.covestro.com/en/brands/platilon (accessed on 30 November 2022).
35. Löher, T.; Seckel, M.; Ostmann, A. Stretchable electronics manufacturing and application. In Proceedings of the 3rd Electronics System Integration Technology Conference (ESTC), Berlin, Germany, 13–16 September 2010; pp. 1–6. [CrossRef]
36. Zoschke, K.; Löher, T.; Kallmayer, C.; Jung, E. Flexible and Stretchable Systems for Healthcare and Mobility. In *Flexible, Wearable, and Stretchable Electronics*, 1st ed.; Katsuyuki, S., Ed.; CRC Press: Boca Raton, FL, USA, 2020; pp. 269–282.
37. Dils, C.; Werft, L.; Walter, H.; Zwanzig, M.; von Krshiwoblozki, M.; Schneider-Ramelow, M. Investigation of the Mechanical and Electrical Properties of Elastic Textile/Polymer Composites for Stretchable Electronics at Quasi-Static or Cyclic Mechanical Loads. *Materials* **2019**, *12*, 3599. [CrossRef]
38. von Krshiwoblozki, M.; Linz, T.; Neudeck, A.; Kallmayer, C. Electronics in Textiles—Adhesive Bonding Technology for Reliably Embedding Electronic Modules into Textile Circuits. *Adv. Sci. Technol.* **2012**, *85*, 1–10. [CrossRef]
39. IPC-TM-650 TEST METHODS MANUAL no. 2.4.9: *Peel Strength, Flexible Printed Wiring Materials*; The Institute for Interconnecting and Packaging Electronic Circuits: Northbrook, IL, USA, 1988.
40. EN 16812:2016; Textiles and Textile Products—Electrically Conductive Textiles—Determination of the Linear Electrical Resistance of Conductive Tracks; German Version. European Committee for Standardization: Brussels, Belgium, 2016.
41. Agcayazi, T.; Chatterjee, K.; Bozkurt, A.; Ghosh, T.K. Flexible Interconnects for Electronic Textiles. *Adv. Mater. Technol.* **2018**, *3*, 1700277. [CrossRef]
42. Biermaier, C.; Bechtold, T.; Pham, T. Towards the Functional Ageing of Electrically Conductive and Sensing Textiles: A Review. *Sensors* **2021**, *21*, 5944. [CrossRef]
43. Persons, A.K.; Ball, J.E.; Freeman, C.; Macias, D.M.; Simpson, C.L.; Smith, B.K.; Burch V., R.F. Fatigue Testing of Wearable Sensing Technologies: Issues and Opportunities. *Materials* **2021**, *14*, 4070. [CrossRef]
44. Decaens, J.; Vermeersch, O. Specific testing for smart textiles. In *Advanced Characterization and Testing of Textiles*, 1st ed.; Dolez, P., Vermeersch, O., Izquierdo, V., Eds.; Elsevier: Cambridge, UK, 2018; pp. 351–374. [CrossRef]
45. Shuvo, I.I.; Decaens, J.; Lachapelle, D.; Dolez, P.I. Smart textiles testing: A roadmap to standardized test methods for safety and quality-control. In *Textiles for Functional Applications*, 1st ed.; Kumar, B., Ed.; IntechOpen: London, UK, 2021; pp. 1–15. Available online: https://www.intechopen.com/chapters/75712 (accessed on 22 December 2022). [CrossRef]
46. IEC 60068-2-14:2009; Environmental Testing—Part 2-14: Tests—Test N: Change of Temperature; German Version. International Electrotechnical Commission: Geneva, Switzerland, 2009.
47. IEC 60068-2-1:2007; Environmental Testing—Part 2-1: Tests—Test A: Cold; German Version. International Electrotechnical Commission: Geneva, Switzerland, 2007.

48. *IEC 60068-2-78:2012*; Environmental Testing—Part 2-78: Tests—Test Cab: Damp Heat, Steady State; German Version. International Electrotechnical Commission: Geneva, Switzerland, 2013.
49. Rotzler, S.; von Krshiwoblozki, M.; Schneider-Ramelow, M. Washability of e-textiles: Current testing practices and the need for standardization. *Text. Res. J.* **2021**, *91*, 2401–2417. [CrossRef]
50. Rotzler, S.; Schneider-Ramelow, M. Washability of E-Textiles: Failure Modes and Influences on Washing Reliability. *Textiles* **2021**, *1*, 37–54. [CrossRef]
51. Rotzler, S.; Schneider-Ramelow, M. Development of a Testing Protocol to Assess the Washability of E-Textiles. *Solid State Phenom.* **2022**, *333*, 3–10. [CrossRef]

Disclaimer/Publisher's Note: The statements, opinions and data contained in all publications are solely those of the individual author(s) and contributor(s) and not of MDPI and/or the editor(s). MDPI and/or the editor(s) disclaim responsibility for any injury to people or property resulting from any ideas, methods, instructions or products referred to in the content.

Article

Eco-Friendly Natural Thickener (Pectin) Extracted from Fruit Peels for Valuable Utilization in Textile Printing as a Thickening Agent

Sara A. Ebrahim [1], Hanan A. Othman [1], Mohamed M. Mosaad [1] and Ahmed G. Hassabo [2,*]

[1] Textile Printing, Dyeing and Finishing Department, Faculty of Applied Arts, Benha University, Benha P.O. Box 13518, Egypt
[2] National Research Centre (Scopus Affiliation ID 60014618), Textile Research and Technology Institute, Pre-Treatment, and Finishing of Cellulose-Based Textiles Department, 33 El-Behouth St. (Former El-Tahrir Str.), Dokki, Giza P.O. Box 12622, Egypt
* Correspondence: aga.hassabo@hotmail.com or ag.hassabo@nrc.sci.eg; Tel.: +20-110-22-555-13

Abstract: Fruit peels are a rich source of many substances, such as pectin. Extraction of natural thickening agent (pectin) from fruit waste such as (orange and pomegranate peels) is an environmentally friendly alternative to commercial thickeners and is cheap for use in the printing of natural and synthetic fabrics, especially polyester and polyacrylic fabrics. Hexamine was used to treat the extracted pectin to make it appropriate for use in an alkali medium for printing cotton fabric. The results showed that the extracted and modified pectin have good rheological properties as well as bacterial resistance. Pectin is suitable for use in an acidic medium. All the printed samples with pectin and its modified synthetic dyes (reactive, acid, and disperse) exhibited good fastness towards washing and wet and dry rubbing. The light fastness of printed textiles was excellent (7), which is more than using alginate as a thickener (5). In both acidic and alkaline perspiration, the perspiration fastness characteristic revealed 3–4 to 4–5 color differences. Colorfastness to rubbing was tested in both dry and wet conditions, and it was revealed that dry rubbing had the same effect as wet rubbing. Printed textiles using pectin or modified pectin as thickeners exhibit antibacterial activity. Physical and mechanical properties of all printed fabrics such as (tensile strength, elongation, and surface roughness) were enhanced.

Keywords: pectin; hexamine; plant peel waste; thickener; printing

Citation: Ebrahim, S.A.; Othman, H.A.; Mosaad, M.M.; Hassabo, A.G. Eco-Friendly Natural Thickener (Pectin) Extracted from Fruit Peels for Valuable Utilization in Textile Printing as a Thickening Agent. *Textiles* **2023**, *3*, 26–49. https://doi.org/10.3390/textiles3010003

Academic Editor: Laurent Dufossé

Received: 16 November 2022
Revised: 3 January 2023
Accepted: 7 January 2023
Published: 11 January 2023

Copyright: © 2023 by the authors. Licensee MDPI, Basel, Switzerland. This article is an open access article distributed under the terms and conditions of the Creative Commons Attribution (CC BY) license (https://creativecommons.org/licenses/by/4.0/).

1. Introduction

Textile printing is becoming a well-known technology in the textile wet-processing industries for all fibers, textiles, and garments. Textile printing seems to be the most versatile and widely used way of imparting color and pattern to textile fibers. Textile printing is a significant method of producing ornamental textile fabric. Color is obtained using dyes or pigments in printing paste. Printing is a type of dyeing in which colors are applied to specific parts of the cloth rather than the entire fabric. The dyes and other auxiliaries are glued with a natural or synthetic thickening agent to confine the coloring materials to the design area. A successful print requires precise color, sharpness of mark, levelness, excellent hand, and efficient dye use: all of these characteristics are affected by the type of thickener employed [1–7].

Thickening agents are an important component of any printing process. Thickeners are high-molecular-weight viscous chemicals that form a sticky paste with water, imparting stickiness and plasticity to the printing paste. These thickeners help to keep the design outlines from spreading even under high pressure. The primary function of thickeners in the textile industry is to retain or stick dye particles to the targeted portions of the fabric until the dye has transferred to the fabric surface and its fixation has been completed. A thickener adds viscosity to printing pastes, prevents early reactions between the print paste's chemicals, and aids in the seizing of the print paste's constituents on textiles. The thickener must be stable and suitable with the dyes and dyeing auxiliaries that are being employed [8–11].

Natural thickeners are polysaccharides derived from nature, such as plant exudates, seaweeds, seeds, and roots, which are commonly utilized. Some of them seem to be appropriate for printing with a specific color category, but they should be chemically adapted to satisfy the standards for printing [1,9].

Pectin is a complex polysaccharide combination that accounts for around one-third of the dry cell-wall material in higher plants. These chemicals are present in much lower concentrations in grass cell walls. Pectin concentrations are greatest in the middle lamella of the cell wall, with a steady decline as one moves through the main wall into the plasma membrane. Although pectin is found in almost all plant tissues, the number of sources from which pectin may be commercially manufactured is quite restricted. Because pectin's capacity to form gel is dependent on the molecular size and degree of esterification (DE), pectin from different sources does not have the same gelling ability owing to variances in these characteristics [12–15].

As a result, detecting a substantial amount of pectin in fruit is insufficient to certify that fruit is a source of commercial pectin. Now, commercial pectin is virtually entirely generated from citrus peel or apple pomace, both of which are byproducts of juice (or cider) production. On a dry-matter basis, apple pomace comprises 10–15% pectin. Citrus peel contains 20–30% vitamin C. Citrus and apple pectin are identical in terms of application. Citrus pectin is pale cream or light tan in color, but apple pectin is usually deeper [16].

Pectin is used in the cosmetics, pharmaceutical, and food industries to stabilize acidified milk drinks or juice and as a gelling or thickening agent. Pectin has also been the subject of special attention from nutritionists. It is used as dietary fiber and exerts physiological effects on the intestinal tract by increasing the transit time and the absorption of glucose [17].

This study aims to use a natural thickening agent extracted from fruit peel (orange and pomegranate peels) in printing paste. Then the extracted pectin from both orange and pomegranate peel was modified with hexamine to improve its ability to store and against high pH to be able to be used as a thickener in the printing paste of cotton fabric.

The rheological properties of extracted pectin from orange and pomegranate peels and modified pectin were evaluated, including the influence of pH. The effect of prepared pectin thickeners on the color strength, fastness, mechanical, and physical qualities of the different textiles was also analyzed.

During this study, the different fabrics (cotton, wool, acrylic, and polyester) were printed using the appropriate dyes for the fibers (reactive, acid, and disperse dyes).

2. Materials and Methods

2.1. Material

Cotton (100%; 158 g/m^2), polyester (100%; 160 g/m^2), polyacrylic (100%; 160 g/m^2), and wool (100%; 210 g/m^2) fabrics were purchased from Misr El-Mahala Co., El-Mahala, Egypt. The natural gel was extracted from the citrus peel (orange and pomegranate peels collected from the local market) to use as an eco-friendly thickener. Alginate was purchased from Fluka BioChemica GmbH Co. and DELL thickener P (as a polyacrylate inverse nonionic emulsion), produced by Delta for Chemical Industries, Egypt.

C.I. Reactive blue 19 as a reactive dye, Telon Blue BRL as an acid dye, and Dianix Blue SE-2R as a dispersed dye were kindly supplied by Dystar Co., Cairo, Egypt. Hostapal CV, an anionic textile auxiliary based on alkyl aryl polyglycol ether, was used as a detergent. The sodium bicarbonate, acetic acid, ethyl alcohol, and urea used were laboratory-grade chemicals.

2.2. Method

2.2.1. Extraction of Pectin from Fruit Peels

Pomegranate or orange peels were split/cut into four parts, and the peels were removed (a soft white substance inside the skin of citrus fruits). The peels were cut further into smaller pieces for easy drying and then washed with a large volume of water to remove glycosides and the bitter taste of the peels and to remove the residues of the pesticide spray. Afterward, they were air-dried for 24 h [18–20].

The peel pieces (30 g) were weighed and put into a 1000 mL beaker containing 450 mL distilled water, and 2.6 mL hydrochloric acid was used to achieve a pH of 1.27. After that, the fruit peel was boiled for an hour. The residues were then removed from the extracts by filtering them through a fine mesh. The fine mesh was rinsed with 250 mL of boiling water, and the combined filtrate was allowed to cool to 25 °C to prevent the heat destruction of the pectin. The extracted pectin was precipitated by adding 200 mL of 95% isopropanol to 100 mL of extracted pectin and thoroughly swirling for 30 min to allow the pectin to rise to the surface. The gelatinous pectin flocculants were subsequently removed by filtering. The pectin extract was refined by washing it in 200 mL of isopropanol and pressing it on nylon fabric to remove the HCl and universal salt residues. The pectin that resulted was measured, crushed into little pieces, and air-dried.

Finally, the dried pectin was ground into tiny parts with a pestle and mortar and measured with a digital weighing balance [20,21] Figures 1 and 2 illustrate a schematic diagram and photo for the preparation of pectin from fruit peel.

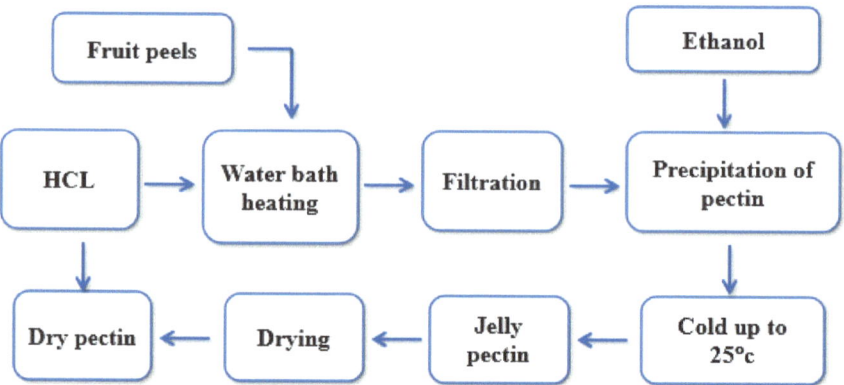

Figure 1. Scheme diagram for extraction of pectin from the fruit peels (orange and pomegranate).

Figure 2. Photo of extracted pectin from fruit peels. (**A**) Total fruit peels, (**B**) inside white part in fruit peels, (**C**) precipitated and filtrated pectin, (**D**) jelly pectin.

2.2.2. Modification of Pectin with Hexamine

Hexamine compound was used to modify the extracted pectin as follows: 5 g extracted pectin was dissolved in 100 mL distilled water under stirring until it formed a clear solution. Then, 30 g hexamine was added to the solution under constant stirring; the temperature was adjusted to 30 °C, and the mixture was kept under stirring for 48 h. The mixture was then dried overnight in a vacuum oven at 30 °C. The resulting modified pectin was used as resulted without further purification

2.2.3. Preparation of Thickening Agents

Different thickening agents (alginate, DELL P, extracted pectin, and modified pectin) were prepared in various concentrations (1, 2, 3, 4, 5%) to investigate their rheological behavior during this work. Alginate and DELL P thickeners were used for comparison with the prepared thickeners.

2.2.4. Preparation of Printing Paste

The printing paste was prepared using a different thickener, as described in the previous recipes. To prepare the printing paste, sodium carbonate, urea, glycerin, dispersing agent, citric acid, and dyes were added to the thickeners. Before applying the paste to the textiles, it was kept for 2–3 h after mixing (to allow the paste to relax after mixing and cooling to the internal temperature). The samples were then printed using a manual silkscreen with the printing produced.

The following recipe listed in Table 1 shows the recipe used to prepare the printing paste for cotton fabric using C.I. Reactive blue 19 as a reactive dye, for wool fabric using a Telon Blue BRL as an acid dye, and for polyester and acrylic fabrics using Dianix Blue SE-2R as a dispersed dye

Table 1. The recipe used to prepare the printing paste.

Type of Dye	Reactive Dye (Reactive Blue 19)	Acid Dye (Telon Blue BRL)	Disperse Dye (Dianix Blue SE-2R)
Amount of dye	3 g	3 g	3 g
Thickener	70 g	70 g	70 g
Urea	15 g	15 g	15 g
Sodium carbonate	till pH = 6 with pectin and pH 8 with alginate and modified pectin	till pH = 5.5	till pH = 6
Glycerin	2 g	2 g	2 g
Dispersing agent			2 g
Sodium dihydrogen phosphate			1 g
Water	Y g	Y g	Y g
Total	100 g	100 g	100 g

2.2.5. Application of Printing Paste to Textile Fabrics

Flat-screen printing was used during this research. The printed cotton fabric was dried at 100 °C for 3 min before being thermofixed at 140 °C for 5 min. The printed fabrics were then rinsed with cold water for about 15 min before being washed in warm water (50 °C) with a nonionic detergent for about 15 min and then dried at 85 °C for 5 min. The printed wool fabric was dried at 100 °C for 7 min before being thermofixed at 180 °C for 5 min. The printed fabrics were then rinsed with cold water for about 15 min before being washed in warm water (50 °C) with a nonionic detergent for about 15 min and then dried at 85 °C for 5 min. The printed polyester and acrylic fabrics were dried for 3 min at 100 °C and thermofixed for 3 min at 180 °C. The printed fabrics were then rinsed with cold water for about 15 min before being washed in warm water (50 °C) with a nonionic detergent for about 15 min, rinsed well, and dried at 85 °C for 5 min. The printed fabrics were subsequently immersed in a reducing bath with a 1:50 liquor ratio containing 2 g/L sodium hydrosulfite, 2 g/L sodium hydroxide, and 2 g/L wetting agent for 10 min at 60–70 °C. The textiles were then thoroughly washed with cold water, neutralized with 1 g/L acetic acid at 40 °C for 5 min, and dried at 85 °C for 5 min.

2.3. Analysis and Measurements

2.3.1. Characterization of Extracted Pectin

Fat Content

With 5 g dry peels powder and Soxhlet's equipment, the fat content was calculated using the ether extraction technique according to AOAC method no. 945.44 [22]. The extraction was then dried in a water bath before being baked in a vacuum oven at 100 °C for 1 h. The extract was then filtered to remove impurities from the fatty sample. The fat content of the sample was calculated as a percentage using the formula below.

$$\text{fat (\%)} = \frac{\text{weight of fat (g)}}{\text{weight of the sample}} \times 100$$

Protein Content

Micro-equipment Kjeldahl's was used to determine the protein content in line with AOAC method number 991.20 [23]. A total of 0.2 g defatted peel powder was digested in a mixture of catalyst (1 g) and 5 mL each of H_2O_2 and concentrated H_2SO_4 Before collecting the ammonia distillate liberated in boric acid, the digested sample was brought to a boil. Hydrochloric acid was used to titrate the distillation until the blue hue was fully eliminated. The equation outlined below was used to determine the protein content.

$$N(\%) = \frac{(S-B) \times N \times 14.007 \times \text{Volume made (mL)}}{\text{Weight of sample (g)} \times \text{Volume taken (mL)} \times 100}$$

$$\text{Protein (\%)} = \text{Nitrogen (\%)} \times 6.25$$

where S = mL of HCl required for sample titration, B = mL of HCl required for blank titration, N = normality of HCl (0.02 N)

Ash Content

Carbonizing a 2 g material at 290 °C for 6 h in a muffle furnace according to AOAC method 923.03 yielded the ash content [24]. The amount of ash in the sample was calculated using the formula below.

$$\text{Ash (\%)} = \frac{\text{Weight of ash (g)}}{\text{Weight of sample (g)}} \times 100$$

Carbohydrate Content

Carbohydrates were determined using the equation shown below:

$$\text{Carbohydrate} = 100 - (\text{moisture} + \text{fat} + \text{protein} + \text{ash})\%$$

The Percentage Yield of Pectin

The total extraction pectin yield (Y_{pec}) was calculated using the equation below:

$$Y_{pec}(\%) = \frac{P}{P_o} \times 100$$

where Y_{pec} (%) is the extracted pectin yield in percent, P is the amount of extracted pectin in grams, and P_o is the initial amount of pomegranate peel (30 g).

Solubility of Dry Pectin in Cold and Hot Water

By combining 0.25 g pectin with 10 mL 95 percent ethanol and 50 mL distilled water, the cold-water solubility of the pectin powder was evaluated. To determine solubility, the combination was forcefully agitated, and any quantity of insoluble material in the solution was weighed. The solution was then heated for 15 min at 85–95 °C to determine the existence of any insoluble material. [25]

Solubility of Pectin in Cold and Hot Alkali (NaOH)

A total of 5 mL of pectin solution was mixed with 1 mL of 0.1 N NaOH and heated at 85–90 °C for 15 min. [20,21]

Sugar and Organic Acids

A total of 1 g of pectin was added in separate 500 mL flasks and moistened with 5 mL ethanol, 100 mL water poured fast, agitated, and left to stand for 10 min; 100 mL ethanol with 0.3 mL hydrochloric acid was added to this solution, quickly combined, and filtered; then 2.5 mL of the filtrate was measured into a conical flask (25 mL), the liquid evaporated on a steam bath, and the residual dried in an oven at 50 °C for 2 h.

Determination of Equivalent Weight of Dried Pectin

In a 250 mL conical flask, 0.5 g of dried pectin was placed, and 5 mL of 95% ethanol was added. One gram of sodium chloride and 100 mL of deionized water were added. Then, six drops of phenol red were added and titrated against 0.1 M sodium hydroxide. The pink hue denoted the titration point. The equivalent weight was calculated using the following equation: [26]

$$EW = \frac{W_s \times 1000}{V_{NaOH} \times M_{NaOH}}$$

where EW = equivalent weight of dried pectin (%), V_{NaOH} = volume of alkali (ml), M_{NaOH} = molarity of alkali, and W_s = weight of sample

Determination of Methoxyl Content of Dried Pectin (MeO%)

The neutralized solution was collected after calculating the equivalent weight, and 25 mL of sodium hydroxide was then added. The blended solution was well agitated and left at room temperature (25 °C) for 30 min. Then, 30 min later, 25 mL of 0.25 M hydrochloric acid was added and titrated against 0.1 M sodium hydroxide. The methoxyl content was calculated using the following equation: [27]

$$MTC = \frac{V_{NaOH} \times M_{NaOH} \times 3.1}{W_s}$$

MTC = methoxyl content of dried pectin (%).
V_{NaOH} = volume of alkali (mL), M_{NaOH} = molarity of alkali, and W_s = weight of the sample.

Measurement of the Ash Content of Dried Pectin

A total of 5 g of dried pectin was placed in a weighted empty crucible. The crucible was moved to a furnace heated at 60 °C to burn out any biological stuff. The carbon scorched and eventually burned away as carbon dioxide, leaving black ash behind. This procedure took 24 h. The crucible was removed from the furnace and cooled in a desiccator. After cooling, the crucible was reweighed. The ash content was calculated using the following equation: [28]

$$AC = \frac{W_a}{W_s} \times 100$$

AC = ash content of dried pectin (%), W_a = weight of ash (g), and W_s = weight of the sample (g).

Determination of the Degree of Esterification of Dried Pectin

At 40 °C, 20 g of dry pectin was moistened with ethanol and soluble in 20 mL of distilled water. After the pectin had completely dissolved, 5 drops of phenolphthalein were added to the solution. Following that, the solution was titrated with 0.5 M sodium hydroxide; in addition, the amount of the sodium hydroxide solution utilized for color change was noted as V_1. Following that, 10 mL of 0.5 M sodium hydroxide was added, and the solution was vigorously agitated and allowed to rest for 15 min. In addition, 10 mL of 0.5 M hydrochloric acid was added, and the solution was agitated till the pink tint was gone. During the last stage, the solution was titrated with 0.5 M sodium hydroxide, and the amount consumed was noted as V_2. The degree of esterification was calculated using the following equation: [29–31]

$$\text{DE (\%)} = \frac{V_2}{V_1 + V_2} \times 100$$

DE = degree of esterification of dried pectin (%), V_1 = initial titration volume (ml), and V_2 = final titration volume (mL).

Determination of Total Anhydrouronic Acid Content of Dried Pectin

Estimating the anhydrouronic acid content (AUA) is critical for determining purity and the degree of esterification (DE), as well as for assessing pectin's physical characteristics. Pectin, a partially esterified polygalacturonide, comprises at least 10% organic material in the form of galactose, arabinose, and other sugars. Using comparable weight, methoxyl content, and ash alkalinity values, anhydrouronic acid was calculated [32].

$$\text{AUA (\%)} = \frac{(V_m + V_e) \times M_{NaOH} \times 176}{W_s \times 1000} \times 100$$

where V_m = mL (titer) of NaOH from methoxyl content determination, V_e = mL (titer) of NaOH from equivalent weight determination, W_s = weight of the sample, and 176 is the molecular weight of anhydrouronic acid.

Determination of Acetyl Value (ACV) of Dried Pectin

With stirring, 0.5 g of dried pectin sample was dissolved in 0.1 M sodium hydroxide solution and left to stand overnight. Distilled water was used to dilute the contents to 50 mL, and an aliquot (20 mL) was placed in the distillation apparatus. A magnesium sulfate–sulfuric acid solution (20 mL) had also been added to the distillation device and distilled, yielding approximately 100 mL of distillate. Utilizing phenol red as an indicator, the distillate was titrated with 0.5 M sodium hydroxide. A blank distillation was performed utilizing 20 mL of the magnesium sulfate–sulfuric acid solution, and the distillate was titrated. The acetyl content was calculated using the following equation: [27]

$$\text{ACV} = \frac{V_{NaOH} \times N_{NaOH} \times 4.3}{W_s}$$

ACV = acetyl value of orange peel dried pectin (%)

Moisture Content Determination

An emptied dry petri dish was baked, chilled in a desiccator, and weighed. A total of 0.5 g of pectin samples was placed in the crucibles in an oven set at 130 °C for one hour, following which the Petri dish was removed, cooled in a desiccator, and weighed.

This procedure was carried out once more. The moisture content was calculated using the following equation: [20,21]

$$\text{Moisture Content (\%)} = \frac{\text{Weight of the Residue}}{\text{Weight of the Sample}} \times 100$$

2.3.2. Rheological Behavior and Power Law

The rheological activity and power low of the thickening agent were studied at $25 \pm 0.1\ °C$ with a coaxial rotary viscometer (HAAK V20), Germany [33,34]. A co-axial rotary viscometer (HAAK V20, Germany) was used to evaluate the thickening agent's rheological behavior at $25\ °C$. The result of shear stress (τ; dyn/cm^2) versus shear rate (γ; s^{-1}) is known in Newton's law as the apparent viscosity (η; cP), as shown in the equation: [35]

$$\eta = \frac{\tau}{\gamma}$$

The power-law equation of Ostwald de Waele is another name for the power-law model.

$$\tau = K\gamma^n$$

where γ represents shear rate, and K represents the consistency coefficient, which specifies the entire viscosity range throughout the region of the current flow curve and is the viscosity or stress at a certain shear rate point. The index of the power law is represented by the n number. The n value for a shear-thinning fluid was larger than 0 and less than 1 (0 n 1), indicating that the closer a sample is to zero, the more shear thinning. As a result, the viscosity may be defined as [36–38]

$$\eta = K\gamma^{n-1}$$

When shown in logarithmic form, log shear stress vs. log shear rate graphs of the top curve for many fluids become linear. The data of shear thinning and shear thickening for fluids is described by the power-law model. As a result, the following equation can be constructed by taking the natural logarithms of both sides from the preceding equation:

$$\log \eta = (n-1)\log \gamma + \log K$$

When plotting log (η) versus log (γ), the result is linear. It is, nonetheless, helpful in evaluating and spotting trends in experimental data. This model is especially valuable since it can provide data with shear rates ranging from 10 to 10^4 s^{-1} [36]. The disadvantage of this approach is that it does not account for shear-thinning fluids with a constant viscosity at low and high shear rates [39,40].

2.3.3. FTIR Measurement

The FTIR tester of the JASCO spectrometer connected with the diamond ATR unit was used to analyze the spectrum of the modified pectin. The measurement was done from 400 to 4000 cm^{-1} by the accomplishment of 128 reads.

2.3.4. Color Measurements

The Hunter Lab Ultra-Scan Pro was used to evaluate the color intensity of printed fabrics at Egypt's National Research Center. The traditional form is represented by the characters K/S. The Kubelka–Munk equation was used to determine the K/S values [41–45].

$$K/S = \frac{(1-R)^2}{2R} - \frac{(1-R_o)^2}{2R_o}$$

where K is the absorption coefficient; S is the dispersion coefficient; R_{kmax} is the fabric's reflectance at its highest wavelength.

2.3.5. Colorfastness Properties

The colorfastness to washing was tested by using a Laudner-Ometer according to the AATCC test procedure 61–2013 [46]. The colorfastness against friction (dry and wet) was tested utilizing a crock meter according to the AATCC test method 8–2016 [47]. AATCC test method 15–2013 was used to measure colorfastness to perspiration (acidic and alkaline) [48]. The colorfastness to light was determined according to the AATCC test method 16.1–2014 [49]. The Gray Scale guideline for the color shift was used to evaluate the printed textiles.

2.3.6. Mechanical Properties of the Treated Fabric

Tensile properties must be measured at 25 °C and 65 percent relative moisture using an FMCW 500 tensile strength machine (Veb Thuringer Industrie Werk Rauenstein 11/2612 Germany) following the ASTM D1682-59T [50]. The crease recovery angle (CRA) was measured using the AATCC 66–2014 [51]. The fabric roughness was determined using the surface roughness instrument SE 1700 following ASTM D 7127–13 [52]. The stiffness or rigidity of printed textiles was tested using only cantilever equipment following ASTM D 1388-14e1 [53,54].

2.3.7. Antibacterial Activity

Antibacterial activity was quantitatively tested against *Staphylococcus aureus* (ATCC 29213) as a Gram-positive bacteria and *Escherichia coli* (ATCC 25922) and *Candida Albicans* (ATCC 10231) as a fungus using the AATCC 100-2004 (bacterial-reduction method) [55], which is a popular methodological model for antimicrobial paste studies [56].

This test cultivates uniform microbes with liquid culturing. The culture is diluted in a nutrient solution that has been sterilized. In sealed containers, the thickening agent is infected with microorganisms for 24 h at 37 °C. Shake for 1 min after incubation to measure bacteria levels. Eventually, the quantity of microbes compared to the starting concentration was, as follows, a percentage reduction of bacteria (R%) [57]

$$\text{Percent reduction of bacteria (R\%)} = \frac{B - A}{B} \times 1000$$

Therefore, A is the number of bacteria collected from an inoculated lab test in a jar throughout the desired length of contact, and B is the number of bacteria recovered from an inoculated measurement specimen in the jar immediately after inoculation (at "0" contact time).

2.3.8. Handle and Sharpness

Touch and eye observation were used to evaluate the textile handling and sharpness of the printed region. The cloth was examined by three specialists, and the average of their ratings was recorded. The textile handling was rated as smooth (S) or harsh (H), and the sharp outline of the printed region was rated as sharp (Sh) or not sharp (Ns).

3. Results

3.1. Characterization of Fruit Peels

On a dry basis, moisture content of 9.32 and 9.11%, carbohydrate content of 79.49 and 80.29%, crude protein content of 5.54 and 5.23%, fat content of 2.11 and 2.13%, and ash content of 3.54 and 3.24% were found in dried orange and pomegranate peel powders, respectively.

Due to its gelling characteristic, which is suited for the manufacture of low-calorie and dietetic meals, there has lately been growing interest in the synthesis of pectin [58].

3.2. Characterization of Extracted Pectin

The physical analysis results for extracted pectin from both orange and pomegranate peels are listed in Table 2. The total extraction yield (Y_{pec}) in wet processing from both orange and pomegranate peels was 16.3 and 8.14%, respectively (see Table 3), which is as reported by researchers [32,58–62]. Previous research has found that severe extraction conditions improve pectin extraction yield from banana peel, apple pomace, and lemon by-product [63,64].

Table 2. Physical analysis results for extracted pectin from both orange and pomegranate peels.

Parameter	Orange Peel	Pomegranate Peel
Color	Light yellow	Dark brown
Solubility in cold water	Dissolved slightly and forms a suspension after vigorous shaking	
Solubility at 85–90 °C for 15 min	The mixture dissolves.	
Solubility of pectin suspension in cold alkali	The pectin suspension forms a yellow precipitate.	
Solubility of pectin suspension in hot alkali	The pectin suspension dissolved and turned milky.	
Sugar and organic acid	36.1%	37.2%

The extraction of pectin from orange, passion fruit, dragon fruit, and citric waste peels at a correspondingly increased extraction temperature and at low pH results in increased pectin production. Low pH indicates increased acid concentration (acidic), and a higher acidity level boosts the extraction yields of different kinds of pectin and proto-pectin [65].

This is due to the fractionation of glycosidic bonds in neutral polysaccharides that are much more sensitive to pH than the connection between two galacturonic acids, resulting in the breakdown of neutral sugar side chains.

Furthermore, a high concentration of hydrogen ions in the solvent triggers protopectin hydrolysis at low pH, causing the repression of the ionization of the hydrated carboxylate groups by turning them into hydrated carboxylic acid groups [63,66]. This loss of carboxylate groups reduces the repulsion of polysaccharide molecules, which increases pectin gelation, resulting in much more precipitated pectin at lower pH [67].

In general, pectin is differentiated by the degree of methylation, which is a crucial factor in pectin function regulation. The degree of esterification for dried pectin (DE) from both orange and pomegranate peels was 63.8 and 63.2%, respectively (see Table 3). The degree of esterification is a key molecular marker for pectin categorization because it reflects the extent to which carboxyl groups in pectin molecules exist as methyl ester. The pectin produced is classified as high methoxyl pectin (HMP), since it has a high proportion of esterification, more than 50%. Pectin with a significant level of esterification can form gel fast at high temperatures, resulting in a more effective influence on the lipid profile [68]. The degree of esterification, on the other hand, merely shows the ratio of methanol-esterified carboxyl groups to free carboxyl groups, while the methoxyl rate relates to the number of methoxyl groups in a sample [69].

Because the degree of methyl esterification (DM) for both orange and pomegranate peels were 6.07 and 5.32%, respectively (see Table 3), predominantly high methoxyl pectin was recovered. Lower DM levels were obtained as a result of extraction as corroborating results reported for pectin extraction from banana peels and durian rinds [63]. This is attributed to the de-esterification of the polygalacturonic acid chain being aided by harsher extraction conditions [63,70,71]. Furthermore, stronger treatments may have extracted more firmly bonded pectin, potentially with a lower DM content [72].

Kanmani et al. discovered that the methoxyl content of pectin ranges between 0.2 and 12% based on the source and technique of extraction. Because the value achieved was less than 7%, the dried pectin has a low ester characteristic, implying that it is good in terms of quality. Pectin with a low methoxyl concentration creates a thermo-irreversible gel, meaning it remains gelled even when heated to temperatures that would ordinarily dissolve it [60,73].

The equivalent weights for both orange and pomegranate peels were 614.7 and 523.21 g, respectively (see Table 3). Pectin has a noticeable physical feature called equivalent weight, which is the most significant factor in defining pectin's functional behavior. Individual pectin gelling capacities are extremely strongly related to comparable weight. A higher equivalent weight has a stronger gel-forming impact. A low equivalent weight indicates a higher partial breakdown of the pectin, which is undesirable [74].

The total anhydrouronic acid level of dried pectins for both orange and pomegranate peels were 72.67 and 66.32%, respectively (see Table 3). Pectin must include at least 65% galacturonic acid, according to the Food Chemical Codex (FCC), Food and Agriculture Organization (FAO), and European Union (EU) [75]. The amount of anhydrouronic acid in pectin affects its gelling ability. The high value obtained indicates that the extracted pectin has little protein.

The dried pectin from both orange and pomegranate peels has an acetyl value of 0.39 and 0.35%, respectively (see Table 3). The gelling ability of pectin also decreases as the degree of acetylation increases. The presence of an acetyl group in pectin reduces gel formation. Because of the low value attained, pectin is an effective gelling agent.

Table 3. Quantitative analysis results for extracted pectin from both orange and pomegranate peels.

Parameters	Orange Peel		Pomegranate Peel	
	Actual Result	Reference Result	Actual Result	Reference Result
Degree of esterification	63.85	65.49 ± 0.57% [14]	63.21	61.67% [58]
Equivalent weight	614.73	599.74% [27]	523.21	
Methoxyl content	6.07	6.23% [27]	5.32	
Total anhydrouronic acid	72.67	70.9% [27]	66.32	6.63 ± 0.40 [76]
Acetyl value	0.39	0.4% [27]	0.35	
Ash content	35.88	35% [32]	51.2	4.97 ± 0.22 [76]
Moisture content	92.28	95.25% [32]	66.92	67.26 ± 0.23 [76]
The percentage yield of pectin on a wet basis	16.32	15.92% [32]	8.14	7.84% [58]
The percentage yield of pectin on a dry basis	1.64	1.68 [32]	1.21	

3.3. Rheological Properties of Extracted Pectin

A fluid's "viscosity" is a measurement of its resistance to flow. Newtonian (or simple flow) and non-Newtonian complex flow are both possible in fluids. Newtonian flow is an ideal state in which viscous liquids obey Newton's rule of viscous flow, which says that the flow rate is proportional to applied shearing stress. Most solvents, such as water, alcohol, benzene, real solutions, and extremely dilute colloidal solutions, are found in Newtonian fluids. Shear rate (S) and shear stress (F) in Newtonian fluids have a straight-line connection and are stated in terms of absolute viscosity (η):

$$F = \eta S$$

Non-Newtonian fluids are fluids that do not obey Newton's law of flow. Because shear rate and shear stress in these fluids do not have a linear relationship, they are typically described in terms of apparent viscosity (η_{app}).

The rheological properties of five different concentrations of extraction pectin from orange and pomegranate peels (1, 2, 3, 4, and 5%) were investigated. Figure 3 depicts the effect of shear rate on shear stress and viscosity of extracted pectin, demonstrating that the flow of prepared thickeners is characterized by the formation of a hysteresis loop that starts at 0 and goes up to 40 before reversing and ending at a zero-share rate. It is also possible to assume that the thickener, regardless of concentration, has a pseudo-plastically non-Newtonian shear-thinning flow, implying significant thixotropy. The time required to fix the thickener's distorted internal structures is linked to its thixotropic activity. The degree of thixotropy is defined as the area between curves up and down [77,78].

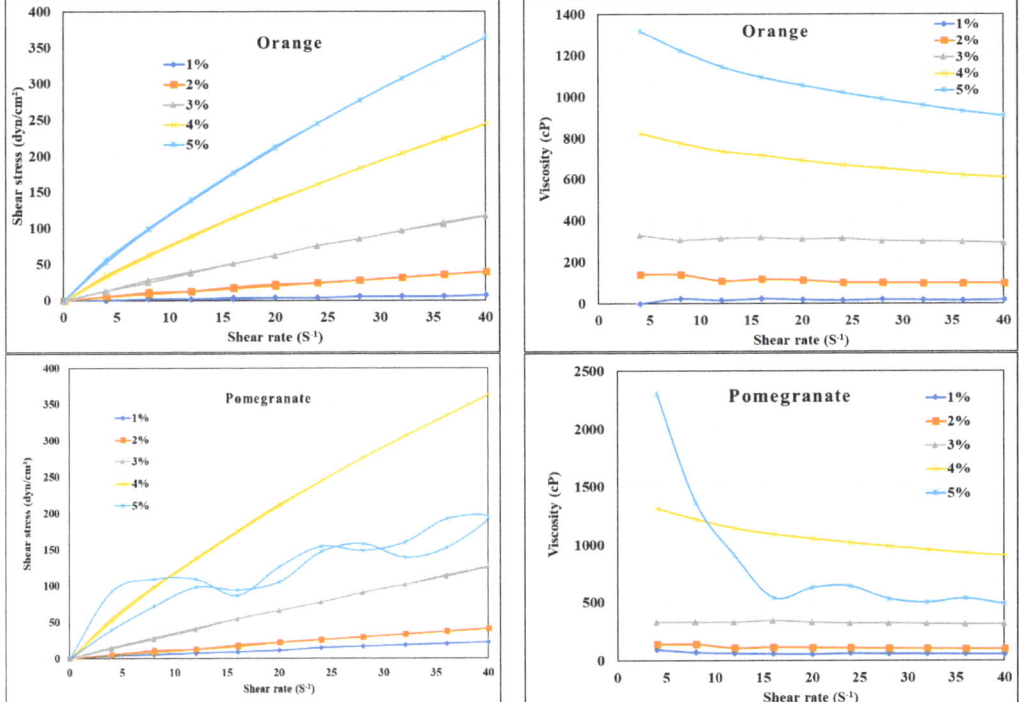

Figure 3. The effect of shear rate on shear stress and apparent viscosity for various pectin concentrations (from orange or pomegranate peels extract) as a thickening agent at $25 \pm 1\ °C$.

As a result, the lower the thixotropy, the more elastic the thickening (i.e., the capacity for the internal structure to recover after releasing the shear). As a result, the thickening is thought to be more elastic due to the lower thixotropy (which is between the up and down curves).

Table 4 shows the area between curves for various extracted pectin concentrations from both orange and pomegranate peels, revealing that 4 percent of pectin has a lower area, while lower concentrations have a practically identical area from up and down the curve, indicating improved elasticity and thickening effectiveness. In addition, the concentration of pectin increasing by more than 4% led to irregular and unstable behavior in both shear stress and viscosity, which proves that 4% of pectin is the most favorable to use as a thickening agent.

3.4. Characterization of Modified Pectin

As pectin has acidic pH and increasing its pH led to destroying its viscosity, which makes it unsuitable for use as a thickening agent in the printing paste with high pH (especially for cotton fabric), a modification for pectin occurred to improve its rheological behavior, as well as to use it as a thickening agent in printing paste by improving its ability to work at high pH (9).

Hexamine, also called hexamethylenetetramine, is a heterocyclic organic compound. Extracted pectin from both orange and pomegranate peel was modified with hexamine to improve its ability to keep and to protect against high pH to be able to be used as a thickener in the printing paste of cotton fabric. Figure 4 illustrates the suggested mechanism for the modification of pectin

Table 4. Shear rate vs. shear stress curves' area between up and down curves using pectin as a thickening agent (from orange or pomegranate peel extract).

Thickener	Source of Pectin	The Area between Up and Down Curves
1% pectin	Orange peels	1.891
	Pomegranate peels	1.845
2% pectin	Orange peels	5.626
	Pomegranate peels	3.781
3% pectin	Orange peels	7.553
	Pomegranate peels	7.477
4% pectin	Orange peels	3.329
	Pomegranate peels	11.182
5% pectin	Orange peels	11.182
	Pomegranate peels	207.48

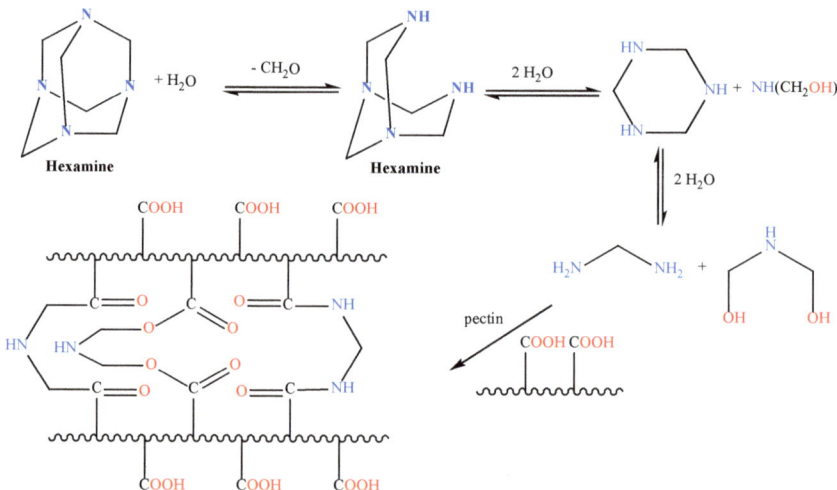

Figure 4. Suggesting mechanism for modification of pectin.

The infrared spectra for pectin, hexamine, and modified pectin are illustrated in Figure 5 and the presence of the common peaks for both original raw materials (pectin and hexamine) is displayed.

The major peaks of extracted pectin are the hydroxyl group that appeared at 3410 cm^{-1} for pectin. The aliphatic C–H groups esterified carboxyl groups, and free carboxyl groups were shown in 2928, 1748, and 1640 cm^{-1} for pectin [79]. The peak at 1443 cm^{-1} describes the presence of a CH methyl group for extracted pectin and standard pectin. In addition, peaks at 1160 and 1154 cm^{-1} indicate the presence of C–O bonds in alcohols, esters, and carboxylic acids [80].

The IR spectrum of the hexamine compound showed absorption bands at 650, 795, 990, 1225, 1370, 1445, 1674, 2866, and 2912 cm^{-1} which were similar to those reported according to the literature [81]. Of these, the bands at 1225 and 990 cm^{-1} were assigned to the C–N stretching modes. In this regard, the modified pectin has the same bands likely attributed to the C–N stretching, which confirmed the modification process.

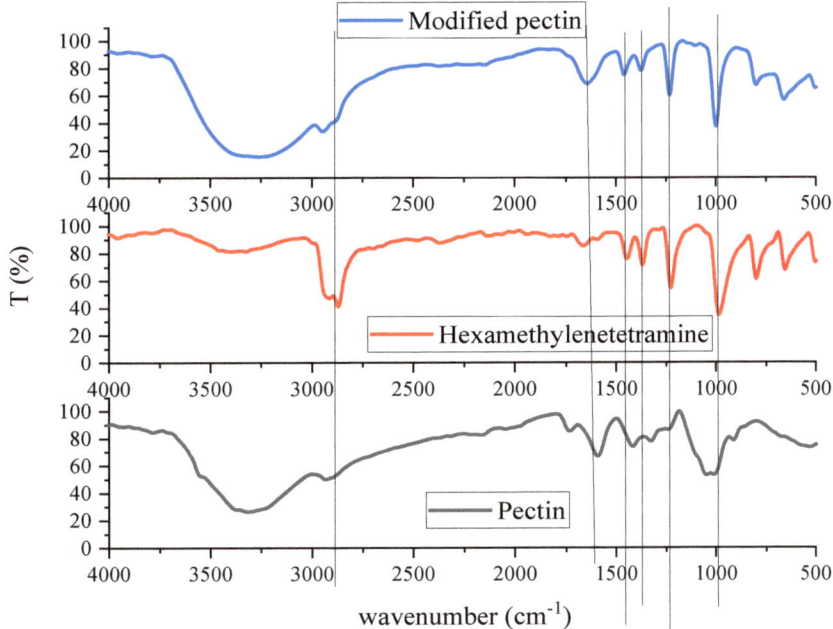

Figure 5. FT-IR spectra of pectin, hexamine, and modified pectin.

In addition, modified pectin has two vibration bands located at 2912 and 1225 cm^{-1} assigned, respectively, to the (CH) and (CN) vibration, and two bands located at 1445 and 1370 cm^{-1}, characteristics of the $\delta_{as}(CH_2)$ and $\delta_s(CH_2)$ deformation of the hexamethylene–tetramine ligand [82].

Effect of Storing Time on Pectin and Modified Pectin Performance

Figure 6 depicts the dehydration mechanisms that cause polysaccharides to become (a) longer and (b) cross-linked and branching

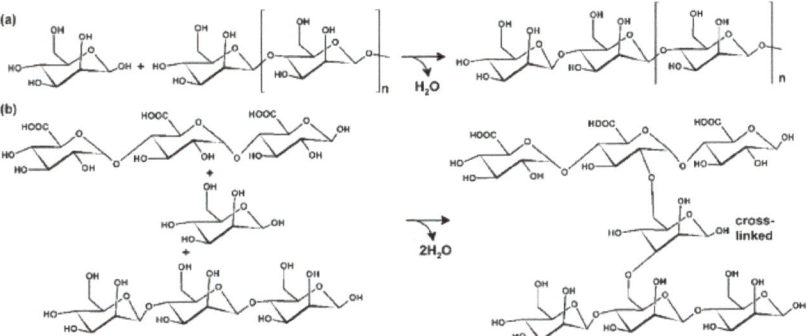

Figure 6. Diagrams depicting the dehydration mechanisms that cause polysaccharides to become (a) longer and (b) cross-linked and branching.

The storage-time investigation revealed that storage time has a significant impact on the color of gels (4%) from both extracted pectins, which are altered by each constituent. When compared to those stored at room temperature (25 °C), the pectin and its modified gel from both fruit peels held at 4 °C were considerably not changed. This might be due to the enzymatic activity being activated at ambient temperature, causing color deterioration [34,83]. When the gels were kept at 4 °C, the greatest color activity occurred without alteration. Furthermore, as demonstrated in Table 5, the modification of both pectins maintained the gel's final color consistent even after 15 days of storage.

Table 5. Effect of storage time on the visual color of pectin and modified pectin.

Storing time (days)	(a) The Visual Color							
	Pectin from orange peel (4%)				Pectin from pomegranate peel (4%)			
	Pectin		Modified Pectin		Pectin		Modified Pectin	
	4 °C	25 °C	4 °C	25 °C	4 °C	25 °C	4 °C	25 °C
0–3	Transparent		Transparent		Yellowish		Yellowish	
5	Transparent	Yellowish	Transparent	Yellowish	Yellowish	Fungi spots		
7	Yellowish	Fungi spots	Transparent	Yellowish	Yellowish	Fungi layer	Yellowish	
15	Yellowish	Fungi layer	Transparent	Yellowish	Yellowish	Fungi layer		

The pH of gels (4%) from both extracted pectins held at ambient temperature ranged between 3.57 and 4.12 and 3.71 and 4.54, which were lower than the pH of the pectin gels ranging from 3.97–4.12 and 4.38–4.54 maintained at 4 °C, for the extracted pectin from orange and pomegranate peels, respectively. This is because the enzymatic activity produces acid, which lowers the pH at ambient temperature. At both storage temperatures, the modified gel is more stable than the pectin gel solely. The pH value of the pectin and its modification following storage are shown in Table 6.

Table 6. Effect of storage time on the pH of pectin and modified pectin.

Storing time (days)	(b) pH							
	Pectin from orange peel (4%)				Pectin from pomegranate peel (4%)			
	Pectin		Modified Pectin		Pectin		Modified Pectin	
	4 °C	25 °C	4 °C	25 °C	4 °C	25 °C	4 °C	25 °C
0	4.12	4.12	9.11	9.11	4.54	4.54	9.01	9.01
1	4.10	4.05	9.06	8.96	4.51	4.47	8.96	8.87
3	4.07	3.97	9.00	8.82	4.49	4.38	8.90	8.72
5	4.05	3.87	8.95	8.68	4.46	4.26	8.85	8.58
7	4.02	3.73	8.89	8.54	4.43	4.11	8.80	8.45
10	4.00	3.57	8.84	8.40	4.41	3.93	8.74	8.31
15	3.97	3.37	8.79	8.27	4.38	3.71	8.69	8.18

The viscosity of the pectin gel increased after being kept at 4 °C for 15 days. The reason for this is that the gel is thicker, and evaporation reduces the amount of water stored in the gel [83]. Furthermore, after 15 days of storage, the pectin structure remained stiff, which is due to the presence of hydrogen bonding, which can create a cross-linking connection that affects its viscosity (see Table 7). Furthermore, gel storage at room temperature (25 °C) was designed to produce an enzyme that attacks the pectin structure and the hydrogen bonding and causes a decrease in the molecular weight as well as the viscosity (see Table 7). The dissociation of the polymer chain of carbohydrates is used in the liquefaction gel-forming a reactive functional group (see Figure 6) [84].

Table 7. Effect of storage time on the apparent viscosity of pectin and modified pectin.

	(c) Apparent Viscosity															
Storing time (days)	Pectin from orange peel (4%)								Pectin from pomegranate peel (4%)							
	Pectin				Modified Pectin				Pectin				Modified Pectin			
	4 °C		25 °C		4 °C		25 °C		4 °C		25 °C		4 °C		25 °C	
Share rate (S^{-1})	4	20	4	20	4	20	4	20	4	20	4	20	4	20	4	20
0	2309.45	924.33	2309.45	924.33	2739.24	990.42	2739.24	990.42	2049.41	820.25	2049.41	820.25	2430.80	878.90	2430.80	878.90
1	2314.50	943.84	1910.12	888.25	2746.81	1011.32	2740.21	1142.36	2053.89	837.56	1695.04	788.23	2437.52	897.45	2431.66	1013.73
3	2319.55	963.68	1623.35	666.74	2754.39	1023.56	2755.36	1148.68	2058.37	855.17	1440.56	591.67	2444.24	908.31	2445.11	1019.33
5	2324.60	983.51	1463.44	443.75	2761.96	1035.80	2770.51	1154.99	2062.85	872.77	1298.66	393.78	2450.96	919.17	2458.55	1024.94
7	2329.65	1003.35	1303.11	321.56	2769.53	1048.04	2785.66	1161.31	2067.33	890.37	1156.38	285.35	2457.68	930.03	2471.99	1030.54
10	2334.70	1023.19	1143.78	320.35	2777.10	1060.28	2800.81	1167.62	2071.81	907.98	1014.99	284.28	2464.40	940.89	2485.44	1036.15
15	2339.75	1043.03	983.91	319.16	2784.68	1072.52	2815.96	1173.94	2076.29	925.58	873.12	283.22	2471.12	951.75	2498.88	1041.75

Table 7 shows the viscosity values of pectin and its modified gel at 4 and 20 share rates during 15 days of storage at 4 °C and room temperature (25 °C). Because it prevents the water molecule from evaporating as a result of its network trapping water molecules, the modified pectin did not affect the viscosity by increasing the storage times of the two rates under investigation (4 and 20) even in both storage conditions (cold (4 °C) or room temperature (25 °C)).

After 15 days of storage at room temperature (25 °C), prepared pectin and its modified gels extracted from orange or pomegranate peels were evaluated for antibacterial activity against Escherichia coli, Staphylococcus aureus, and Candida albicans using the counting process. The percentage of antibacterial decrease in the presence of the produced pectin gels is shown in Table 8.

Table 8. Effect of storage time on the bacteria reduction % of pectin and modified pectin.

	(d) Bacteria Reduction %											
Storing time (days)	Pectin from orange peel (4%)						Pectin from pomegranate peel (4%)					
	Pectin			Modified Pectin			Pectin			Modified Pectin		
	E. coli	S. aureus	C. albicans	E. coli	S. aureus	C. albicans	E. coli	S. aureus	C. albicans	E. coli	S. aureus	C. albicans
0	88.32	87.62	85.62	92.84	92.11	90	94.06	93.31	91.18	91.43	90.71	88.64
5	85.62	84.94	83	90	89.29	87.25	91.18	90.46	88.4	88.64	87.94	85.93
10	83.74	83.08	81.18	88.03	87.33	85.34	89.18	88.47	86.45	86.69	86	84.04
15	81.31	80.67	78.82	85.47	84.8	82.86	86.59	85.91	83.95	84.18	83.51	81.6

The bacteria reduction % for produced pectin gels had the same antibacterial and antifungal effects against both bacteria and fungi. With increasing storage duration, the bacterium-reduction percent of produced gels was decreased in microbial resistance. These data suggested that extending the storage duration affected the produced gels' microbiological resistance. Additional proof indicates that both modified pectin gels increase the microbiological characteristics more than pectin gel alone after increasing the storage duration.

3.5. Characterization of Printed Fabrics

Because the goal of this study was to evaluate and find the best possible printed material to achieve darker prints with better overall fastness properties using new eco-friendly biological thickener extracted from fruit peels, extracted pectin and pectin modified with hexamine ere used in the printing of different fabrics (namely, cotton, wool, acrylic, and polyester) with appropriate dyes for each textile material.

3.5.1. Color Strength and Fastness Properties

Color strength (K/S) and various fastness properties, such as light, washing, perspiration, and rubbing fastness, were evaluated for all printed textiles (cotton, wool, acrylic, and polyester fabrics) with appropriate dye for each fabric using alginate, DELL P, pectin, and modified pectin as thickeners, and the results are listed in Table 9.

Table 9. Color strength (K/S) and fastness properties of different printed fabrics using a different thickening agent.

Fabric	Dye	Thickening Agent (5%)	K/S	Washing			Rubbing		Fastness Properties						Light	Handling	Sharpness
									Perspiration								
									Acidic			Alkaline					
				Alt.	SC	SW	Dry	wet	Alt.	SC	SW	Alt.	SC	SW			
Cotton	Reactive dye	Alginate	8.61	4	3	3	3	2	3	3	3	3	3	3	5	S	Sh
		Pectin	8.83	4–5	3–4	3–4	4	3	3–4	4	4	4	4	4	7	S	Sh
		Modified Pectin	9.75	4–5	4	4	4	4	4–5	4–5	4–5	4–5	4–5	4–5	7	S	Sh
Wool	Acid dye	Alginate	4.19	4	3	3	3	3	3	3	3	3	3	3	7	S	Sh
		Pectin	8.77	4–5	3–4	3–4	4	3–4	4	4	4	4	4	4	7	S	Sh
		Modified Pectin	9.35	4–5	4	4	4–5	4–5	4–5	4–5	4–5	4–5	4–5	4–5	7	S	Sh
Acrylic	Disperse Dye	Alginate	5.15	3–4	3–4	3–4	3–4	3–4	3–4	3–4	3–4	3–4	3–4	3–4	6–7	S	Ns
		DELL	6.87	4–5	4	4	4–5	4–5	4–5	4–5	4–5	4–5	4–5	4–5	7	S	Sh
		Pectin	7.69	4–5	3–4	3–4	4	3–4	4	4	4	4	4	4	7	S	Sh
		Modified Pectin	8.35	4–5	4	4	4–5	4–5	4–5	4–5	4–5	4–5	4–5	4–5	7	S	Sh
Polyester	Disperse Dye	Alginate	4.11	4–5	4–5	4	4–5	4–5	4–5	4	4–5	4–5	4	4–5	5	S	Ns
		DELL	5.02	4–5	4	4	4–5	4–5	4–5	4–5	4–5	4–5	4–5	4–5	7	S	Sh
		Pectin	5.12	4–5	3–4	3–4	4	3–4	4	4	4	4	4	4	7	S	Sh
		Modified Pectin	6.47	4–5	4	4	4–5	4–5	4–5	4–5	4–5	4–5	4–5	4–5	7	S	Sh

The lightfastness of printed textiles containing each of the pectin's gels as thickeners were excellent (6), better than using alginate as a thickener (4). The washing fastness for all printed fabrics was good (4–5). The perspiration fastness property showed 3–4 to 4–5 color differences in both acidic and alkaline perspiration. The colorfastness to rubbing was tested in both dry and wet procession, and it was discovered that dry rubbing had a similar impact as wet rubbing.

Furthermore, for printed cotton and wool fabrics with reactive dye, both pectin thickeners have great color strength and fastness properties, as well as a sharp outline and soft handling in comparison to the printed cotton fabric using alginate as thickeners. Furthermore, modified pectin imparts higher color strength and fastness properties than pectin only. In addition, compared to printed fabrics using all examined thickeners, the printed fabric using alginate as a thickening agent delivered a fair color value.

Furthermore, all thickeners based on aloe vera gel have great color strength and fastness properties with sharp edges and gentle handling for printed polyester or acrylic fabrics with dispersed dye. In addition, compared to printed fabrics using alginate as a thickener, the printed polyester or acrylic fabrics using DELL P as a thickening agent delivered excellent color value.

Pectin and its modified thickeners may be utilized for printing cotton, wool, acrylic, and polyester textiles under varied pH values, depending on the printing condition of each printed fabric, based on their overall performance. These results correspond to the thickener performance. In addition, modified pectin shows better results compared with pectin only.

3.5.2. Mechanical and Physical Properties

The mechanical and physical characteristics of printed textiles, such as tensile strength, elongation at a break, bending length, crease recovery angle, and surface roughness, have been examined, and the findings are displayed in Table 10.

Table 10. Physical and mechanical properties of printed fabrics using a prepared thickening agent.

Fabric	Dye	Thickening Agent (5%)	Physical and Mechanical Properties				
			Tensile Strength (N/mm^2)	Elongation at a Break (%)	Bending Length (cm)	Crease Recovery Angle (Warp + Weft) (°)	Surface Roughness
Cotton	Reactive dye	Alginate	12.2	6.4	3.8	210.8	17.6
		Pectin	13.3	8.8	4.0	202.6	16.7
		Modified Pectin	13.6	8.9	4.1	201.6	16.8
Wool	Acid dye	Alginate	20.9	9.7	4.1	206.7	18.3
		Pectin	22.0	12.1	4.4	207.7	17.6
		Modified Pectin	22.3	12.2	4.5	206.7	17.6
Polyester	Disperse Dye	Alginate	21.0	10.3	4.4	204.6	18.3
		DELL	22.1	12.6	4.6	209.7	17.5
		Pectin	22.4	12.7	4.7	208.7	17.5
		Modified Pectin	22.7	13.2	4.7	208.7	17.4
Acrylic	Disperse Dye	Alginate	23.2	13.3	3.7	200.0	15.5
		DELL	24.3	15.6	4.0	205.0	14.9
		Pectin	24.6	15.7	4.0	203.0	14.9
		Modified Pectin	24.9	16.2	4.1	205.0	14.8

The use of pectin-based thickeners increased the tensile strength and elongation at a break of the printing materials, as seen by the data Table 10. Rather than using pectin only as a thickener in the printing paste, the modified pectin helps to create and alter the thin film in the microstructure of the fabric, which fills gaps on the surface of the printed fabrics and improves both tensile strength and elongation at a break [85].

The bending length presence of printed textiles with pectin-based thickener showed superior values compared to the printed fabrics with alginate as a thickener, according to the data Table 10. Furthermore, the use of modified pectin as a thickener in the printing of all the textiles tested, with all the essential colors, had a better bending length value than the use of alginate. This behavior might be attributed to the presence of amine groups in the coating layer that forms as a result of the printing paste. The use of a novel thickener in the printing of various textiles has helped in the stiffness of the printed fabrics, according to these studies.

Further investigation into the angle of the crease recovery of printed textiles in both the warp and weft directions revealed that all printed materials have comparable angles of crease recovery. These findings show that the new thickeners did not affect the angle of the crease recovery of printed textiles in a variety of printing paste formulas and pH levels.

3.5.3. Antimicrobial Activity of Printed Fabrics

Using three types of microbes, Gram-positive bacteria (Staphylococcus aureus), Gram-negative bacteria (Escherichia coli), and fungus (Candida albicans), the antimicrobial activities of printed fabrics using different thickening agents (alginate, DELL P, pectin, and modified pectin) and suitable dyes have been quantitatively demonstrated.

Table 11 shows the antibacterial properties of printed materials. Printed textiles that use pectin or modified thickeners have better antibacterial activity than printed fabrics that use alginate. Due to differences in the composition of the cell walls of both the examined bacterium strains, printed textiles are more effective against Gram-positive bacteria than against Gram-negative bacteria because they decrease ergosterol, a critical component of the fungal cell membrane [86–88].

Table 11. Microbial reduction % of printed fabrics with different thickening agents before and after 10 washing cycles.

Fabric	Dye	Thickening Agent	Microbial Reduction %					
			E. coli (ATCC 25922)		*S. aureus* (ATCC 29213)		*C. albicans* (ATCC 10231)	
			Before Washing	After Washing	Before Washing	After Washing	Before Washing	After Washing
Cotton	Reactive dye	Alginate	47.695	44.18	59.115	48.74	38.47	35.525
		Pectin	54.75	41.38	66.79	56.965	48.965	46.3525
		Modified Pectin	78.84	74.15	84.855	79.945	75.945	74.6375
Wool	Acid dye	Alginate	64.605	59.06	72.165	64.715	55.59	53.085
		Pectin	78.23	74.1	81.515	75.72	70.97	69.44
		Modified Pectin	93.445	90.07	96.235	94.465	92.965	92.5175
Acrylic	Disperse Dye	Alginate	58.67	53.44	66.91	58.84	49.64	47.07
		DELL	67.31	63.87	75.585	68.83	63.33	61.535
		Pectin	79.86	72.35	84.375	80.69	77.69	76.71
		Modified Pectin	81.17	75.32	88.125	83.27	79.315	78.2525
Polyester	Disperse Dye	Alginate	59.115	53.53	66.695	59.15	50.305	47.8425
		DELL	67.815	61.96	75.34	69.2	64.2	62.565
		Pectin	80.875	72.53	83.885	81.43	79.43	78.775
		Modified Pectin	81.755	74.31	86.39	83.15	80.515	79.8075

This test cultivates the uniform microbe in liquid culturing. The culture is diluted in a nutrient solution that has been sterilized. In sealed containers, the thickening agent is infected with microorganisms for 24 h at 37 °C. Shake for 1 min after incubation to measure bacteria levels. Eventually, the quantity of microbes compared to the starting concentration was as follows as a reduction percentage of bacteria (R%).

The antibacterial behavior of all printed textiles using modified pectin as a thickener was found to be superior to that of printed fabrics using alginate, DELL, or pectin only as a thickener.

After washing, the percentage of antibacterial reduction dropped. The percentage reduction decreased depending on the thickeners. Even after 10 washing cycles, the printed materials had good antibacterial capability against the tested microorganisms due to their longevity.

4. Conclusions

Pectin is a naturally occurring biopolymer that may be employed as a thickening agent in textiles because of its chemistry and ability to produce gels. Pectin may be extracted in an easy and safe method for the environment.

Pectin from orange and pomegranate peels was extracted by cutting, removing the white portion connected to the outer peel of plants, drying them, then pouring 30 g/L and 2.6 mL/L of HCL to set the pH at 2, boiling for an hour, filtering, precipitating with alcohol, filtering, and drying.

The contents of fats, protein, ash, carbohydrates, and sugars; the equivalent weight of dried pectin and methoxyl groups; acetyl value; moisture; organic acids; ability to dissolve in cold/hot water; ability to dissolve in cold/hot alkali; and the degree of esterification were measured. The effect of shear rate on shear stress to evaluate the viscosity was also measured.

The pectin was modified with hexamine and characterized by the IR spectrum. The IR spectrum of the hexamine compound showed absorption bands at 650, 795, 990, 1225, 1370, 1445, 1674, 2866, and 2912 cm^{-1}, which are similar to those reported according to the literature. The rheological properties of pectin extracted from orange peel and modified have the properties of pseudo-plastic pastes, while the pectin extracted from pomegranate peel has the properties of a non-Newtonian thixotropic paste. Thickeners bear a pH of up to 6 in the acidic direction and decompose at pH 7.5 in the alkaline direction. The pectin was modified by adding hexamine to change the pH degrees in the basic direction; the rheological properties of thickeners did not change during a storage period of up to 15 days.

Pectin and its modification were used for printing, and they were compared to other thickeners such as alginates for natural textiles and Dell P as manufactured thickeners for industrial fabrics with various dyes. After the printing process, they were dried and then fixed. The printed fabrics were characterized using color strength (K/S), fastness properties (washing, dry and wet rubbing, acid and alkaline perspiration, light), mechanical properties (surface roughness, tensile strength, elongation at the break, crease recovery angle), and anti-bacterial protection (Gram-positive, Gram-negative, fungi).

The results showed an increase in the color intensity for all materials printed with pectin and modified compared to the standard sample printed with sodium alginate and Dell p and increasing in K/S for printed cotton fabrics (3%), (13%) for wool (109%), (123%) for polyester (49%), (64%), (12%), (22%) and for polyacrylic (25%), (57%), (2%), (29%) sequentially. The improvement in fastness properties (washing, dry and wet rubbing, acidic and alkaline perspiration, light) was from satisfactory to good to excellent. There was also improvement in fabric roughness, tensile strength, elongation at the break, and crease recovery. Printed fabrics with pectin and modified pectin gave high resistance to Gram-negative and Gram-positive bacteria, as well as fungi.

Through the printing process, natural thickeners (extracted pectin from fruit peels) and modified pectin were added to various textiles (cotton, wool, acrylic, and polyester). All the printed samples with pectin and its modified synthetic dyes (reactive, acid, and disperse) exhibited good fastness towards washing and wet and dry rubbing. The light fastness of printed textiles was excellent (7), better than when using alginate as a thickener (5). All printed textiles had good washing fastness (4–5). In both acidic and alkaline perspiration, the perspiration fastness characteristic revealed 3–4 to 4–5 color differences. Colorfastness to rubbing was tested in both dry and wet conditions, and it was revealed that dry rubbing had the same effect as wet rubbing. Printed textiles using pectin or modified pectin as thickeners exhibit antibacterial activity.

Author Contributions: Conceptualization, A.G.H. and H.A.O.; methodology, S.A.E.; software, A.G.H. and S.A.E.; validation, A.G.H.; formal analysis, A.G.H. and S.A.E.; investigation, A.G.H., H.A.O., M.M.M. and S.A.E.; resources and data curation, M.M.M. and S.A.E.; writing—original draft preparation, A.G.H., M.M.M. and S.A.E.; writing—review and editing, A.G.H., H.A.O. and S.A.E.; visualization and supervision, A.G.H. and H.A.O.; project administration, S.A.E.; funding acquisition, S.A.E. All authors have read and agreed to the published version of the manuscript.

Funding: This research received no external funding.

Institutional Review Board Statement: Not applicable.

Informed Consent Statement: Not applicable.

Data Availability Statement: All needed data are presented in this study.

Acknowledgments: The authors gratefully acknowledge the faculty of Applied Arts, at Benha University. Furthermore, the authors gratefully acknowledge the Central Labs Services (CLS) and Centre of Excellence for Innovative Textiles Technology (CEITT) in Textile Research and Technology Institute (TRTI), National Research Centre (NRC) for the facilities provided.

Conflicts of Interest: The authors declare no conflict of interest.

References

1. Chaudhary, H.; Singh, V. Eco-friendly tamarind kernel thickener for the printing of polyester using disperse dyes. *Fibers Polym.* **2019**, *19*, 2514–2523. [CrossRef]
2. Clarke, W.; Miles, L.W.C. Synthetic thickening agents for textile printing. *Rev. Prog. Color.* **1983**, *13*, 27–31. [CrossRef]
3. Madhu, C.R.; Patel, D.M.C. Reactive dye printing on wool with natural and synthetic thickeners. *Int. Res. J. Eng. Technol. (IRJET)* **2016**, *3*, 1236–1238.
4. Ahmed, N.S.E.; El-Shishtawy, R.M. The use of new technologies in coloration of textile fibers. *J. Mater. Sci.* **2009**, *45*, 1143–1153. [CrossRef]
5. Abou Taleb, M.; Haggag, K.; B Mostafa, T.; El-Kheir, A.; El-Sayed, H. Preparation, characterization, and utilization of the suspended-keratin based binder in pigment printing of man-made fibres. *Egypt. J. Chem.* **2017**, *60*, 2–6. [CrossRef]
6. Harlapur, S.F.; Airani, N.R.; Gobbi, S.S. Appliance of natural gums as thickeners in the process of cotton printing. *Adv. Res. Text. Eng.* **2020**, *5*, 1048–1051.
7. El-Shishtawy, R.M.; Ahmed, N.; Nassar, S. Novel green coloration of cotton fabric. Part ii: Effect of different print paste formulations on the printability of bio-mordanted fabric with madder natural dye. *Egypt. J. Chem.* **2020**, *63*, 6–7. [CrossRef]
8. Teli, M.D.; Shanbag, V.; Kulkami, P.R.; Singhal, R.S. Amaranthus paniculates (rajgeera) starch as a thickener in the printing of textiles. *Carbohydr. Polym.* **1996**, *31*, 119–122. [CrossRef]
9. Hamdy, D.M.; Hassabo, A.G.; Osman, H.A. Recent use of natural thickeners in the printing process (a mini review). *J. Text. Color. Polym. Sci.* **2021**, *18*, 75–81.
10. Kibria, M.G.; Rahman, F.; Chowdhury, D.; Uddin, M.N. Effects of printing with different thickeners on cotton fabric with reactive dyes. *IOSR J. Polym. Text. Eng. (IOSR-JPTE)* **2018**, *5*, 5–10.
11. Harlapur, S.F.; Airani, N.; Gobbi, S.S. Natural gums as thickeners in the process of cotton fabric printing. *J. Fash. Technol. Text. Eng.* **2020**, *8*, 186–188.
12. Raj, A.A.S.; Rubila, S.R.J.; Ranganathan, T.V. A review on pectin: Chemistry due to general properties of pectin and its pharmaceutical uses. *Open Access Sci. Rep.* **2012**, *1*, 1–4.
13. Jarvis, M.C. Structure and properties of pectin gels in plant cell walls. *Plant Cell Environ.* **1984**, *7*, 153–164.
14. Sayah, M.Y.; Chabir, R.; Benyahia, H.; Rodi Kandri, Y.; Ouazzani Chahdi, F.; Touzani, H.; Errachidi, F. Yield, esterification degree and molecular weight evaluation of pectins isolated from orange and grapefruit peels under different conditions. *PLoS ONE* **2016**, *11*, e0161751. [CrossRef] [PubMed]
15. Srivastava, P.; Malviya, R. Sources of pectin, extraction and its applications in pharmaceutical industry–An overview. *Indian J. Nat. Prod. Resour.* **2011**, *2*, 10–18.
16. Talmadge, K.W.; Keegstra, K.; Bauer, W.D.; Albersheim, P. The structure of plant cell walls. *Plant Physiol.* **1973**, *51*, 158–173. [CrossRef]
17. Williams, P.A. Chapter 1. Natural polymers: Introduction and overview. *RSC Polym. Chem. Ser.* **2011**, 1–14.
18. May, C.D. Industrial pectins: Sources, production and applications. *Carbohydr. Polym.* **1990**, *121*, 79–99. [CrossRef]
19. Pandharipande, S.; Makode, H. Separation of oil and pectin from orange peel and study of effect of ph of extracting medium on the yield of pectin. *J. Eng. Res. Stud.* **2012**, *3*, 6–9.
20. Aina, V.O.; Barau, M.M.; Mamman, O.A.; Zakari, A.; Haruna, H.; Umar, M.H.; Abba, Y.B. Extraction and characterization of pectin from peels of lemon (citrus limon), grape fruit (citrus paradisi) and sweet orange (citrus sinensis). *Br. J. Pharmacol. Toxicol.* **2012**, *3*, 259–262.
21. Bhavya, D.K.; Suraksha, R. Value added products from agriculture: Extraction of pectin from agro waste product musa acuminata and citrus fruit. *Res. J. Agric. For. Sci.* **2015**, *3*, 13–18.
22. *AOAC 945.44-1945*; Fat in Fig Bars and Raisin-Filled Crackers Ether Extraction Method. Association of Official Analytical Chemists: Washington, DC, USA, 2000.
23. *AOAC 991.20-1994(1996)*; Nitrogen (Total) In Milk. Kjeldahl Methods. Association of Official Analytical Chemists: Washington, DC, USA, 2000.
24. *AOAC 923.03-1923*; Ash of Flour. Direct Method. Association of Official Analytical Chemists: Washington, DC, USA, 2000.
25. Fishman, M.L.; Chau, H.K.; Hoagland, P.D.; Hotchkiss, A.T. Microwave-assisted extraction of lime pectin. *Food Hydrocoll.* **2006**, *20*, 1170–1177. [CrossRef]
26. Devi, W.E.; Shukla, R.N.; Bala, K.L.; Kumar, A.; Mishra, A.; Yadav, K.C. Extraction of pectin from citrus fruit peel and its utilization in preparation of jelly. *Int. J. Eng. Res. Technol.* **2014**, *3*, 1925–1932.
27. Fakayode, O.A.; Abobi, K.E. Optimization of oil and pectin extraction from orange (citrus sinensis) peels: A response surface approach. *J. Anal. Sci. Technol.* **2018**, *9*, 20. [CrossRef]
28. Chanthaphon, S.; Chanthachum, S.; Hongpattarakere, T. Antimicrobial activities of essential oils and crude extracts from tropical citrus spp. Against food-related microorganisms. *Songklanakarin J. Sci. Technol.* **2008**, *30*, 125–131.
29. Morris, E.R.; Foster, T.J.; Harding, S.E. The effect of the degree of esterification on the hydrodynamic properties of citrus pectin. *Food Hydrocoll.* **2000**, *14*, 227–235. [CrossRef]
30. Harker, J.H.; Backhurst, J.R.; Richardson, J.F. *Chemical Engineering Volume 2*; Elsevier: New York, NY, USA, 1978.
31. Liew, S.Q.; Chin, N.L.; Yusof, Y.A. Extraction and characterization of pectin from passion fruit peels. *Agric. Agric. Sci. Procedia* **2014**, *2*, 231–236. [CrossRef]

32. Twinomuhwezi, H.; Awuchi, C.G.; Kahunde, D. Extraction and characterization of pectin from orange (citrus sinensis), lemon (citrus limon) and tangerine (citrus tangerina). *Am. J. Phys. Sci.* **2020**, *1*, 17–30.
33. Saad, F.; Hassabo, A.; Othman, H.; Mosaad, M.M.; Mohamed, A.L. A valuable observation on thickeners for valuable utilisation in the printing of different textile fabrics. *Egypt. J. Chem.* **2022**, *65*, 431–448. [CrossRef]
34. Saad, F.; Hassabo, A.G.; Othman, H.A.; Mosaad, M.M.; Mohamed, A.L. Improving the performance of flax seed gum using metal oxides for using as a thickening agent in printing paste of different textile fabrics. *Egypt. J. Chem.* **2021**, *64*, 4937–4954. [CrossRef]
35. Barnes, H.A. Thixotropic—A review. *J. Non-Newton. Fluid Mech.* **1997**, *70*, 1–33. [CrossRef]
36. Evans, J.; Beddow, J. Characterisation of particle morphology and rheological behaviour in solder paste. *IEEE Trans. Compon. Hybrids Manuf. Technol.* **1987**, *10*, 224–231. [CrossRef]
37. Bullard, J.W.; Pauli, A.T.; Garboczi, E.J.; Martys, N.S. Comparison of viscosity-concentration relationships for emulsion. *J. Colloid Interface Sci.* **2009**, *330*, 186–193. [CrossRef]
38. McLelland, A.R.A.; Henderson, N.G.; Atkinson, H.V.; Kirkwood, D.H. Anomalous rheological behavior of semi-solid alloy slurries at low shear rates. *Mater. Sci. Eng. A* **1997**, *232*, 110–118. [CrossRef]
39. Durairaj, R.; Man, L.W.; Ramesh, S. Rheological characterisation and empirical modelling of lead-free solder pastes and isotropic conductive adhesive pastes. *J. ASTM Int.* **2010**, *7*, 1–13. [CrossRef]
40. Pospischil, M.; Zengerle, K.; Specht, J.; Birkle, G.; Koltay, P.; Zengerle, R.; Henning, A.; Neidert, M.; Mohr, C.; Clement, F.; et al. Investigations of thick-film-paste rheology for dispensing applications. *Energy Procedia* **2011**, *8*, 449–454. [CrossRef]
41. Kubelka, P.; Munk, F. Ein beitrag zur optik der farbanstriche. *Z. Tech. Phys.* **1931**, *12*, 593.
42. Mehta, K.T.; Bhavsar, M.C.; Vora, P.M.; Shah, H.S. Estimation of the kubelka—Munk scattering coefficient from single particle scattering parameters. *Dye. Pigment.* **1984**, *5*, 329–340. [CrossRef]
43. Waly, A.; Marie, M.M.; Abou-Zeid, N.Y.; El-Sheikh, M.A.; Mohamed, A.L. Process of single—Bath dyeing, finishing and flam—Retarding of cellulosic textiles in presence of reactive tertiary amines. In Proceedings of the 3rd International Conference of Textile Research Division, NRC; Textile Processing: State of the Art & Future Developments, Cairo, Egypt, 2–4 April 2006; pp. 529–543.
44. Waly, A.; Marie, M.M.; Abou-Zeid, N.Y.; El-Sheikh, M.A.; Mohamed, A.L. Flame retarding, easy care finishing and dyeing of cellulosic textiles in one bath. *Egypt. J. Text. Polym. Sci. Technol.* **2008**, *12*, 101–131.
45. Mohamed, A.L.; Hassabo, A.G. Cellulosic fabric treated with hyperbranched polyethyleneimine derivatives for improving antibacterial, dyeing, ph and thermo-responsive performance. *Int. J. Biol. Macromol.* **2021**, *170*, 479–489. [CrossRef]
46. AATCC Test Method (61-2013). Color fastness to laundering: Accelerated. In *Technical Manual Method*; American Association of Textile Chemists and Colorists: Durham County, NC, USA, 2017; p. 108.
47. AATCC Test Method (8-2016). Colorfastness to crocking, crockmeter method. In *Technical Manual Method*; American Association of Textile Chemists and Colorists: Durham County, NC, USA, 2018; Volume 86, pp. 17–19.
48. AATCC Test Method (15-2013). Colour fastness to perspiration. In *Technical Manual Method*; American Association of Textile Chemists and Colorists: Durham County, NC, USA, 2017; Volume 86, pp. 30–32.
49. AATCC Test Method (16.1-2014). Colour fastness to light: Outdoor. In *Technical Manual Method*; American Association of Textile Chemists and Colorists: Durham County, NC, USA, 2015; Volume 16.1, pp. 33–48.
50. *ASTM Standard Test Method (D5035-2011 (Reapproved 2019))*; Standard Test Method for Breaking Force and Elongation of Textile Fabrics (Strip Method). ASTM International: West Conshohocken, PA, USA, 2019.
51. AATCC Test Method (66-2014). Wrinkle recovery of fabric: Recovery angle method. In *Technical Manual Method*; American Association of Textile Chemists and Colorists: Durham County, NC, USA, 2017; pp. 113–116.
52. *ASTM Standard Test Method (D7127-13)*; Standard Test Method for Measurement of Surface Roughness of Abrasive Blast Cleaned Metal Surfaces Using a Portable Stylus Instrument. ASTM International: West Conshohocken, PA, USA, 2016.
53. *ASTM Standard Test Method (D1388-14e1)*; Standard Test Methods for Stiffness of Fabrics. ASTM International: West Conshohocken, PA, USA, 2016.
54. Hassabo, A.G.; Shaarawy, S.; Mohamed, A.L.; Hebiesh, A. Multifarious cellulosic through innovation of highly sustainable composites based on moringa and other natural precursors. *Int. J. Biol. Macromol.* **2020**, *165*, 141–155. [CrossRef] [PubMed]
55. AATCC Test Method (100-2019). Assessment of antimicrobial finishes on textile materials. In *Technical Manual Method*; American Association of Textile Chemists and Colorists: Durham County, NC, USA, 2019; Volume 68.
56. Khattab, T.A.; Mohamed, A.L.; Hassabo, A.G. Development of durable superhydrophobic cotton fabrics coated with silicone/stearic acid using different cross-linkers. *Mater. Chem. Phys.* **2020**, *249*, 122981. [CrossRef]
57. Pargai, D.; Jahan, S.; Gahlot, M. *Functional Properties of Natural Dyed Textiles*; IntechOpen: London, UK, 2020.
58. Pereira, P.H.; Oliveira, T.I.; Rosa, M.F.; Cavalcante, F.L.; Moates, G.K.; Wellner, N.; Waldron, K.W.; Azeredo, H.M. Pectin extraction from pomegranate peels with citric acid. *Int. J. Biol. Macromol.* **2016**, *88*, 373–379. [CrossRef]
59. El-Nawawi, S.A.; Shehata, F.R. Extraction of pectin from egyptian orange peel. Factors affecting the extraction. *Biol. Wastes* **1987**, *20*, 281–290. [CrossRef]
60. Kanmani, P. Extraction and analysis of pectin from citrus peels: Augmenting the yield from citrus limon using statistical experimental design. *Iran. J. Energy Environ.* **2014**, *5*, 303–312. [CrossRef]
61. Moorthy, I.G.; Maran, J.P.; Surya, S.M.; Naganyashree, S.; Shivamathi, C.S. Response surface optimization of ultrasound assisted extraction of pectin from pomegranate peel. *Int. J. Biol. Macromol.* **2015**, *72*, 1323–1328. [CrossRef] [PubMed]

62. Hashmi, S.H.; Ghatge, P.; Machewad, G.M.; Pawar, S. Studies on extraction of essential oil and pectin from sweet orange. *J. Food Process. Technol.* **2012**, *1*, 1–3. [CrossRef]
63. Happi Emaga, T.; Ronkart, S.N.; Robert, C.; Wathelet, B.; Paquot, M. Characterisation of pectins extracted from banana peels (musa aaa) under different conditions using an experimental design. *Food Chem.* **2008**, *108*, 463–471. [CrossRef]
64. Garna, H.; Mabon, N.; Robert, C.; Cornet, C.; Nott, K.; Legros, H.; Wathelet, B.; Paquot, M. Effect of extraction conditions on the yield and purity of apple pomace pectin precipitated but not washed by alcohol. *J. Food Sci.* **2007**, *72*, C001–C009. [CrossRef] [PubMed]
65. Putnik, P.; Kovačević, D.B.; Jambrak, A.R.; Barba, F.J.; Cravotto, G.; Binello, A.; Lorenzo, J.M.; Shpigelman, A. Innovative "green" and novel strategies for the extraction of bioactive added value compounds from citruswastes—A review. *Molecules* **2017**, *22*, 680. [CrossRef]
66. Sereewatthanawut, I.; Prapintip, S.; Watchiraruji, K.; Goto, M.; Sasaki, M.; Shotipruk, A. Extraction of protein and amino acids from deoiled rice bran by subcritical water hydrolysis. *Bioresour. Technol.* **2008**, *99*, 555–561. [CrossRef] [PubMed]
67. Maxwell, E.G.; Belshaw, N.J.; Waldron, K.W.; Morris, V.J. Pectin e an emerging new bioactive food polysaccharide. *Trends Food Sci. Technol.* **2012**, *24*, 64–73. [CrossRef]
68. Brouns, F.; Theuwissen, E.; Adam, A.; Bell, M.; Berger, A.; Mensink, R. Cholesterol-lowering properties of different pectin types in mildly hyper-cholesterolemic men and women. *Eur. J. Clin. Nutr.* **2012**, *66*, 591–599. [CrossRef]
69. Dominiak, M.; Søndergaard, K.M.; Wichmann, J.; Vidal-Melgosa, S.; Willats, W.G.T.; Meyer, A.S.; Mikkelsen, J.D. Application of enzymes for efficient extraction, modification, and development of functional properties of lime pectin. *Food Hydrocoll.* **2014**, *40*, 273–282. [CrossRef]
70. Mort, A.J.; Qiu, F.; Maness, N. Determination of the pattern of methyl esterification in pectin. Distribution of continuous nonesterified residues. *Carbohydr. Res.* **1993**, *247*, 21–35. [CrossRef]
71. Masmoudi, M.; Besbes, S.; Chaabouni, M.; Robert, C.; Paquot, M.; Blecker, C.; Attia, H. Optimization of pectin extraction from lemon by-product with acidified date juice using response surface methodology. *Carbohydr. Polym.* **2008**, *74*, 185–192. [CrossRef]
72. Deroeck, A.; Sila, D.; Duvetter, T.; Vanloey, A.; Hendrickx, M. Effect of high pressure/high temperature processing on cell wall pectic substances in relation to firmness of carrot tissue. *Food Chem.* **2008**, *107*, 1225–1235. [CrossRef]
73. Yapo, B.M.; Koffi, K.L. Extraction and characterization of highly gelling low methoxy pectin from cashew apple pomace. *Foods* **2014**, *3*, 1–12. [CrossRef]
74. Yadav, S.D.; Bankar, N.S.; Waghmare, N.N.; Shete, D.C. Extraction and characterization of pectin from sweet lime. *Int. Conf. Multidiscip. Res. Pract.* **2017**, 58–63.
75. Willats, W.G.T.; Knox, P.; Mikkelsen, J.D. Pectin: New insights into an old polymer are starting to gel. *Trends Food Sci. Technol.* **2006**, *17*, 97–104. [CrossRef]
76. Abid, M.; Cheikhrouhou, S.; Renard, C.M.; Bureau, S.; Cuvelier, G.; Attia, H.; Ayadi, M.A. Characterization of pectins extracted from pomegranate peel and their gelling properties. *Food Chem.* **2017**, *215*, 318–325. [CrossRef] [PubMed]
77. Ragheb, A.; Abd El-Thalouth, I.; El-Sayad, H.; Hebeish, A. Preparation and characterization of carboxymethylcellulose from jute wastes. *Indian J. Fibre Text. Res.* **1991**, *12*, 263.
78. Hebeish, A.; Ibrahim, N.A.; Abo Shosha, M.H.; Fahmy, H.M. Rheological behavior of some polymeric sizing agents alone and in admixtures. *Polym. Plast. Technol. Eng.* **1996**, *35*, 517–543. [CrossRef]
79. Sato, M.d.F.; Rigoni, D.C.; Canteri, M.H.G.; Petkowicz, C.L.d.O.; Nogueira, A.; Wosiacki, G. Chemical and instrumental characterization of pectin from dried pomace of eleven apple cultivars. *Acta Sci. Agron.* **2011**, *33*, 383–389.
80. Wathoni, N.; Yuan Shan, C.; Yi Shan, W.; Rostinawati, T.; Indradi, R.B.; Pratiwi, R.; Muchtaridi, M. Characterization and antioxidant activity of pectin from indonesian mangosteen (*Garcinia mangostana* L.) rind. *Heliyon* **2019**, *5*, e02299. [CrossRef]
81. Gerasimova, A.V.; Alekhina, O.V.; García-Cruz, L.; Iniesta, J.; Melezhik, A.V.; Tkachev, A.G. Polycondensation of hexamethylenetetramine in anhydrous acid media as a new approach to carbyne-like materials and its application as dispersant of carbon materials. *C* **2019**, *5*, 54. [CrossRef]
82. Ezzayani, K.; Khelifa, A.B.; Saint-Aman, E.; Loiseau, F.; Nasri, H. Complex of hexamethylenetetramine with magnesium-tetraphenylporphyrin: Synthesis, structure, spectroscopic characterizations and electrochemical properties. *J. Mol. Struct.* **2017**, *1137*, 412–418. [CrossRef]
83. Suriati, L.; Utama, I.M.S.; Harjosuwono, B.A.; Gunam, I.B.W. Stability aloe vera gel as edible coating. *IOP Conf. Ser. Earth Environ. Sci.* **2020**, *411*, 012053. [CrossRef]
84. Suriati, L.; Mangku, I.G.P.; Rudianta, I.N. The characteristics of aloe vera gel as an edible coating. *IOP Conf. Ser. Earth Environ. Sci.* **2018**, *207*, 012051. [CrossRef]
85. Hebeish, A.; Shaarawy, S.; Hassabo, A.G.; El-Shafei, A. Eco-friendly multifinishing of cotton through inclusion of motmorillonite/chitosan hybrid nanocomposite. *Pharma Chem.* **2016**, *8*, 259–271.
86. Biswas, B.; Rogers, K.; McLaughlin, F.; Daniels, D.; Yadav, A. Antimicrobial activities of leaf extracts of guava (*Psidium guajava* L.) on two gram-negative and gram-positive bacteria. *Int. J. Microbiol.* **2013**, *2013*, 746165. [CrossRef] [PubMed]

87. Metwally, A.M.; Omar, A.A.; Harraz, F.M.; El Sohafy, S.M. Phytochemical investigation and antimicrobial activity of *Psidium guajava* L. Leaves. *Pharm. Mag.* **2010**, *6*, 212–218.
88. Dhiman, A.; Nanda, A.; Ahmad, S.; Narasimhan, B. In vitro antimicrobial activity of methanolic leaf extract of *Psidium guajava* L. *J. Pharm. Bioallied Sci.* **2011**, *3*, 226–237. [CrossRef] [PubMed]

Disclaimer/Publisher's Note: The statements, opinions and data contained in all publications are solely those of the individual author(s) and contributor(s) and not of MDPI and/or the editor(s). MDPI and/or the editor(s) disclaim responsibility for any injury to people or property resulting from any ideas, methods, instructions or products referred to in the content.

Article

A New Perspective on the Textile and Apparel Industry in the Digital Transformation Era

Waleed Hassan Akhtar [1,*], Chihiro Watanabe [1,2], Yuji Tou [3] and Pekka Neittaanmäki [1]

1. Faculty of Information Technology, University of Jyväskylä, 40014 Jyväskylä, Finland
2. International Institute for Applied Systems Analysis (IIASA), 2361 Laxenburg, Austria
3. Department of Industrial Engineering and Management, Tokyo Institute of Technology, Tokyo 152-8550, Japan
* Correspondence: waleed.akhtr@gmail.com

Abstract: The textile and apparel (fashion) industry has been influenced by developments in societal socio-cultural and economic structures. Due to a change in people's preferences from economic functionality to supra-functionality beyond economic value, the fashion industry is at the forefront of digitalization. The growing digitalization in the fashion industry corresponds to digital fashion, which can satisfy the rapid shift in consumers' preferences. This paper explores the evolving concept of innovations in digital fashion in the textile and apparel industry. Specifically, it centers on the evaluation of Amazon's digital fashion initiatives, which have made the platform the United States' top fashion retailer. An analysis of the business model of Amazon's digital fashion business showed that with the advancements in artificial intelligence (AI) powered by advanced Amazon Web Services (AWS), Amazon has introduced novel digital solutions for the fashion industry, such as advanced digital fashions (ADFs), on-demand manufacturing, neo-luxury, and, ultimately, cloud-based digital fashion platforms, that is, a supra-omnichannel, where all stakeholders are integrated, and their activities are visible in real time. This can be attributed to the learning orchestration externality strategy. This study concludes that with the advancement of digital innovations, Amazon has fused a self-propagating function that advances digital solutions. This study shows that Amazon is the largest R&D company. Its R&D process is based on users' knowledge gained by their participation through AWS-driven ICT tools. This promotes a culture of experimentation in the development of user-driven innovations. Such innovations have further advanced the functionality of AWS in data analysis and business solutions. This dynamism promotes the development of soft innovation resources and revenue streams. These endeavors are demonstrated in a model, and their reliability is validated through an empirical analysis focused on the emergence of ADF solutions. Therefore, based on an analysis of the development trajectories of Amazon's digital fashion technologies, such as ADFs, on-demand manufacturing, and neo-luxury, insightful suggestions and a framework for solutions beyond e-commerce are provided.

Keywords: Amazon; textile and apparel; fashion; advanced digital fashions; supra-omnichannel; non-contact society; beyond e-commerce

Citation: Akhtar, W.H.; Watanabe, C.; Tou, Y.; Neittaanmäki, P. A New Perspective on the Textile and Apparel Industry in the Digital Transformation Era. *Textiles* **2022**, *2*, 633–656. https://doi.org/10.3390/textiles2040037

Academic Editor: Laurent Dufossé

Received: 24 September 2022
Accepted: 7 November 2022
Published: 5 December 2022

Publisher's Note: MDPI stays neutral with regard to jurisdictional claims in published maps and institutional affiliations.

Copyright: © 2022 by the authors. Licensee MDPI, Basel, Switzerland. This article is an open access article distributed under the terms and conditions of the Creative Commons Attribution (CC BY) license (https://creativecommons.org/licenses/by/4.0/).

1. Introduction

The textile and apparel (fashion) industry has social, cultural, and economic significance in many societies. Nevertheless, the word fashion has different meanings ranging from the way of doing things to textiles and apparel. Among others, Hansen's [1] study found that fashion implies discourses representing the developments in arts, social structure, and culture. In Western society, it is highly associated with "style", "dress", or "clothes" [2,3]. As these features make fashion a meaningful phenomenon, in this study, the fashion industry corresponds to fashion-driven textiles and apparel as well as other fashion-related products. It includes a wide range of business networks ranging from raw materials production, design, manufacturing, and retail [4,5]. These advancements improve fashion, and apparel accelerates this development [6].

In response to shifting people's preferences towards suprafunctionality beyond economic value [7–10], and a prevalence of a non-contact society, the digital economy has accelerated the development of digital solutions that transform the traditional fashion industry [11–13].

In the digital economy, the traditional fashion industry is at the center of global dynamic change [11,12] driving its volatility, velocity, variety, complexity, and dynamism [14,15] which necessitate digital solutions.

The United States is one of the leaders in new technological innovations in the global fashion industry. For example, among technology giants (GAFAM), Amazon is a leader in digital services provision. Its heavy R&D investments enabled it to develop a novel R&D-based disruptive business model that converts its investments and R&D activities into a new concept of R&D that serves as a locomotive for innovations in Amazon's businesses ranging from Amazon's brick-and-mortar to e-commerce-based businesses [4,16,17].

Previous studies have examined the identical features of fashion such as cultural aspects, supply chain management, designing, manufacturing, marketing, and technological developments in the fashion industry towards sustainability (e.g., [18–24]), and also Amazon's R&D system from the viewpoints of technology operation strategy as well as financial management system (e.g., [25–27]), no one has analyzed their co-evolutionary progression leading to further development of solutions provided by the digitalization of the fashion industry and Amazon's R&D-driven, customer-centric virtuous cycle toward stakeholder capitalization. Nevertheless, consideration of practical beyond e-commerce solutions, supporting tools such as high-performance computing, and its impact on the global fashion industry has also not been actively pursued. Finally, the lack of conceptualization on practical solutions that encounter fashion-driven luxury brands' internet dilemma that hinders the development of the fashion industry in the digital era, which is critical to transforming the global fashion industry to survive and compete in a non-contact society.

At the same time, the significance of stakeholder capitalism is gaining momentum in a newly emerged non-contact society [28]. It enables companies to protect and satisfy stakeholders' concerns by engaging them to create a shared and sustained value. This can be achieved by corresponding to their changing preferences. Thus, while analyzing the co-evolutionary paths of Amazon's business model and the fashion industry, an approach toward stakeholders' capitalization is attempted.

Amazon develops its business empire by undertaking frontier innovation and companywide experimentation based on heavy investments in "technology and content" that generates a big data collection system enabling it to harness the power of users which functions as a virtuous cycle leading to the transformation of "routine or periodic alterations" into "significant improvement" during the R&D process [27]. For example, recently, a few studies have suggested that Amazon's strength lies in artificial intelligence, whereas Amazon web services (AWS) are a locomotive for AI-driven innovations reflecting sociocultural, economic, and technological changes in society [29–31].

The new socio-economic conditions, such as the emergence of a noncontact society during COVID-19, urged the digital transformation of the fashion industry [32]. Digital transformation indicates company-wide changes that result in the emergence of new business models [33]. The digital transformation of the fashion industry refers to "the reception of the digital environment by the industry" [34], whereas digital fashion represents an overlying area between fashion and ICT tools [35].

In this study, advanced digital fashions (ADFs) are referred to as fast fashion products which are developed by studying the effects of prior digital innovations and are powered by Amazon's recommendation engines. Their enabling tools and preceding innovations are illustrated in Table 1. They represent Amazon's fashion catalog, influencer styles, Amazon's labels, and third-party fashion lines, whereas supra-omnichannel is a cloud-based fashion platform that emerged by the co-evolution of ADFs, ODM, and neo-luxury, all stakeholders are combined in a cloud-based fashion platform so that their activities are visible in real-time.

Table 1. Functions and enabling core technologies for ADFs development.

ADFs	Functionality	Preceding Innovations	Core Technology
Prime Wardrobe (2017)	Try at home before purchase service, customers can receive up to 15 items at home and pay only for the selected outfit. Sizing and returns are the biggest online shopping barriers; this feature has removed such barriers.	Endless.com, my habit.com, Body Labs, 3D body modeling AI (ML), IoT, VR/AR, and mobile devices.	AI-based matching recommendations, Amazon fashion catalog, recommender system
AI Algo. fashion designer (2017)	The algorithm learns about a particular style of fashion from the web, and social media (images, videos), and generates new items in similar styles.	Body labs based on ADFs technologies	ML and DL-based Generative Adversarial Networks (GAN)
EchoLook (2017)	Smart speaker, voice assistant, and hands-free camera. It was introduced to train Alexa in becoming a fashion advisor.	Echo Look (2017) has emerged from the classic Echo (2014) device and is similar to Echo Dot (2016). Its features are based on previous, Outfit Compare (share photos), and Style Check (second opinion)	Based on AI, Echo Look (2017) incorporates CV, NLP, and ML.
AR Mirror (2018)	A mirror-based display system that enables users to interact with virtual objects. The blended AR systems combine images reflected by a mirror with the images that are transmitted from the screen behind the mirror.	Further development of Echo Look. BodyLabs software, and Lab 123 hardware.	Two-way mirror with electronic display, depth-sensing camera, projectors, CV algorithms, blended reality
Personal Shopper (2019)	Customized clothing box. Incorporates curation function, consumers co-create with Amazon's designers. This curation function satisfies customers' personalized requirements.	Further evolution of Prime Wardrobe, sophisticated curation ability accumulated through the series of ADFs development.	With the addition of the Personal Shopper (2019) to Prime Wardrobe (2017), the company uses ML and personalized recommendation algorithms along with personal human stylists.
Style Snap (2019)	Fashion recommendations are based on user-submitted photos in real-time. Connects high-profile social media fashion influencers to Amazon fashion and customers.	Amazon Associates, Amazon influencers.	CV and DL identify apparel in a photo. DL classifies the apparel items in the image.
The Drop (2019)	Social media fashion influencers and Amazon fashion designers co-create limited edition apparel available only for 30 h.	Amazon influencers program, Influencers drove the fashion line.	SM, BDA

Sources: Authors' elaboration based on [36–41].

Applying techno-economic analysis and in-depth literature reviews, an empirical co-evolutional analysis was conducted on Amazon's recent AI-oriented R&D-driven developments in fashion, i.e., the introduction of advanced digital fashions (*ADFs*) leading to the development of supra-omnichannel, and the contribution of digitalization towards sustainable fashion. Thus, this paper investigates the development status of the textile and apparel industry in the digital transformation era, starting from an analysis of Amazon's fashion business and presents the technical frameworks in developing e-commerce-based ADFs, on-demand manufacturing (ODM), neo-luxury, and construction of the supra-omnichannel. It also discusses the future development of Amazon beyond e-commerce endeavors leading to a metaverse society [32,42].

Section 2 introduces fashion as a representation of social life and the historic development of the fashion industry, and Section 3 introduces the contributing factors for growth in Amazon's fashion business such as Amazon web services, artificial intelligence, and the emergence of soft innovation resources. Section 4 introduces the Amazon's learning orchestration externality in developing digital fashion business that covers the concept of learning orchestration externality in the development of ADFs, ODM, and with their dual co-evolution emergence of the supra-omnichannel. Section 5 introduces a framework encompassing learning orchestration externality beyond e-commerce, and the conclusion part summarizes the development of this paper and, finally, provides direction for future research.

2. Fashion as a Representation of Social Life

2.1. Global Fashion Industry

The contemporary fashion industry is based on social and cultural phenomena referred to as the "fashion system". The fashion system is a highly influential force that encompasses art, design, manufacturing, branding, and retail. For example, art and design contribute to the formation of fashion trends in society [43,44] as illustrated in Figure 1.

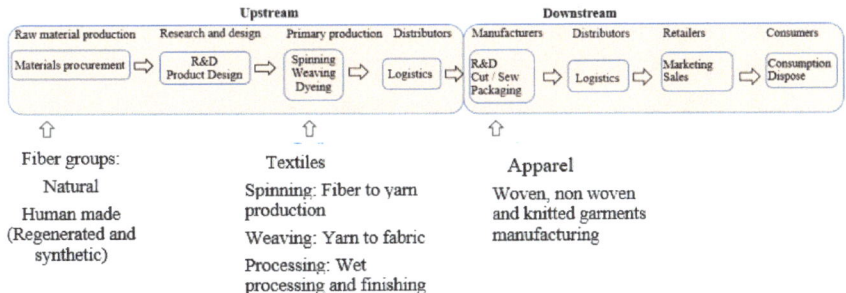

Figure 1. Value chain structure of the fashion industry.

In recent years, the fashion industry has seen a paradigm shift from traditional retailer-driven manufacturing to modern (fast fashion) consumer-driven manufacturing [45]. This industry is well-known for its unpredictable demand–supply relationships, long production lead times, short fashion cycles, choice of raw materials, seasonal demands, and heterogeneity [46,47].

The contemporary fashion value chain has transformed from forecast-based bulk production to season-based assortments. Apparel retailers can receive and deliver orders with shorter lead times. Today, industrial trends such as fast fashion and fashion mass customization are all supported by quick response and delivery speed [48].

Given the economic benefits for developed and developing economies, several constraints, such as shorter life cycles, tariff barriers, speed to market, and seasonality, have obstructed the balanced development of the industry. However, in addition to recent policies for removing trade impediments, the fashion industry is advancing toward digital solutions, for instance, real-time supply chain visibility, supply chain optimization, on-demand cloud manufacturing, stock-level optimization, and adjustment with demand planning. The most common type of digital platform is a digital marketplace, and the hyper-personalized solutions provided by e-commerce solutions eliminate geographic constraints among the stakeholders in the value chain. Thus, digital solutions can transform the contemporary value chain of the fashion industry into a disruptive platform [4].

2.2. Historical Development of the Fashion Industry

The fashion industry is a highly globalized industry, and its value chain is spread over different countries. For example, a fashion brand in the United States might source raw materials in China, have the apparel produced in Vietnam, and ship it to warehouses in Europe and the U.S. for distribution to retail outlets globally. This is mainly due to the quota system, rising materials and labor costs, as well as environmental regulations. However, the fashion industry is transitioning from mass production in standard-size systems to consumer-driven personalized manufacturing. This fragmented value chain and employment shift is not only responsible for a significant share of world economic output but has also gained momentum worldwide [9,49] in response to the shift in consumers' preferences from economic functionality to supra-functionality beyond economic value, encompassing social, cultural, emotional, and aspirational value, as illustrated in Figures 2 and 3.

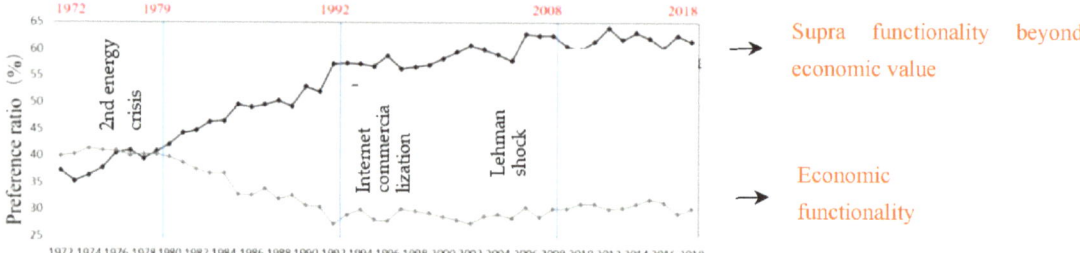

Figure 2. The trend in the shift of people's preferences in Japan (1972–2018). Source: National Survey of Lifestyle Preferences [50].

For example, in aging societies, consumers tend to buy products that correspond to both their functional and supra-functional (emotional, social, and cultural) requirements. Consumers are more demanding and prefer products that are best suited to their lifestyles. A psychological barrier develops when a product or service is unable to satisfy an individual user. This barrier hinders customers from developing relationships with those products and services; as a result, products are abandoned [7,51].

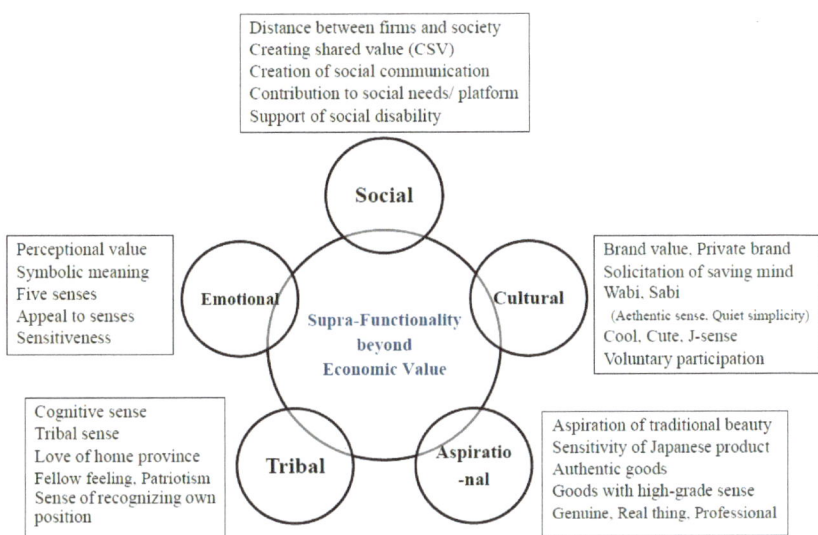

Figure 3. Concept of supra-functionality beyond economic value. Sources: [7–9].

3. Contributing Factors for Growth in Amazon's Fashion Business

3.1. Growth of the Fashion Industry

The recent COVID-19 pandemic exerted a shock on the global economy. However, despite this crisis, the fashion industry has been an engine of economic growth. Historically, the fashion industry, including textiles and apparel, has been a method for industrializing. For example, the market size of the global fashion industry improved substantially from USD 1.05 trillion in 2011 to USD 1.25 trillion in 2015, USD 1.40 trillion in 2017, and USD 1.65 trillion in 2020 (40% is shared by the EU and the US) [52]. It developed more rapidly than the global economy: in contrast to the average growth rate of 2.70% in the global gross domestic product (GDP) from 2011 to 2015, the global apparel market increased by 4.70% in the same period. The fast fashion industry established a larger increase (10.0%) during this period, as shown in Figure 4. The GDP elasticity to fashion ε_{FG} (1% increase in GDP increases ε_{FG} % increase in fashion) is more than double the GDP elasticity of textile and apparel, as shown in the right of Figure 4.

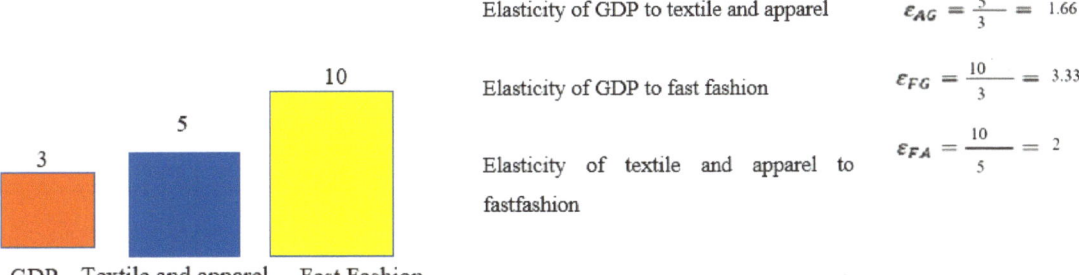

Figure 4. Comparison of the average growth rate of global GDP, total textile and apparel, and fast fashion (2011–2015).—%p.a. Source: Authors' elaboration based on [53,54].

3.2. Fashion Industry in the Digital Economy

Fashion represents people's lifestyles. It is a means of communication among people with common preferences, trends, and traditions that collectively form and represent the taste and lifestyle of that society. Such fashionability, including tastes, preferences, and lifestyles, is converted into art and fashion concepts and then translated into fashion products [55].

In the fashion industry, traditional approaches to the production and selling of fashion products are being challenged [56]. The key problems are longer production lead times, fragmented supply chains, time to market, and increased fashion seasons encompassing volume, variety, and velocity [43]. With these features, supply systems in the fashion industry that are fast-moving and trend-driven are categorized by shorter lead times in terms of responsiveness, production, and high fashionability [57]. This indicates that flexibility is critical in fast fashion to ensure the rapid delivery of trendy products. As fast fashion is unpredictable, the implementation of a lean manufacturing system including just-in-time, agile supply chains, and quick responses is expected to reduce the processes involved in the buying cycle and lead times for getting new fashion products into stores [14].

The advancement of the digital economy has accelerated the demands described above. At the same time, it provides the fashion industry with a new solution, a digital solution. As suggested by Sun and Zhao [58], advancements in digital technology, from AI, robotics, and agile and on-demand manufacturing to the virtual dressing room, e-commerce, and social media, are becoming key growth drivers in the fashion industry. In addition, these advancements have emerged in new environments, shifting to a sharing economy and a circular economy, which is driving the fashion industry to change to a disruptive business model.

Advancements in ICT have accelerated digital innovations in the fashion industry. For instance, cyber-physical systems, the Internet of Things, personalization, customization, AI, and high-performance computing are accelerating digital innovations. Fashion brands use these innovations to improve customer experiences. For example, AI is used to offer personalized recommendations and curation. Augmented reality and virtual reality are used before buying in a simulated environment [59]. Therefore, confronting the demands of fast fashion and incorporating technological breakthroughs, particularly digital solutions utilizing the dramatic advancement of digital innovation, is expected. Noteworthy endeavors at the forefront of the fashion industry are described in Table 2.

Table 2. Digital innovations supporting the fashion industry toward advanced digital fashions.

Artificial intelligence (AI)	Fashion design, real-time, recommendation, forecasting, and trend analysis
Machine learning (ML)	Product development, demand forecasting, complex data analysis
Virtual reality/Augmented reality (VR/AR)	Creates virtual world, 3D body scanning, customer experience monitoring, virtual stores, and metaverse society
Big data analysis (BDA)	Enables real-time personalization based on purchase history and preferences
Social media	Explores influencers to enhance curation function.
On-demand manufacturing	Satisfies every individual customer's needs, automation
AWS	Locomotive for innovations by providing cloud computing platforms.
IoT	Enable wearables, optimize product assortment and customize recommendations.

3.3. Customer-Centric R&D-Driven Advancement in the Digital Economy

According to Jeff Bezos, "Our success at Amazon is a function of how many experiments we do per year, per month, per week, per day" [60]. In the digital economy, companies have easy access to customer data generated with every digital interaction. Such big data are extremely useful in driving insights that are used for experimentation.

Amazon's business model is based on customer-centric R&D-driven developments. Company-wide experimentation and R&D have been key to Amazon growing its empire. This is achieved by continuous interaction with users based on an architecture of participation, and an advanced assimilation capacity based on rapidly increasing R&D investment.

For instance, Amazon's R&D investments increased from USD 16 billion in 2016 to USD 56 billion in 2021 [61]. The significant increase in Amazon's R&D investment in the digital economy suggests the possibility of a structural change in the concept of R&D, similar to its output. Amazon's concept of R&D refers to technology and content. This indicates thoughtful insights into the R&D model in the digital economy [16,62,63]. The substantial increase in R&D enabled Amazon to become the world's top R&D firm in 2017, with a skyrocketing increase in its market capitalization, making it nearly the world's biggest company, as shown in Figures 5 and 6.

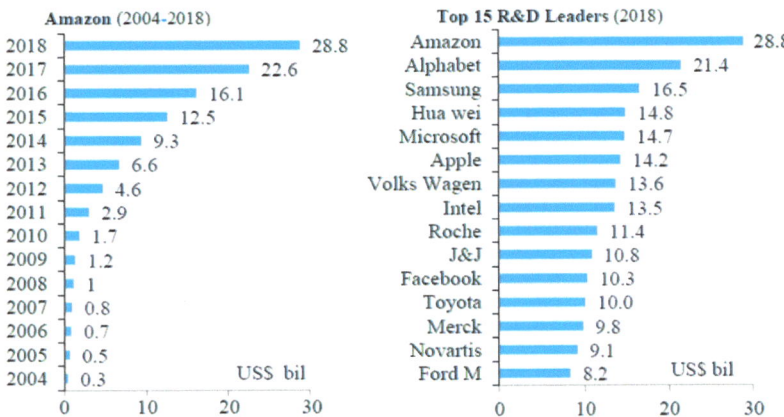

Figure 5. Amazon's conspicuous jump to become the world's top R&D leader—R&D investment. Sources: [64,65].

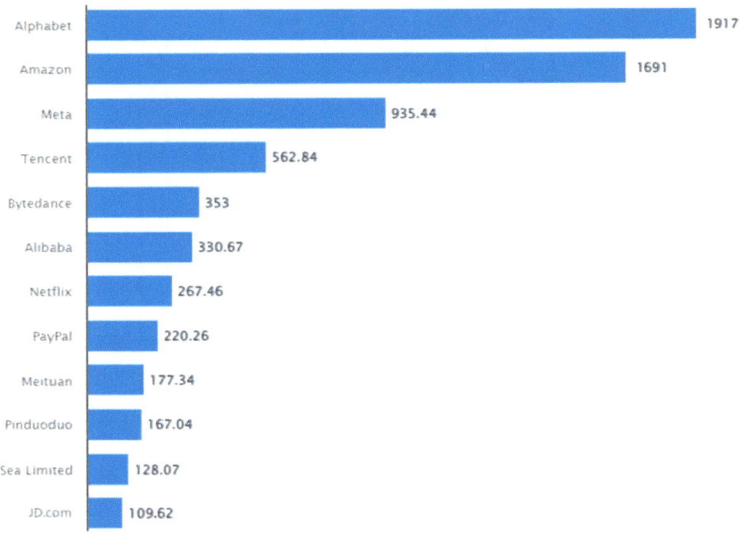

Figure 6. The market capitalization of the largest internet companies worldwide (June 2022). Source: [66].

Amazon also absorbs soft innovation resources (SIRs) from external sources and incorporates them into its business model, which converts "routine or periodic alterations" of business activities into "significant improvement" [32,67]. Thus, user-driven innovation accelerates the co-emergence of SIRs enabled by the advancement of the internet and communication technologies in the digital economy.

SIRs trigger a self-propagating function to satisfy changing customers' preferences beyond economic value. Amazon succeeded due to its customer obsession approach, talented employees, AI/AWS-powered products and services, and timely decision-making, as shown in Figure 7.

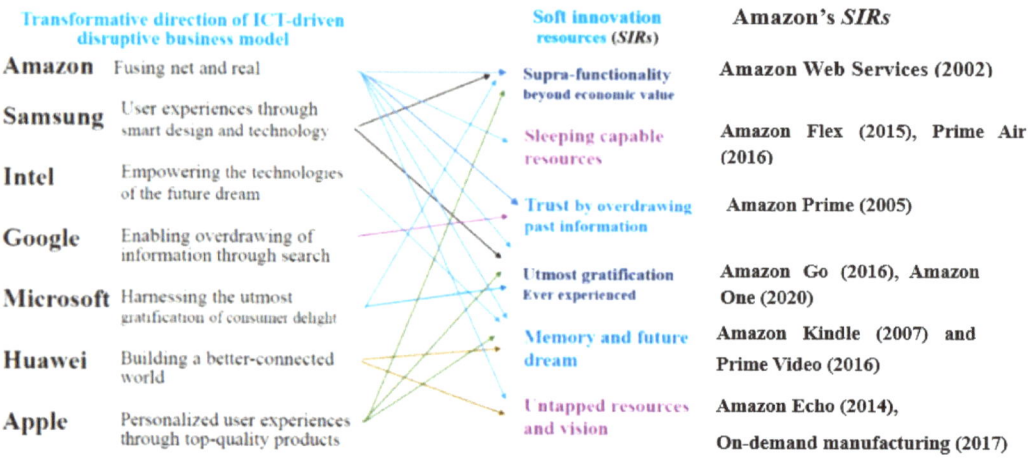

Figure 7. Soft innovation resources developed by top ICT leaders and corresponding examples in Amazon.

Amazon's endeavor to develop SIRs is shown in Figure 7. The SIRs comprise Amazon's initiation of user-driven innovation, such as (i) a shift in preferences toward supra-functionality (e.g., AWS in 2002), (ii) sleeping resources (e.g., Amazon Flex in 2015 and Prime Air in 2016), (iii) drawing on previous information and fostering trust (e.g., Amazon Prime in 2005), (iv) providing the most gratification ever experienced (e.g., Amazon Go in 2016 and Amazon One in 2020), (v) memory and future dreams (e.g., Amazon Kindle in 2007 and Prime Video in 2016), and (vi) untapped resources and vision (e.g., Amazon Echo in 2014 and ODM in 2017).

These SIRs are in line with customers' changing preferences, and Amazon's R&D investments in the development of the digital fashion business are associated with the integration of these resources. Fashion with artistic and functional features accelerates developments in SIRs, ranging from supra-functionality to untapped resources and vision. These SIRs, comprising aesthetic features, lead to advances in the fashion industry.

Thus, Amazon endeavors to create co-evolution between the development of SIRs and fashion advancement. Amazon has been developing a digital fashion business powered by AWS that acts as a locomotive for innovation and a carrier of digital solutions, which is a highly profitable category for the company [52]. Innovations in Amazon's digital fashion business create synergies with each other in an ecosystem rather than behaving individually. Inspired by the digital innovations illustrated in Figure 8, Amazon emerged as the second-largest retailer of apparel in the US, with a 7.9% share (after Walmart, 8.6%) in 2017, from a 3.7% share in 2016 [32]. Thus, Amazon has been expanding its highly profitable fashion-driven apparel business.

Fashion, including apparel and footwear, became Amazon's best-selling segment in 2018–2019, from fourth in 2017–2018, surpassing books, beauty, and electronics, and Amazon silently became the leading apparel retailer in the U.S. in 2019 [68]. In line with this growth and significant improvements in private-label fashion, Amazon attempted to move from selling basic apparel as traditional value to more fashionable higher-value categories.

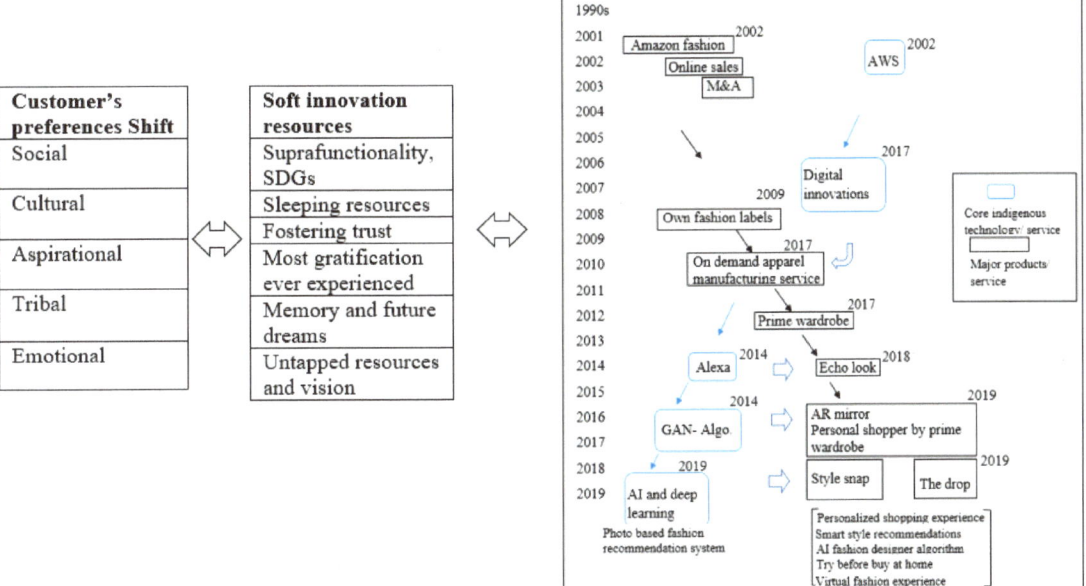

Figure 8. Co-evolution between SIRs-induced innovation and Amazon's digital fashion business.

This was not Amazon's first attempt to break into the luxury fashion market. The company tried similar moves in 2007 with endless.com and in 2012, when endless.com was renamed Amazon Fashion [69]. The increasing share of apparel sales in all of Amazon's online sales, as well as in the U.S. market, is shown in Figure 9. Among other business lines in 2018–2019, Amazon fashion was the most profitable business, with a 9.5% share in 2019. Amazon emerged as a leader in fashion sales in 2019. Since then, the company has maintained its leadership, with a 56.7% share of the U.S. market and a 14.6% share in online sales during 2021–2022.

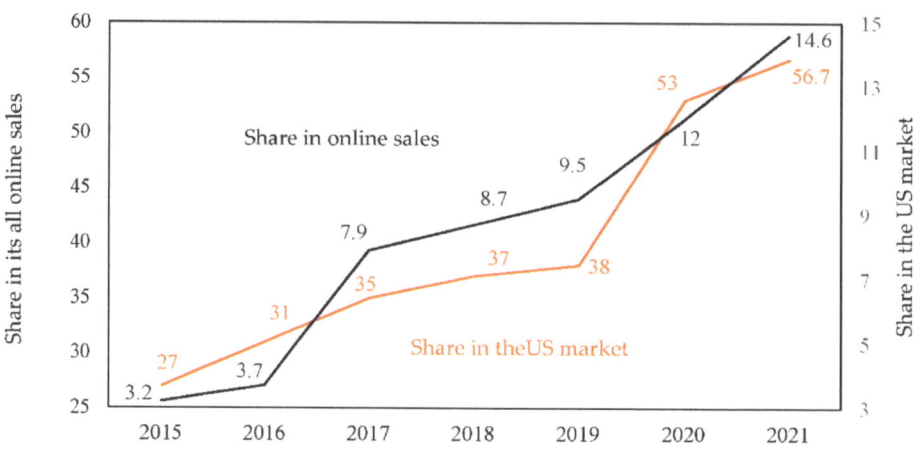

Figure 9. Trends in Amazon's apparel sales share (2015 to 2021): percentage. Sources: [70–76].

4. Amazon's Learning Orchestration Externality in Developing Digital Fashion Business

4.1. Lessons from the Past Experiences

Since 2002, Amazon has made several abortive attempts to capture the high-end fashion market. Its mission to capture this market has faced several historical hindrances, as customers have not trusted buying apparel online out of a desire to try on the items first, and Amazon was not perceived as a "cool" brand.

Amazon debuted endless.com in 2007 followed by myhabit.com in 2011, when the e-tail model was in its early stages. Both were discontinued. Part of the problem was the e-commerce brand's image in the luxury fashion industry. Moreover, luxury brands have been reluctant to adopt digitalization and an e-commerce presence. The prevalence of the "internet dilemma" in which luxury brands have been reluctant to adopt advanced digital solutions and e-commerce impeded Amazon from attracting them [4,77]. These challenges turned into learning experiences that accelerated Amazon's endeavor to develop its fashion empire with its own labels, powered by in-house AWS technology [78,79].

4.2. Creation of Advanced Digital Fashions (ADFs)

Developments in ICT are changing the buying habits of young luxury fashion consumers [4,80]. To capture this segment of the digital economy, Amazon undertook the following initiatives:

1. Entered the virtual assistant (Alexa) business with the introduction of Amazon Echo (2014). It trained Alexa to be a fashion advisor.
2. Activated its AWS team followed by Body Labs in the development of AI-led advanced digital fashions solutions (2017–2020).
3. The acquisition of brick-and-mortar Whole Foods (2017) led to the introduction of Amazon Go (2018) technology in Whole Foods stores. This is a unique venture because Amazon's other products and services are online and AWS-based.
4. Activated Prime Video in the acquisition of the rights to the Lord of the Rings series (2017).

These AI–AWS ventures do not act as stand-alone businesses; they create synergies for innovations. For example, they enabled Amazon to advance its fashion business and laid the foundations for e-commerce-based advanced digital fashions (2017–2020) and the brick-and-mortar fashion store Amazon Style (2022). This was an attempt to capture young affluent consumers, as acquiring a hot digitally native vertical brand is essential for shedding the company's "uncool" image [4].

By developing core AI-, IoT-, virtual- and augmented-reality-based digital tools, and mobile devices as reviewed in Section 3.2 and Table 2, Amazon has presented numerous innovations intended to advance its fashion-driven apparel business by using its big data collection system, user-driven innovation, and advanced logistics system, as illustrated in Table 1.

One of the earliest digital solutions in the ADFs series was Prime Wardrobe (2017), a subscription clothing box service that allows customers to try at home before buying. It was followed by Personal Shopper by Prime Wardrobe (2019). The Personal Shopper service is based on a co-creation and curation function in which customers and Amazon's designers create fashion items. It debuted on the Alexa-powered device Echo Look (2018), which contains a "hands-free camera and style assistant." The addition of a camera on the Echo device enables it to record and comment on its owner's clothing choices using a combination of machine learning and human stylist feedback. Echo Look provides recommendations that drive revenue to Amazon fashion. Specifically, Amazon created an AI algorithm for its private-label brands to design clothes (2018) by analyzing images, copying them as new items, and patenting AR Mirror (2019) that shows customers wearing virtual clothes in virtual locations. In addition, Amazon launched The Drop (2019), which sells limited-edition items designed by influencers. The detailed development paths of these ADFs are shown in Figure 10.

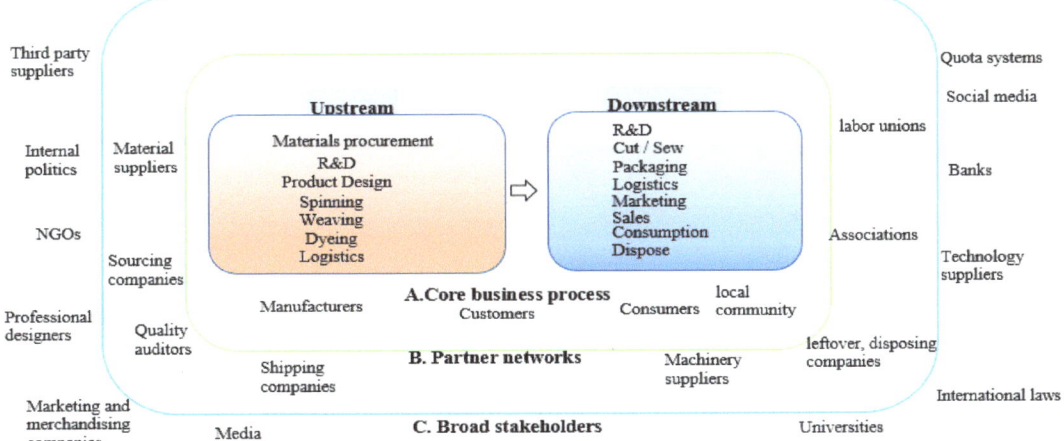

Figure 10. Relative importance of stakeholders in the Amazon fashion business. Source; Authors.

The Drop (2019) service is based on internal and external SIRs that combine broad designer and fashion influencer channels, including entertainment, social media, blogs, videos, webinars, and fashion shows. Moreover, the SIRs-driven The Drop (2019) is a platform for limited-edition designer collections. In this innovative service, broad stakeholders, such as external brands and designers, present their collections for an extremely short period. However, the incorporation of the curation function and ODM broadened the scope of stakeholders' participation more than that of stakeholders in the traditional fashion industry, as shown in Figure 10.

These broad external stakeholders provide Amazon with unstructured data in the form of documents, posts, audio/video, and reviews. Amazon's R&D hub has developed an AI-based algorithm that uses generative adversarial networks to extract meaningful content from the data. It can create a new design by analyzing trendy images on the web. This whole process can be defined as learning effects.

4.3. Learning Orchestration towards Advanced Digital Fashions (ADFs)

A firm's innovation culture is based on its higher capacity to assimilate internal and external knowledge [8]. The capacity to integrate knowledge is a function of the richness of the pre-existing knowledge structure. This indicates that learning is cumulative, and that learning performance is highest when the substance of learning is relevant to what is already known. Moreover, preceding knowledge allows for the integration and exploration of new knowledge [8].

Amazon follows a learning orchestration externality strategy and effectively utilizes the learning effects of similar challenges in three pillars (learning by orchestration), as illustrated in Figure 10:

(i) Customer-centric R&D-driven advancement.

Amazon emerged as a customer-centric company where R&D is the core of its business model.

(ii) Frontier innovation and companywide experimentation.

Amazon's founder Jeff Bezos has always stressed company-wide experimentation. It has become Amazon's culture of innovation. It enabled the growth of Amazon's empire and subsequent big data collection system.

(iii) User-driven innovation.

Amazon demonstrates communication with users for user-driven innovation based on the architecture of participation and a high level of assimilation capacity based on a significant increase in R&D spending [81]. Amazon's development of ADFs is a typical case, as illustrated in Figure 11.

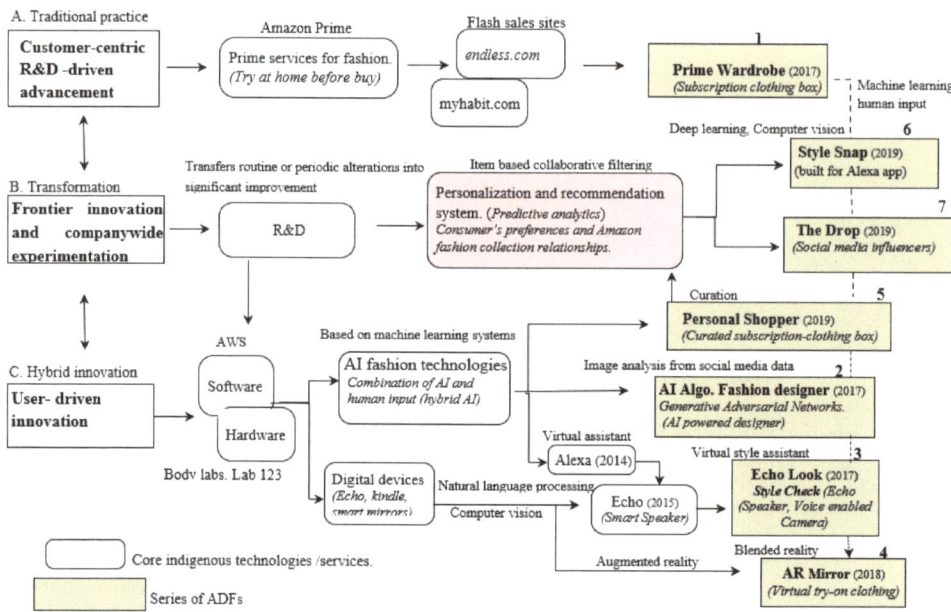

Figure 11. Amazon's ADFs development by learning from preceding innovations. Source: Authors' elaboration based on [10].

A detailed version of this process flow can be found in the authors' previous study [10]. This process flow indicates that Amazon initiated a series of ADFs successively by deploying an orchestration strategy for innovations by learning effectively from previous innovations.

According to [10], the Prime Wardrobe service was created with the mission of changing Amazon's brand image from a basic apparel retailer to a cool fashion brand. This service constructed the foundation of the business model to understand customers' needs and product and style preferences, as well as the measurement of personal data.

Prime Wardrobe was upgraded to Personal Shopper by Prime Wardrobe in 2019. It incorporates a curation function that lets customers co-create with Amazon's designers. At the same time, Amazon developed the AI algo. fashion designer that creates fashion styles without human involvement. This algorithm enhances video and imagery-based fashion services, such as AI-powered Echo Look and its derivatives. Amazon also introduced an ML-based Style Snap service to find similar styles from customers' provided photos. Echo Look's voice-enabled selfies and short videos trained Alexa to become a fashion advisor. Given the potential cost-efficiency benefits in the fashion supply chain, its functionality was transferred to mobile apps. The device was discontinued due to privacy, trust, and ethical issues. If a customer is not satisfied with the algorithmic recommendations provided by Echo Look, it then suggests ordering specific styles from Amazon's fashion catalog. This suggests that Amazon's ODM will soon be able to produce hyper-customized outfits. Moreover, instead of uploading photos to Style Snap and then waiting for recommendations, customers can use Amazon's patented AR Mirror, which provides real-time recommendations. The Style Snap service provides Amazon with customers' behavioral data and encourages collaboration with fashion influencers; ultimately, the scope of external stakeholders is broadened. This has led to the development of The Drop service. It provides limited-edition styles curated by celebrity fashion influencers on social media. Notable fashion influencers present trend-led limited collections for 30 hours. Items recommended by The Drop are from either Amazon's fashion catalog or designers' creations

5. Creation of Supra-Omnichannel during COVID-19

Amazon has transformed the traditional fashion value chain, as illustrated in Figure 1, into a supra-omnichannel. Due to the COVID-19 pandemic, there has been a major drop in textile and apparel sales, and Amazon has been undergoing a digital solution-oriented transformation [82].

During this crisis, Amazon accelerated its strategic actions in developing more advanced digital solutions to support the declining fashion business. Amazon has been growing its fashion empire by introducing a series of ADFs comprising physical and digital commodities. This service increased Amazon's omnichannel dependence based on seamless switching by utilizing its innovative assets. Capturing an e-commerce-based luxury fashion market has always been Amazon's long-awaited vision. To solve this challenge, the company has endeavored to shed its uncool brand image and advanced the AI-driven curation function by introducing an e-commerce-based series of ADFs. Other services, Style Snap and The Drop, enabled Amazon to collaborate with external resources such as global influencers to co-design trendy fashion manufactured in line with demand.

In 2020, Amazon introduced Luxury Stores, a unique digital platform for luxury and high-end fashion. This was in response to the decline in the luxury business during the COVID-19 pandemic and the increasingly non-contact society, which necessitated the addition of extra channels for luxury brands [10,62]. At the same time, luxury brands are confronting "the Internet dilemma" [52], which is the reluctance to incorporate ICT and e-commerce technologies into their business models [83]. The internet dilemma impedes luxury brands, such as the luxury giant LVMH (Kering and Hermes) Group, from collaborating with Amazon [84]. Despite these challenges, the Luxury Stores initiative is gaining momentum; thus far, more than 50 luxury brands have collaborated with Amazon, as illustrated in Figure 12.

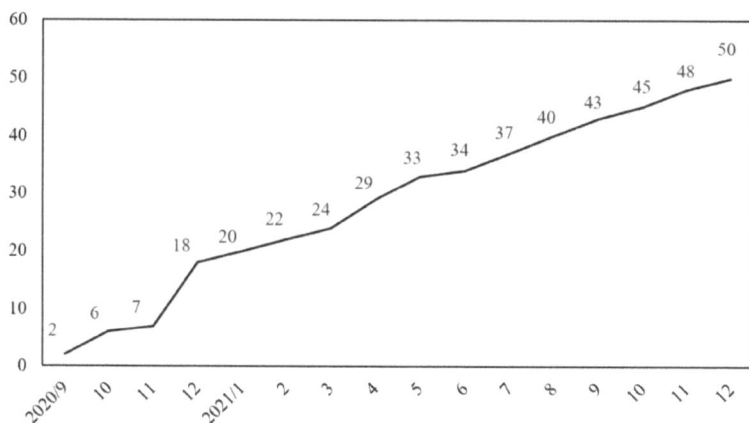

Figure 12. The trend in the cumulative number of luxury brands joining Amazon.

The Luxury Stores platform provides the following advantages to Amazon and its stakeholders:

(1) The millions of Amazon Prime members provide feedback. This initiates an iterative process in which users' feedback is used for significant improvements, experimentation, and innovations.
(2) In contrast to Amazon's traditional business model the company provides freedom to luxury brands in controlling inventory, pricing, and distribution. Amazon provides digital tools and customer data for creating and personalizing content for each brand's identity.
(3) This digital store-within-a-store concept and freedom to control can build trust between Amazon and luxury brands to solve the historic internet dilemma.
(4) Amazon provides luxury brand customers with an opportunity for free shipping and returns.

Amazon reduces luxury brands' dependency on brick-and-mortar stores by providing an independent digital space.

In the same light of solving the internet dilemma, Amazon received a patent for ODM in 2017 that enables luxury brands to consolidate their supply chains. This challenge can be expected to be solved by synchronizing ADFs, Luxury Stores, and ODM. This endeavor suggests a solution to the historic internet dilemma as well as the shift from multichannel and cross-channel to omnichannel [52].

The mission to be a fashion-driven apparel leader has led Amazon to focus on following a three-dimension approach consisting of (i) involving customers in the co-creation of their preferred styles, (ii) improving brand image with the curation function, and (iii) capturing the luxury fashion market with digital services. Co-evolution and synchronization of these three initiatives are expected to lead to ODM for the fashion business, together with shedding Amazon's uncool brand image and diving into the luxury fashion business, as illustrated in Figure 13.

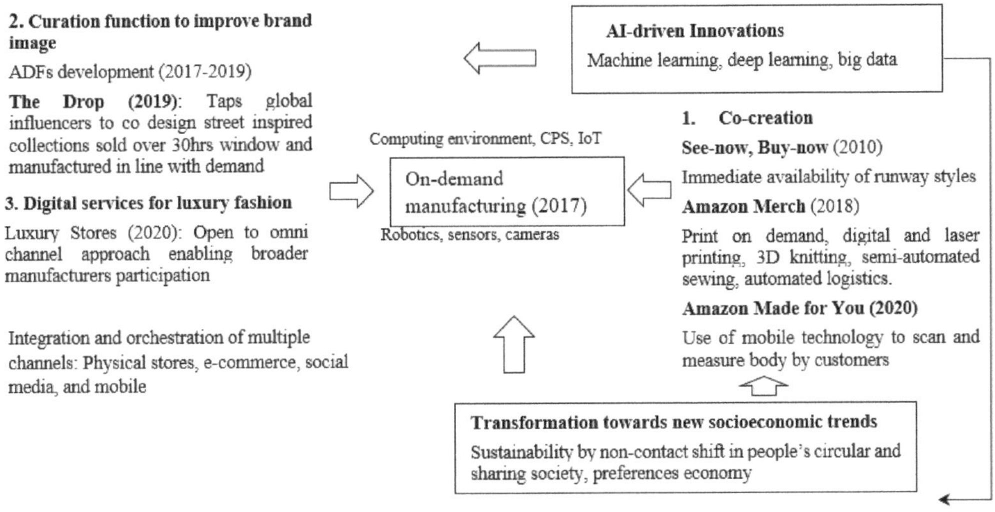

2. Curation function to improve brand image

ADFs development (2017-2019)

The Drop (2019): Taps global influencers to co design street inspired collections sold over 30hrs window and manufactured in line with demand

3. Digital services for luxury fashion

Luxury Stores (2020): Open to omni channel approach enabling broader manufacturers participation

Integration and orchestration of multiple channels: Physical stores, e-commerce, social media, and mobile

AI-driven Innovations
Machine learning, deep learning, big data

Computing environment, CPS, IoT

On-demand manufacturing (2017)
Robotics, sensors, cameras

1. Co-creation

See-now, Buy-now (2010)
Immediate availability of runway styles

Amazon Merch (2018)
Print on demand, digital and laser printing, 3D knitting, semi-automated sewing, automated logistics.

Amazon Made for You (2020)
Use of mobile technology to scan and measure body by customers

Transformation towards new socioeconomic trends
Sustainability by non-contact shift in people's circular and sharing society, preferences economy

Figure 13. Amazon's initiatives toward on-demand manufacturing in the luxury fashion business.

5.1. Involving Customers in the Co-Creation of Their Preferred Styles

Amazon started inducing customers' preferences with co-creation initiatives such as See Now Buy Now (2010), Amazon Merch (2018), and Amazon Made for you (2020). This corresponds with the increasing trend of prosumers (consumers as producers) in response to the increasing anger of consumers at remaining non-producers, in contrast to their desire to enjoy an exciting story with their initiatives as heroes/heroines of a drama [9,85].

5.2. Improving Brand Image by Curation Function

Amazon introduced the curation function to shed its uncool brand image in the fashion industry. It improved customers' abilities in fashion co-designing by developing a series of ADFs leading to collaboration with global social media fashion influencers to co-design the most fashionable collections sold and manufactured in line with demand.

This development of a series of ADFs corresponds to the shift from multichannel and cross-channel to omnichannel [52].

5.3. Capturing the Luxury Fashion Market with Digital Services

Amazon's core multichannel fulfillment service enabled it to delve deeper into the luxury fashion business. For example, the Luxury Stores initiative allowed luxury brands to make decisions about their inventory, selection, timing, and pricing. Oscar de la Renta, Elie Saab, and Altuzarra were early partners of Luxury Stores.

This challenge started with the co-existence of luxury brands' traditional channels and Amazon's channels. However, the goal is to transform this co-existence into co-evolution, as illustrated in Figure 14.

With this co-evolution, Amazon can provide a solution to the previously impossible conundrum "the Internet dilemma" [77,86]. Thus, a possible solution is a cloud-based fashion platform that combines ADFs, luxury brands, and all sales channels. This also transforms the traditional value chain of the fashion industry into a digital platform that combines stakeholders and consumers at one point. With a cloud-based fashion platform, the digitalization of the upstream will provide real-time information on changing customer preferences and enable downstream industries to use digital solutions. Additionally, increasing environmental consciousness enables stakeholders to adopt green practices when the value chain of fashion is visible, and customers know the origin of their fashion products.

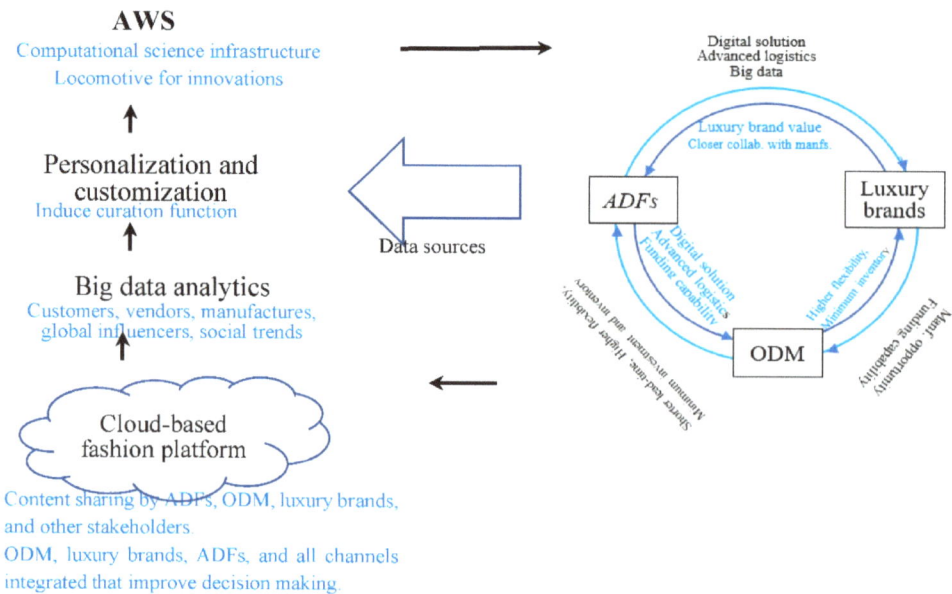

Figure 14. Dual co-evolution Among ADFs, luxury brands, ODM co-evolution, and cloud-based fashion platform advancement.

A cloud-based fashion platform enables personalization and customization with an advanced curation function by way of seamless switching on an on-demand basis. Consequently, big data on customers, vendors, manufacturers, global influencers, and social trends can be analyzed, which further improves AWS's functionality with learning orchestration externality. AWS, as a computational science infrastructure, grows and expands by learning digital advancements initiated by preceding endeavors [32,87]. AWS provides solutions to big data analytic requirements, so that companies focus on business problems instead of managing these tools. Advanced AWS, in turn, further accelerates co-evolution among ADFs, luxury brands, and ODM. Activated co-evolution has led to a cloud-based fashion platform, resulting in a virtuous cycle. Thus, dual co-evolution occurs among ADFs, luxury brands, and ODM, and a cloud-based fashion platform.

6. Learning Orchestration beyond E-commerce

In the previous sections, Amazon's ambitious goals of becoming a digital fashion leader based on digital innovations in fashion were shown and discussed. Considerable R&D investment, ADFs, luxury brands, and ODM emerged. Moreover, their dual co-evolution resulted in a cloud-based fashion platform representing a supra-omnichannel that combines all the stakeholders.

Amazon's new CEO, Andy Jassy, suggested that "customers will eventually do their shopping from their *Fire TV, Omni, or Fire TV 4-Series*. It's part of the company's effort to shift you away from tapping on apps, an experience that will soon feel outdated" [88]. E-commerce apps could be replaced by voice (Alexa), buying while viewing the same fashion show on TV, and digital–physical shopping, such as Amazon Style (2022). Thus, based on Andy Jassy's suggestion, it is crucial to investigate a digital solution that leads beyond e-commerce endeavors. Our research beyond e-commerce focuses on the emergence of Amazon's innovations during the COVID-19 pandemic by following learning orchestration externality strategies, such as the following:

1. Amazon Aware (2022) represents a circular and sharing economy in a carbon-neutral society.

2. Amazon Style (2022) represents on-demand manufacturing
3. Luxury Stores (2020) represent co-creation and customization.
4. Making the Cut (2020) for sociocultural engagement.
5. Amazon Braket (2019) for quantum computing.

When learning orchestration externality is applied, the emergence of beyond e-commerce solutions can be expected, as illustrated in Figure 15.

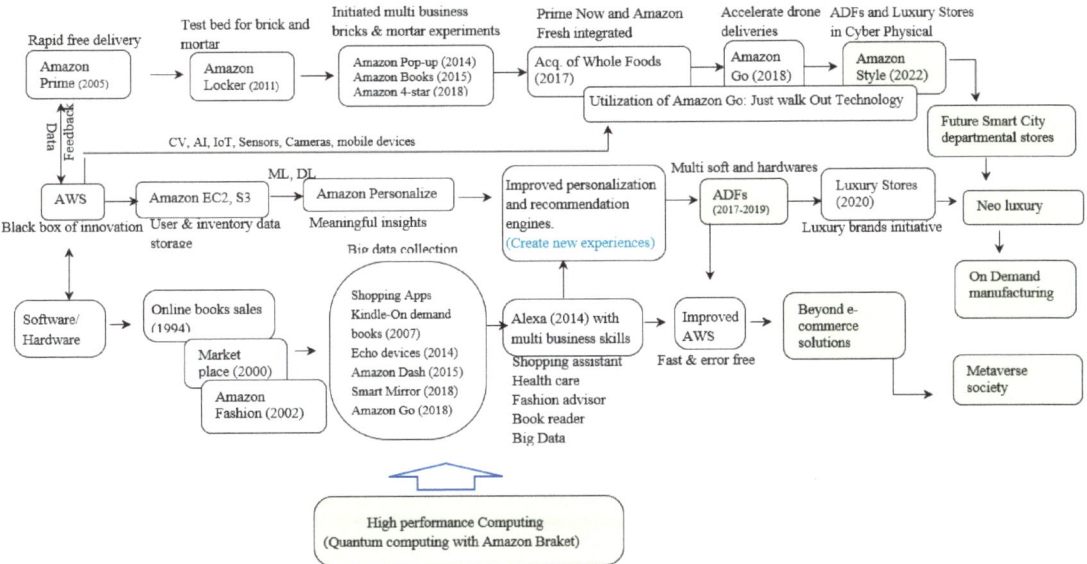

Figure 15. Learning orchestration beyond e-commerce.

6.1. Amazon Aware

Amazon introduced a new fashion brand Amazon Aware (2022). All items including fashion are certified as Climate Pledge Friendly. Since Amazon Aware (2022) is its private label, it is expected that fashion items representing this label will be available at Amazon Style (2022). Both, ODM and Amazon Aware (2022) also correspond to sharing and circular economy [89].

6.2. Amazon Style

Its preceding ADFs (programs, algorithms, devices) and in-store technologies collect insights on customers' digital shopping preferences as well as their brick-and-mortar behaviors. Such first-party data enable Amazon in creating feedback (preferences, reviews, ratings, suggestions) channels in finding what customers value and enable it in creating highly curated and personalized products, recommending them to every visiting customer in real-time. This data leads to effective, personalization, experimentation, and optimization. Amazon's ability to adjust its offerings (curation) based on customer preference data leads to on-demand satisfaction. Since customers expect responses to their demands quickly, Amazon Style is designed to provide on-demand satisfaction to every visiting customer by using enabling technologies consisting of the mobile app, QR codes, cameras, screens, and sensors [90].

6.3. Luxury Stores

Luxury Stores was initiated with Oscar de la Renta (September 2020), followed by Roland Mouret (September 2020), Altuzarra (October 2020), Cle de Peau (October 2020), Car Shoe (October 2020), and Revive Skin Care (October 2020). Since 2020, Amazon endeavored to expand Luxury Stores. In solving the internet dilemma, the success of this endeavor depends on the construction between the traditional business model in brands and Amazon's efforts for a more sophisticated omnichannel approach [91].

6.4. Making the Cut

"Making the Cut" (reality TV fashion show) is a transition from traditional e-commerce to TV commerce, which allows entertainment and shopping at the same time. While using Amazon's devices, customers can purchase their desired styles as soon as a TV episode ends. Along with dramatic features, "Making the Cut" demonstrates a glimpse into what could prove to be the future of shoppable videos. Thus, "[t]ying the content with the opportunity to purchase" as soon as the episode ends is a novel method that goes beyond traditional e-commerce [92].

6.5. Amazon Braket

The emergence of a non-contact society during the COVID-19 pandemic increased people's dependence on ICT, social media, and mobile devices. This increased dependence has changed customers' buying habits. With growing big data, traditional computing impedes efficient solutions to large-scale complex problems [93].

With the amalgamation of the IoT and AI approaches, it is assumed that great insights beyond mere knowledge are achieved. However, as indicated by several studies, there are several data challenges with classical AI approaches. First, most big data are unstructured [94,95]. There are also a lack of personalized data [13] and uneven data flow [96].

To gain meaningful insights (wise decision-making) from the growing big data generated by ADFs, ODM, Echo devices, and other IoT products, more advanced optimization solutions that also require high-performance computing are required. Moreover, classical computing becomes inefficient over time [97]. This suggests the integration of quantum computing with classical AI, IoT, and data analytics. Compared to classical algorithms for computation, quantum algorithms are expected to solve a set of challenges, including computational optimization for the information and natural sciences, with improved efficiency. To gain strategic advantages, technology giants such as Amazon [98] and government organizations are heavily investing in the research and development of these systems.

7. Conclusions and Suggestions

This paper presented the contribution of Amazon's preceding innovations (learning orchestration strategy) that led the company to become a largest digital fashion retailer in e-commerce, brick-and-mortar, and future endeavors beyond e-commerce business. Over the years, Amazon's considerable R&D investments have contributed significantly to the development of the disruptive business model. This achievement is also associated with a virtuous cycle of user-driven innovation, AWS, and SIRs that activate the self-propagation function. Amazon's complex customer-centric R&D process transforms routine changes into numerous significant improvements. These practices led to the advent of digital fashion solutions that turned the COVID-19 pandemic into a springboard for innovations in Amazon's fashion empire.

Fashion is a mode of self-expression that reflects changes in aesthetic, economic, political, cultural, and social life. These changes, in turn, change fashion, and apparel boosts this change. Thus, in response to a shift in people's preferences, the fashion industry has been gaining momentum worldwide. At the same time, digital solutions in terms of communication, devices, services, and e-commerce in fashion are gaining momentum. An analysis of Amazon's endeavor to develop advanced digital fashion with aggressive AI-oriented R&D, an empirical co-evolutional analysis of the development trajectories of Amazon's ADFs, and the fashion industry with special attention to the role of AI advancement was conducted.

The findings include the following:

(1) Amazon's innovations are transforming the traditional value chain of the fashion industry into a platform that harnesses data directly from consumers to develop more customer-centric products and services.
(2) The recent COVID-19 pandemic contributed as a springboard for innovations.
(3) The fashion industry must accelerate digital innovations through emerging tools, such as AI, cloud computing technology, AR/VR, blockchain, etc. These digital technologies can transform the traditional fashion industry into a digital platform industry. For example, Amazon's fashion business secured a timely digital solution by developing a series of ADFs, a supra-omnichannel, and ODM based on the digital tools described above.
(4) The advancement of AWS, ADFs, and ODM led to the development of Luxury Stores in 2020, which emerged as neo-luxury. Amazon's enthusiastic efforts to become an AI giant enabled this success. The Luxury Stores initiative has the potential to solve luxury brands' historic e-commerce dilemma.
(5) The activation of dual co-evolution among ADFs, luxury brands, and ODM is driven by advancements in AWS/AI that contribute to the development of the supra-omnichannel. This incorporates a generative function and evolves a cloud-based fashion platform that integrates internal and external stakeholders. The fashion value chain can be synchronized with ODM in real time, and stakeholders and customers can communicate within the system.
(6) Amazon has been advancing AWS as an innovative, advanced composite cloud infrastructure. This infrastructure incorporates a generative function and develops a cloud-based fashion platform by integrating all stakeholders into one place. These developments have enabled Amazon to gain the outcomes of learning orchestration externalities.

Future research could investigate Amazon's latest innovations during the COVID-19 pandemic, leading beyond e-commerce endeavors in a non-contact society. It would also be interesting to examine the contribution of Amazon Braket, a quantum computing platform, to solve non-e-commerce challenges, such as a carbon-neutral society, an age of meaning, on-demand satisfaction, shopping by amusement, and immersive technologies that correspond to a non-contact society. It is also important to examine the role of advanced preceding innovations such as ADFs, ODM, neo-luxury, and supra-omnichannel that emerged with the dual co-evolution of ADFs, ODM, and neo-luxury beyond e-commerce endeavors.

Thus, the emergence of a non-contact society has created more demand for digital solutions for the fashion industry due to lockdowns, store closures, and social distancing, etc. Future research should focus on advanced digital solutions to develop functioning beyond an e-commerce non-contact society, such as a metaverse society [32,42].

Author Contributions: Conceptualization, W.H.A. and C.W.; methodology, W.H.A. and C.W.; software, Y.T. and W.H.A.; validation, W.H.A., C.W. and P.N.; formal analysis, W.H.A.; investigation, W.H.A. and Y.T.; resources, W.H.A. and C.W.; data curation, W.H.A. and C.W.; writing—original draft preparation, W.H.A.; writing—review and editing, W.H.A. and P.N.; visualization, C.W. and W.H.A.; supervision, C.W.; project administration, P.N. and C.W. All authors have read and agreed to the published version of the manuscript.

Funding: This research received no external funding.

Institutional Review Board Statement: Not applicable.

Informed Consent Statement: Not applicable.

Data Availability Statement: The data presented in this study are available on request from the corresponding author.

Acknowledgments: The research leading to these results was funded by a grant provided by the Jenny and Antti Wihuri Foundation.

Conflicts of Interest: The authors declare no conflict of interest.

References

1. Aaronson, S.A. America's Uneven Approach to AI and Its Consequence. Institute for International Economic Policy Working Paper Series Elliott School of International Affairs The George Washington University. 2020. Available online: https://www2.gwu.edu/~{}iiep/assets/docs/papers/2020WP/AaronsonIIEP2020-7.pdf (accessed on 19 September 2020).
2. Adejeest, D.-A. Amazon's U.S. Market Share of Clothing Soars to 14.6 Percent. 2022. Available online: https://fashionunited.com/news/retail/amazon-s-u-s-marketshare-of-clothing-soars-to-14-6-percent/2022031546520 (accessed on 20 September 2022).
3. Amazon. Amazon Com. Inc. Annual Report 2017. Amazon.com, Inc., Seattle. 2018. Available online: http://www.annualreports.com/Company/amazoncom-inc (accessed on 10 September 2020).
4. Amazon. Amazon.com, Inc. Annual Report 2018. Amazon.com, Inc., Seattle. 2019. Available online: https://ir.aboutamazon.com/static-files/0f9e36b1-7e1e-4b52-be17-145dc9d8b5ec (accessed on 13 September 2022).
5. Amazon.com. Luxury Stores. 2022. Available online: https://www.amazon.com/stores/luxury/page/6D2D5FDF-3E7A-42C7-A1A1-F978792B5E4D (accessed on 18 September 2022).
6. Anderson-Connell, L.J.; Ulrich, P.V.; Brannon, E.L. A Consumer-driven Model for Mass Customization in the Apparel Market. *J. Fash. Mark. Manag.* **2002**, *6*, 240–258. [CrossRef]
7. Arrigo, E. Global Sourcing in Fast Fashion Retailers: Sourcing Locations and Sustainability Considerations. *Sustainability* **2020**, *12*, 508. [CrossRef]
8. Backs, S.; Jahnke, H.; Lüpke, L.; Stücken, M.; Stummer, C. Traditional Versus Fast Fashion Supply Chains in the Apparel Industry: An Agent-based Simulation Approach. *Ann. Oper. Res.* **2021**, *305*, 487–512. [CrossRef]
9. Baker, J.; Ashill, N.; Amer, N.; Diab, E. The Internet Dilemma: An Exploratory Study of Luxury Firms' Usage of Internet-based Technologies. *J. Retail. Consum. Serv.* **2018**, *41*, 37–47. [CrossRef]
10. Barnard, M. *Fashion as Communication*; Routledge: London, UK, 2013.
11. Bebchuk, L.A.; Kastiel, K.; Tallarita, R. Stakeholder Capitalism in the Time of COVID. *Forthcom. Yale J. Regul.* **2023**, *40*. [CrossRef]
12. Berg, A. How Current Global Trends Are Disrupting the Fashion Industry. McKinsey & Company. 2022. Available online: https://www.mckinsey.com/industries/retail/our-insights/how-current-global-trends-are-disrupting-the-fashion-industry (accessed on 8 September 2022).
13. Bloomberg. *2018 Global Innovation 1000 Study*; Bloomberg: New York, NY, USA, 2018.
14. Boyle, A. Amazon's Blended-Reality Mirror Shows You Wearing Virtual Clothes in Virtual Locales. Geekwire. 2018. Available online: https://www.geekwire.com/2018/amazon-patents-blended-reality-mirrorshows-wearing-virtual-clothes-virtual-locales/ (accessed on 9 September 2022).
15. Brown, R. Top Quantum Techniques to Optimize That Complex Last Mile. QCI. 2022. Available online: https://www.quantumcomputinginc.com/blog/last-mile/ (accessed on 22 September 2022).
16. Burns, L.V.; Mullet, K.K.; Bryant, N.O. *The Business of Fashion: Designing, Manufacturing, and Marketing*, 4th ed.; Fairchild Books, Inc.: New York, NY, USA, 2011. Available online: https://salon.thefamily.co/11-notes-on-amazon-part-1-cf49d610f195 (accessed on 6 May 2022).
17. Cholachatpinyo, A.; Fletcher, B.; Padgett, I.; Crocker, M. A Conceptual Model of the Fashion Process-part 1: The Fashion Transformation Process Model. *J. Fash. Mark. Manag.* **2002**, *6*, 11–23. [CrossRef]
18. Christopher, M.; Lowson, R.; Peck, H. Creating Agile Supply Chains in the Fashion Industry. *Int. J. Retail Distrib. Manag.* **2004**, *32*, 367–376. [CrossRef]
19. Chuprina, N.V.; Krotova, T.F.; Pashkevich, K.L.; Kara-Vasylieva, T.V.; Kolosnichenko, M.V. Formation of Fashion System in the XX-the beginning of the XXI century. *Vlak. A Text. (Fibres Text.)* **2020**, *27*, 48–57.
20. Čiarnienė, R.; Vienažindienė, M. Management of Contemporary Fashion Industry: Characteristics and Challenges. *Procedia-Soc. Behav. Sci.* **2014**, *156*, 63–68. [CrossRef]
21. Clark, E. Amazon Shuttering MyHabbit.com. WWD. 2016. Available online: https://wwd.com/business-news/technology/amazon-closing-myhabit-com-10415899/ (accessed on 16 September 2022).
22. Clement, J. Market Value of the Largest Internet Companies Worldwide 2022. Statista. 2022. Available online: https://www.statista.com/statistics/277483/market-value-of-the-largest-internet-companies-worldwide/ (accessed on 13 September 2022).
23. Clifford, S. Amazon Leaps into High End of the Fashion Pool. The Newyork Times. 2017. Available online: https://www.nytimes.com/2012/05/08/business/amazon-plans-its-next-conquest-your-closet.html (accessed on 14 September 2022).
24. CNN. Amazon's New Line Is All about Sustainable Essentials. 2022. Available online: https://edition.cnn.com/cnn-underscored/home/amazon-aware-launch (accessed on 21 September 2022).
25. Cohen, W.M.; Levinthal, D.A. Absorptive Capacity: A New Perspective on Learning and Innovation. *Adm. Sci. Q.* **1990**, *35*, 128–152. [CrossRef]
26. Coppolla, D. Annual Technology and Content Expenses of Amazon from 2016 to 2021. 2022. Available online: https://www.statista.com/statistics/991947/amazons-annual-technology-and-content-expenses/ (accessed on 12 September 2022).
27. Dickerson, K.G. *Textiles and Apparel in the Global Economy*, 3rd ed.; Prentice hall: London, UK, 1999.
28. Doeringer, P.; Crean, S. Can Fast Fashion Save the U.S. Apparel Industry? *Socio-Econ. Rev.* **2006**, *4*, 353–377. [CrossRef]
29. Dubey, A.; Bhardwaj, N.; Abhinav, K.; Kuriakose, S.M.; Jain, S.; Arora, V. AI Assisted Apparel Design. *arXiv* **2020**, arXiv:2007.04950.
30. Financial Times. Can Amazon Upend the Luxury Sector? 2020. Available online: https://www.ft.com/content/bcc14d0a-a6a5-40bb-b5dc-333e6f3d7775 (accessed on 18 September 2022).

31. Forte, D. Amazon Is America's Most Shopped Retailer in Apparel. Multichannel Merchant. 2019. Available online: https://multichannelmerchant.com/ecommerce/amazon-americas-shopped-retailer-apparel/#:~{}:text=Amazon%20has%20jumped%20from%20a,from%20around%2060%25%20last%20year (accessed on 20 September 2022).
32. Fraser, S.; Oberlack, U.; Wright, E. Trends and Tradition: Negotiating Different Cultural Models in Relation to Sustainable Craft and Artisan production. Bangalore, India 29th September to 1st October 2010. 2010. Available online: https://ualresearchonline.arts.ac.uk/id/eprint/4615 (accessed on 23 September 2022).
33. Galloway, S. *The Hidden DNA of Amazon, Apple, Facebook, and Google*; Penguin Random House LLC: New York, NY, USA, 2017.
34. Gereffi, G. Global Value Chains in a Post-Washington Consensus World. *Rev. Int. Political Econ.* **2014**, *21*, 9–37. [CrossRef]
35. Gonzalo, A.; Harreis, H.; Altable, C.; Villepelet, C. Fashion Digital Transformation: Now or Never. Mc Kinsey & Co. 2020. Available online: https://www.mckinsey.com/industries/retail/our-insights/fashions-digital-transformation-now-or-never (accessed on 18 September 2022).
36. Stone, E. *The Dynamics of Fashion*, 3rd ed.; Fairchild Books: New York, NY, USA, 2008.
37. Sull, D.; Turconi, S. Fast Fashion Lessons. *Bus. Strategy Rev.* **2008**, *19*, 5–11. [CrossRef]
38. Sun, L.; Zhao, L. Technology Disruptions: Exploring the Changing Roles of Designers, Makers, and Users in the Fashion Industry. *Int. J. Fash. Des. Technol. Educ.* **2018**, *11*, 362–374. [CrossRef]
39. Thomassey, S.; Zeng, X. Erratum to: Artificial Intelligence for Fashion Industry in the Big Data Era. In *Artificial Intelligence for Fashion Industry in the Big Data Era*; Springer: Singapore, 2018; p. E1.
40. Tou, Y.; Watanabe, C.; Moriya, K.; Naveed, N.; Vurpillat, V.; Neittaanmäki, P. The Transformation of R & D into Neo Open Innovation-a new Concept in R & D Endeavor Triggered by Amazon. *Technol. Soc.* **2019**, *58*, 101141.
41. Vasan, S. Amazon Reimagines Instore Shopping with Amazon Style. 2022. Available online: https://www.aboutamazon.com/news/retail/amazon-reimagines-in-store-shopping-with-amazon-style (accessed on 21 September 2021).
42. Gosh, P. Amazon Is Now America's Biggest Apparel Retailer, Here's Why Walmart Can't Keep Up. 2020. Available online: https://www.forbes.com/sites/palashghosh/2021/03/17/amazon-is-now-americas-biggest-apparel-retailer-heres-why-walmart-cant-keep-up/?sh=7aaf4b9131ce (accessed on 20 September 2022).
43. Hansen, K.T. The World in Dress: Anthropological Perspectives on Clothing, Fashion, and Culture. *Annu. Rev. Anthropol.* **2004**, *33*, 369–392. [CrossRef]
44. Hardt, M.; Chen, X.; Cheng, X.; Donini, M.; Gelman, J.; Gollaprolu, S.; Kenthapadi, K. Amazon Sagemaker Clarify: Machine Learning Bias Detection and Explainability in the Cloud. *arXiv* **2021**, arXiv:2109.03285.
45. Hartmans, A. Amazon's New Echo Device Is a Hands-Free Camera That Helps You Decide What to Wear. Business Insider.com. 2017. Available online: http://uk.businessinsider.com/amazon-look-camera-outfit-analysis-2017- (accessed on 17 September 2022).
46. Hautala, L. New Amazon CEO Andy Jassy Says Voice Is the Future. Tapping on Apps Is 'So Circa 2005'. Cnet. 2021. Available online: https://www.cnet.com/tech/tech-industry/new-amazon-ceo-andy-jassy-says-voice-is-the-future-tapping-on-apps-so-circa-2005/ (accessed on 20 September 2022).
47. Huang, M.H.; Rust, R.T. Engaged to a Robot? The Role of AI in Service. *J. Serv. Res.* **2021**, *24*, 30–41. [CrossRef]
48. Huber, B. Amazons the Drop Will Sell Limited Edition Items Designed by Influencers. Refinery29.com. 2019. Available online: https://www.refinery29.com/en-us/2019/05/233447/amazon-fashion-the-drop-limited-edition-street-style-clothing (accessed on 15 September 2022).
49. Japan Cabinet Office. *National Survey of Lifestyle Preferences*; Japan Cabinet Office: Tokyo, Japan, 2019.
50. Kaiser, S.B. *Fashion and Cultural Studies*; Berg Publishers: Oxford, UK, 2012.
51. Kellie, E. Amazon Takes on Stitch Fix. Women's Wear Daily, Los Angeles. 2019. Available online: https://search.proquest.com/docview/2319664381/AB33481834C54061PQ/4?accountid=11774 (accessed on 15 September 2022).
52. Keyes, D. Amazon Opens Prime Wardrobe to More Shoppers. Business Insider. 2018. Available online: https://static3.businessinsider.com/amazon-opens-prime-wardrobe-to-more-shoppers-2018-4 (accessed on 20 September 2022).
53. Knott, A.M. *How Innovation Really Works: Using the Trillion-Dollar R & D Fix to Drive Growth*; McGraw Hill: New York, NY, USA, 2017.
54. Ko, E.; Costello, J.P.; Taylor, C.R. What is a Luxury Brand? A New Definition and Review of the Literature. *J. Bus. Res.* **2019**, *99*, 405–413. [CrossRef]
55. Kohlbacher, F.; Hang, C.C. Applying the Disruptive Innovation Framework to the Silver Market. *Ageing Int.* **2011**, *36*, 82–101. [CrossRef]
56. Lee, M.R.; Kim, M.S. A Study on the Digitalization of the Fashion Industry. *Int. J. Costume Cult.* **2001**, *4*, 124–137.
57. Leighton, M. Amazon's 'Try Before You Buy' Shopping Service, Prime Wardrobe, Is Free For Prime members and Easy to Use—Here's How it Works. Businessinsider.com. 2020. Available online: https://www.businessinsider.com/what-is-prime-wardrobe?r=US&IR=T (accessed on 12 August 2020).
58. Light, L. Amazon's Prime Time for Luxury. Forbes. 2020. Available online: https://www.forbes.com/sites/larrylight/2020/10/16/amazons-prime-time-for-luxury/ (accessed on 21 September 2021).
59. Major, J.S.; Steele, V. Fashion Industry. Encyclopedia Britannica. 2022. Available online: https://www.britannica.com/art/fashion-industry (accessed on 3 September 2022).

60. Mc Kinsey and Company. Ten Trends for the Fashion Industry to Watch in 2019. Mc Kinsey & Company. 2019. Available online: https://www.mckinsey.com/industries/retail/our-insights/ten-trends-for-the-fashion-industry-to-watch-in-2019 (accessed on 5 September 2022).
61. Mc Kinsey & Co. The Future of Personalization and How to Get Ready for It. 2019. Available online: https://www.mckinsey.com/business-functions/growth-marketing-and-sales/our-insights/the-future-of-personalization-and-how-to-get-ready-for-it (accessed on 16 September 2022).
62. Mc Kinsey & Co. State of Fashion 2022: An Uneven Recovery and New Frontiers. 2022. Available online: https://www.mckinsey.com/industries/retail/our-insights/state-of-fashion (accessed on 13 September 2020).
63. McDonagh, D. Satisfying Needs beyond the Functional: The Changing Needs of the Silver Market Consumer. In Proceedings of the International Symposium on the Silver Market Phenomenon-Business Opportunities and Responsibilities in the Aging Society, Tokyo, Japan, 1 April 2018.
64. Moore, M.E.; Rothenberg, L.; Moser, H. Contingency Factors and Reshoring Drivers in the Textile and Apparel Industry. *J. Manuf. Technol. Manag.* **2018**, *29*, 1025–1041. [CrossRef]
65. Nagurney, A.; Yu, M. Fashion Supply Chain Management through Cost and Time Minimization from a Network Perspective. *Fash. Supply Chain Manag. Ind. Bus. Anal.* **2011**, 1–20. [CrossRef]
66. Nahm, K. Fast Retailing (Part 1): Transforming the Clothes Shopping Experience. AWS. 2021. Available online: https://aws.amazon.com/blogs/industries/fast-retailing-part-1-transforming-the-clothes-shopping-experience/ (accessed on 16 September 2022).
67. Nakano, K. *Apparel Innovators*; Nihon Jitsugyou Syuppansha: Tokyo, Japan, 2020.
68. Noris, A.; Nobile, T.H.; Kalbaska, N.; Cantoni, L. Digital Fashion: A Systematic Literature Review. A Perspective on Marketing and Communication. *J. Glob. Fash. Mark.* **2021**, *12*, 32–46. [CrossRef]
69. Oliveira, M.; Fernandes, T. Luxury Brands and Social Media: Drivers and Outcomes of Consumer Engagement on Instagram. *J. Strateg. Mark.* **2022**, *30*, 389–407. [CrossRef]
70. Polhemus, T.; Proctor, L. *Fashion and Anti-Fashion: An Anthology of Clothing and Adornment*; Cox & Wyman: London, UK, 1978.
71. PYMNTS. Amazon Aims to Enable Purchases from TV Screens Via T-commerce. PYMNTS. 2021. Available online: https://www.pymnts.com/news/retail/2021/amazon-aims-to-enable-purchases-from-tv-screens-via-tcommerce/ (accessed on 21 September 2022).
72. PYMNTS. Amazon, Walmart Battle for the Consumer's Whole Paycheck: Who's Winning by the Numbers. PYMNTS. 2022. Available online: https://www.pymnts.com/whole-paycheck-consumer-spending/2020/amazon-walmart-battle-for-the-consumers-whole-paycheck-whos-winning-by-the-numbers/ (accessed on 20 September 2022).
73. Richter, F. Amazon: Not That Big after All. Statista. 2019. Available online: https://www.statista.com/chart/18755/amazons-estimated-market-share-in-the-united-states/ (accessed on 20 September 2022).
74. Runfola, A.; Guercini, S. Fast Fashion Companies Coping with Internationalization: Driving the Change or Changing the Model? *J. Fash. Mark. Manag.* **2013**, *17*, 190–205.
75. Shen, B.; Zhu, C.; Li, Q.; Wang, X. Green Technology Adoption in Textiles and Apparel Supply Chains with Environmental Taxes. *Int. J. Prod. Res.* **2021**, *59*, 4157–4174. [CrossRef]
76. Singh, G. The Apparel Market Is Growing Faster Than the Global Economy. 2017. Available online: https://fee.org/articles/fast-fashion-has-changed-the-industry-and-the-economy/ (accessed on 10 September 2022).
77. Statista. Top Internet Companies: Global Market Value 2018. Statista, Hamburg. 2019. Available online: https://www.statista.com/statistics/264621/market-value-of-the-top-20-internet-companies-in-japan/ (accessed on 13 September 2022).
78. Statista. Apparel Sales of Amazon as a Percentage of Total Apparel Sales in the United States from 2011 to 2016. 2021. Available online: https://www.statista.com/statistics/755262/us-amazon-share-of-total-apparel-sales-market/ (accessed on 20 September 2022).
79. Statista. Market Growth of the Apparel Industry Worldwide from 2012 to 2020. Statista. 2022. Available online: https://www.statista.com/statistics/727541/apparel-market-growth-global/#:~{}:text=Global%20apparel%20market%20growth%202012%2D2020&text=It%20was%20estimated%20in%202017,6.2%20percent%20expected%20in%202020 (accessed on 6 September 2022).
80. Steele, V. *Encyclopedia of Clothing and Fashion*; Charles Scribner's Sons: New York, NY, USA, 2005; Volume 1.
81. Verhoef, P.C.; Broekhuizen, T.; Bart, Y.; Bhattacharya, A.; Dong, J.Q.; Fabian, N.; Haenlein, M. Digital Transformation: A Multidisciplinary Reflection and Research Agenda. *J. Bus. Res.* **2021**, *122*, 889–901. [CrossRef]
82. Watanabe, C.; Tou, Y. Transformative Direction of R & D: Lessons from Amazon's Endeavor. *Technovation* **2020**, *88*, 102081.
83. Watanabe, C. Innovation-consumption Co-emergence Leads a Resilience Business. *Innov. Supply Chain Manag.* **2013**, *7*, 92–104. [CrossRef]
84. Watanabe, C.; Akhtar, W.; Tou, Y.; Neittaanmäki, P. Amazon's New Supra-omnichannel: Realizing Growing Seamless Switching for Apparel During COVID-19. *Technol. Soc.* **2021**, *66*, 101645. [CrossRef]
85. Watanabe, C.; Akhtar, W.; Tou, Y.; Neittaanmäki, P. Amazon's Initiative Transforming a non-contact Society-Digital Disruption Leads the Way to Stakeholder Capitalization. *Technol. Soc.* **2021**, *65*, 101596. [CrossRef]
86. Watanabe, C.; Akhtar, W.; Tou, Y.; Neittaanmäki, P. A New Perspective of Innovation Toward a Non-contact Society-Amazon's Initiative in Pioneering Growing Seamless Switching. *Technol. Soc.* **2022**, *69*, 101953. [CrossRef]

87. Watanabe, C.; Akhtar, W.; Tou, Y.; Neittaanmäki, P. Fashion-driven Textiles as a Crystal of New Stream for Stakeholder Capitalism: Amazon's Endeavor. *Int. J. Manag. Inf. Technol.* **2020**, *12*, 19–24.
88. Watanabe, C.; Naveed, K.; Zhao, W. New Paradigm of ICT Productivity–Increasing role of Un-captured GDP and Growing Anger of Consumers. *Technol. Soc.* **2015**, *41*, 21–44. [CrossRef]
89. Watanabe, C.; Naveed, N.; Neittaanmäki, P. Digitalized Bioeconomy: Planned Obsolescence-driven Circular Economy Enabled by Co-Evolutionary Coupling. *Technol. Soc.* **2019**, *56*, 8–30. [CrossRef]
90. Watanabe, C.; Zhu, B.; Griffy-Brown, C.; Asgari, B. Global Technology Spillover and its Impact on Industry's R & D Strategies. *Technovation* **2001**, *21*, 281–291.
91. Wichser, J.D.; Hart, C.; Yozzo, J. 2019 U.S. Retail Forcast: An FTI Consulting Report. FTI Consulting. 2019. Available online: https://www.fticonsulting.com/~{}/media/Files/us-files/insights/reports/2019-us-online-retail-forecast.pdf (accessed on 20 September 2022).
92. Williamson, B.; Gulson, K.N.; Perrotta, C.; Witzenberger, K. Amazon and the New Global Connective Architectures of Education Governance. *Harv. Educ. Rev.* **2022**, *92*, 231–256. [CrossRef]
93. Xue, J.; Liang, X.; Xie, T.; Wang, H. See Now, Act Now: How to Interact with Customers to Enhance Social Commerce Engagement. *Inf. Manag.* **2020**, *57*, 103324. [CrossRef]
94. Yenipazarli, A. The Marketplace Dilemma: Selling to the Marketplace vs. Selling on the Marketplace. *Nav. Res. Logist. (NRL)* **2021**, *68*, 761–778. [CrossRef]
95. Yeung, J.; Wong, S.; Tam, A.; So, J. Integrating Machine Learning Technology to Data Analytics for E-commerce on Cloud. In Proceedings of the 2019 Third World Conference on Smart Trends in Systems Security and Sustainability (WorldS4), 30–31 July 2019; IEEE: Piscataway, NJ, USA, 2019; pp. 105–109.
96. Zhang, C.; Dong, M.; Ota, K. Employ AI to Improve AI services: Q-Learning based Holistic Traffic Control for Distributed Co-inference in Deep Learning. *IEEE Trans. Serv. Comput.* **2021**, *15*, 627–639. [CrossRef]
97. Zhang, L.; Qi, Z.; Meng, F. A Review on the Construction of Business Intelligence System Based on Unstructured Image Data. *Procedia Comput. Sci.* **2022**, *199*, 392–398. [CrossRef]
98. Zhuang, Y.T.; Wu, F.; Chen, C.; Pan, Y.H. Challenges and Opportunities: From Big Data to Knowledge in AI 2.0. *Front. Inf. Technol. Electron. Eng.* **2017**, *18*, 3–14. [CrossRef]

Article

An Assessment of Energy and Groundwater Consumption of Textile Dyeing Mills in Bangladesh and Minimization of Environmental Impacts via Long-Term Key Performance Indicators (KPI) Baseline

Abdullah Al Mamun [1], Koushik Kumar Bormon [2], Mst Nigar Sultana Rasu [3], Amit Talukder [4], Charles Freeman [4,*], Reuben Burch [1,5] and Harish Chander [5,6]

1. Department of Industrial and Systems Engineering, Mississippi State University, Starkville, MS 39762, USA
2. Department of Hydro Science and Engineering, Technical University of Dresden, 01062 Dresden, Germany
3. Hohenstein Institute Bangladesh, Dhaka 1213, Bangladesh
4. Department of Human Sciences, Mississippi State University, Starkville, MS 39762, USA
5. Human Factors and Athlete Engineering, Center for Advanced Vehicular Systems, Mississippi State University, Starkville, MS 39762, USA
6. Neuromechanics Laboratory, Department of Kinesiology, Mississippi State University, Starkville, MS 39762, USA
* Correspondence: cf617@msstate.edu

Citation: Mamun, A.A.; Bormon, K.K.; Rasu, M.N.S.; Talukder, A.; Freeman, C.; Burch, R.; Chander, H. An Assessment of Energy and Groundwater Consumption of Textile Dyeing Mills in Bangladesh and Minimization of Environmental Impacts via Long-Term Key Performance Indicators (KPI) Baseline. *Textiles* **2022**, *2*, 511–523. https://doi.org/10.3390/textiles2040029

Academic Editor: Laurent Dufossé

Received: 31 August 2022
Accepted: 26 September 2022
Published: 28 September 2022

Publisher's Note: MDPI stays neutral with regard to jurisdictional claims in published maps and institutional affiliations.

Copyright: © 2022 by the authors. Licensee MDPI, Basel, Switzerland. This article is an open access article distributed under the terms and conditions of the Creative Commons Attribution (CC BY) license (https:// creativecommons.org/licenses/by/ 4.0/).

Abstract: Bangladesh's ready-made garment sectors have evolved to increase market share in the global textile supply chain. Textile sectors heavily rely on energy and groundwater consumption during production; mainly, textile dyeing mills contribute to the carbon footprint and water footprint impact to the environment. Textile dyeing mills have become one of the major industries responsible for the continuous depletion of groundwater levels and severe water pollution to the environment. Reduction of long-term key performance indicators (KPI) can be set to a baseline by reducing energy and groundwater consumption in textile dyeing mills. This study has analyzed the energy and groundwater consumption trend based on 15 textile dyeing mills in Bangladesh in 2019. The average dyed fabric production of 15 textile dyeing mills in 2019 was 7602.88 tons by consuming electricity and groundwater, and discharging treated effluent wastewater to the environment, in the amounts of 17,689.43 MWh, 961.26 million liters, and 640.24 million liters, respectively. The average KPI of treated effluent discharged wastewater was 97.27 L/kg, and energy consumption was 2.58 kWh/kg. Considering yearly 5% reduction strategies of groundwater and energy consumption for each factory could save around 355.43 million liters of water and 6540.68 MWh of electricity in 10 years (equivalent to 4167.08-ton CO_2 emission).

Keywords: effluent treatment; energy and water footprint; groundwater level; key performance indicator; heavy metals discharge

1. Introduction

Energy and water play a vital role in the world's textile supply chain. The product lifecycle of a ready-made garment is related to energy and water consumption (EWC) that comprises several phases: utilization of agricultural machinery driven by fossil fuel and water usage during cotton cultivation; EWC in textile production: spinning, weaving, dyeing/finishing, and apparel manufacturing; logistics and transportation of ready-made garments which contributes to energy consumption; personal use of washing machines that require a significant amount of water and energy [1–4]. The environmental impact of the textile supply chain is widespread; for example, it annually contributes 1.7 billion tons of CO_2 emissions, which is around 10% of global greenhouse gas (GHG) exposure [1]. The textile supply chain also consumes 1.5 trillion liters of water each year, which is responsible for 20% of industrial water pollution [5].

Bangladesh is the second-largest exporter of global ready-made garments (RMG), followed by China [6]. In 2019, the total export value of RMG was 34.13 billion US$, contributing 84.21% of Bangladesh's total export value. In Bangladesh, the RMG sector has evolved to expand its global market share and increase its export value by approximately 63.40% (from 2009 to 2019) [7]. However, the RMG sectors of Bangladesh heavily rely on energy and groundwater consumption during the production process, contributing to carbon footprint (CFP) and wastewater discharge to the environment, respectively. Therefore, EWC in the ready-made garments sector in Bangladesh has become a significant concern for environmental sustainability. However, scarcity of sustainable water may hamper the continuous growth of the RMG sector in Bangladesh, mainly groundwater, the largest and only water source for the entire textile dyeing industry [8]. Due to the self-extraction of unpriced groundwater in most factories, textiles have become one of the major industries responsible for the continuous depletion of groundwater levels and water pollution [9]. In most textile factories, the usage of extracted groundwater is inefficient, and the amount of attenuation is insignificant. In 2015, 1700 textile dyeing mills in Bangladesh consumed approximately 1500 billion liters of groundwater. After groundwater usage by textile dyeing mills, they discharge treated effluent wastewater into the environment, causing extreme water pollution and groundwater depletion [10]. In addition, the surface water of the nearby rivers and water canals has been contaminated by this discharged wastewater with harmful fragments of dyes and chemicals, ultimately affecting aquatic ecology and agriculture.

Over the past two decades, groundwater decline has significantly threatened the area in and around Dhaka city and adjacent industrial zones [11]. The extraction is more than the recharge of aquifers, causing the deterioration of groundwater levels [8,12]. With the depletion of the groundwater level, the energy cost for groundwater extraction will also impact production costs in the RMG sector. Therefore, it is high time to tackle this alarming situation to save our environment and the RMG industry. Addressing this issue, establishing a benchmark of key performance indicators (KPI) of energy, groundwater and treated effluent wastewater based on the amount of dyed fabric production will help the sustainable environment performance index.

1.1. Water and Energy Consumption in Textile Dyeing Mills

By the year 2050, the world's population will increase around 35%, significantly increasing textile production and consumption, driving a significant increase in energy and water consumption, ultimately leading to environmental pollution [4]. In addition, the textile industry requires an intensive amount of water, which significantly strains global water resources. As a result, the textile industry is accounted the worst polluter of clean water, followed by agriculture. At the same time, there are significant concerns about wet textile processing consuming a massive amount of freshwater, discharging wastewater and polluting the ecosystem [13]. For example, in 2016, the Chinese textile industry (consisting of 50,000 textile factories) consumed approximately 3000 billion liters of freshwater [2,6]. According to the Turkish Statistical Institute, the textile industry is responsible for 191.5 billion liters of water consumption, the second-largest industry in the manufacturing sector [14]. On average, approximately 2500–3000 L of water are required to manufacture a cotton t-shirt. Moreover, a substantial amount of water consumption is associated with cotton cultivation, followed by wet processing [2]. In addition, conventional textile dyeing and finishing process require approximately 1.5 million liters of water for every ton of textile processing [13]. Researchers measure specific water consumption (SWC) (treated/groundwater) as usage per mass of the product [9]. For example, various investigations showed that in the wet textile process, on average, 200–400 L of water were consumed for dyeing 1 kg of fabric [15–17]. In the meantime, SWC usage in the Turkish textile sector varies from 20 L/kg to 230 L/kg [18]. Therefore, SWC can help promote water footprint awareness and set a benchmark index regarding environmental sustainability. Similarly, yarn spinning, and wet processing consume significant electricity from the national grid and captive power generators, using fossil fuel and natural gas for the textile industry. Therefore, manufacturers measure specific energy consumption (SEC) in wet textile processing as a ratio of electricity consumption for dyeing 1 kg fabric (kWh/kg) [19]. Generally, the SEC of a textile dyeing mill plays a vital role in monitoring electricity usage versus production calculation. SEC is convertible to a carbon footprint based on country-wise "emission factors." In Bangladesh, 0.64 kgCO_2 contributes to the environment, equivalent to generating 1 kWh of electricity [20]. An investigation found that an average SEC in Turkish textile wet processing required 3.4 kWh/kg dyed fabric [21]. Conserving energy and water consumption will help mitigate air and water pollution, which will also be part of a more environmentally friendly production process.

1.2. Impact of Discharged Wastewater on the Environment

Discharged treated effluent wastewater contains intense color, inorganic finishing agents, surfactant, chlorine compounds, high chemical oxygen demand (COD) and bio-chemical oxygen demand (BOD) amounts, and heavy metals [22]. The investigation also showed that wet textile processes, including bleaching, dyeing, printing, and finishing, use 3600 dyes and 8000 chemicals [23]. Therefore, effluent treatment costs may account for 5% of total production costs [6]. However, the conventional effluent treatment method is unsuitable for purifying many toxic and bio-degradable compounds in wastewater [14,17]. Many of these dyes and chemicals account for the direct and indirect causes of water pollution, soil contamination, and threats to aquatic life [24]. An estimation showed that textile effluent discharge was around 280,000 tons of textile dyes annually around the globe. In addition, the discharged treated effluent wastewater temperature is higher (65 °C) than regular water, reducing the dissolved oxygen level of normal water and leading to an imbalance of biodiversity [24]. Due to this, China is facing one of the worst water pollution scenarios, which has happened because 70% of China's rivers, lakes, and reservoirs have already been contaminated mainly by textile industries [2].

In Bangladesh, a massive amount of discharged wastewater from textiles and effluents has already altered the aquatic ecosystem's chemical and physical properties. This alteration of the typical marine environment has impacted humans, livestock, the fish population, and biodiversity [25,26]. Moreover, untreated textile wastewater also has a severely harmful effect on groundwater quality. The location of textile industries is clustered within a range of 60 km in greater Dhaka industrial zones and their vicinity. The region includes Narayangonj, Gazipur, some of Mymensingh, and Narsingdi, where rivers and water canals near these zones are being polluted by discharged wastewater from textile dyeing industries [27]. Major affected rivers in these textile industrial zones include Buriganga, Shitalakkhya, Turag, and Dhaleshwari (Figure 1) [28]. Addressing this dire condition of water pollution, implementing advanced technologies, and cleaner production strategies may help reduce water consumption and effluent volume from textile dyeing industries.

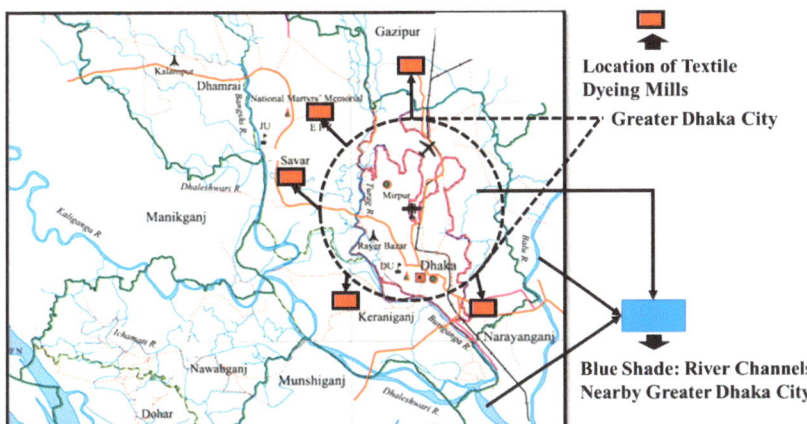

Figure 1. River water pollution by industrial waste in adjacent rivers of Dhaka city. Adapted from [28].

1.3. Groundwater Level Depletion

Water pollution and scarcity of water resources have become severe problems due to a lack of wastewater treatment and water abuse [6]. Freshwater consumption in the wet textile process has also become a significant concern for those countries facing water shortages or those facing it in the near future. For example, textile dyeing mills in Bangladesh utilize a considerable amount of groundwater. As a result, the decrease of groundwater and increasing surface water pollution coincide. Industrial effluents from textile dyeing industries are destroying nearby surface water resources. Currently, water treatment of Shitalakkhya river water by DWASA (Dhaka Water and Sewerage Authority) meets around 22% of the 2.3 billion liters of daily water demand in Dhaka City, with the remaining needs met through underground water resources [29]. Generally, extracted groundwater needs significantly fewer water treatment procedures, while surface water requires various treatment processes that involve substantial investment costs for drinking, domestic and industrial purposes. However, an investigation showed that Dhaka's groundwater level has dropped by 200 feet in the last 50 years, and this trend continues at a high rate [30]. Consequently, large volumes of groundwater extracted by the textile dyeing industries threaten the quality and quantity of drinking water accessible to the residents of Dhaka City.

This study aims to examine the current trend of CFP, effluent discharge (wastewater), and groundwater depletion levels based on textile dyeing mills. Long-term improvement of these trends based on the amount of dyed fabric production will help maintain a sustainable environment in Bangladesh.

2. Materials and Methods

2.1. Study Approach

Figure 2 represents the overall study approach of collecting data, processing data, and analyzing KPI based on production data from the wet processing unit, groundwater extraction volume, amount of discharged wastewater from treated effluent, energy consumption from the national grid, and captive power generation and water & carbon footprint. Data collection, KPI analysis, and recommendations are demonstrated in Sections 2.2, 3.1 and 3.2, respectively. The recommendation section comprises three key points: making a database for future reference, setting goals for a yearly KPI% reduction, and introducing the best available technology to increase productivity and reduce water and energy consumption.

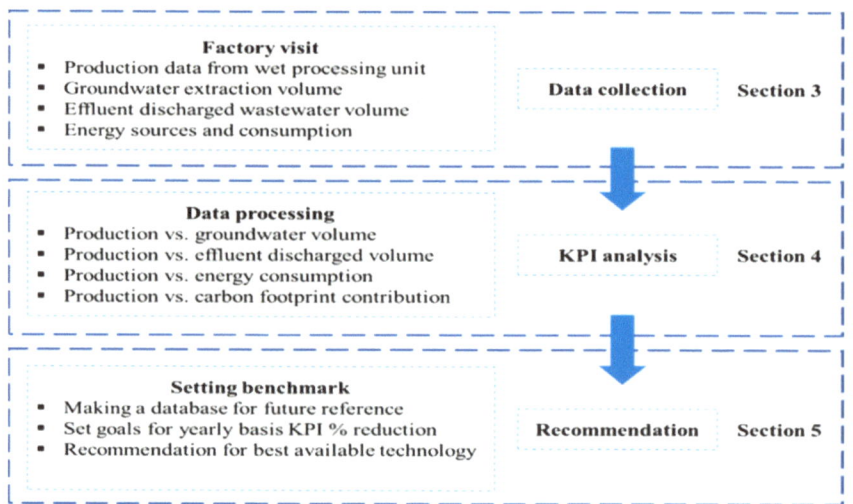

Figure 2. Work procedure of data processing and analyzing KPIs.

2.2. Data Collection

Factory management collected and provided data to the authors from 15 textile factories based in 2019. Data was collected based on dyed fabric amounts from 15 textile factories and associated extracted groundwater, consumed energy, and discharged wastewater. The factory distances from Dhaka city's center range from 20 km to 60 km (Figure 3). In addition, researchers collected secondary data from journal papers, survey reports, international conference papers, newspapers, and textile magazine articles to corroborate data from the 15 participant sites.

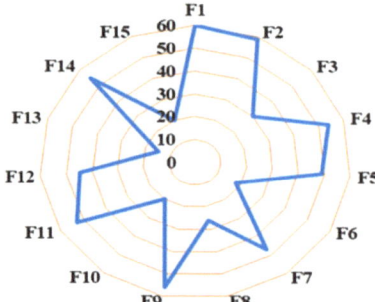

Figure 3. List of factory distances from the center of Dhaka city.

Groundwater Demand and Wastewater Discharge into the Environment

Equations (1)–(3) demonstrate the KPI of groundwater extraction, effluent discharged wastewater, and water loss in the process, respectively. The annual KPI of groundwater is a ratio between the extracted groundwater amount (L) and total dyed fabric amount (kg). Similarly, the yearly KPI of wastewater is the ratio between total treated effluent discharged wastewater and the total dyed fabric amount (kg) in the same year.

$$\text{KPI}_{\text{Groundwater}} = \frac{\sum \text{Extracted groundwater (Liter)}}{\sum \text{Dyed fabric amount (kg)}} \quad (1)$$

$$\text{KPI}_{\text{Wastewater}} = \frac{\sum \text{Discharged wastewater (Liter)}}{\sum \text{Dyed fabric amount (kg)}} \qquad (2)$$

Equation (3) shows that the KPI of water loss in the process is measured based on Equations (1) and (2) calculates the difference between groundwater and wastewater KPIs.

$$\text{KPI}_{\text{Water loss in the process}} = \text{KPI}_{\text{Groundwater}} - \text{KPI}_{\text{Wastewater}} \qquad (3)$$

2.3. Data Collected from Selected Dyeing Mills

Table 1 demonstrates collected data from 15 textile dyeing mills in 2019. Data were collected in each factory based on the annual dyed fabric amount, extracted groundwater volume, the treated effluent wastewater discharge volume, and electricity consumption. Electricity supply from the national grid source and electricity from captive power generation using fossil fuel and natural gas lines determined total electricity consumption. In addition, researchers tracked groundwater extraction and effluent discharged wastewater using a water outlet flow meter.

Table 1. Resource Consumption Data Collected from 15 Textile Dyeing Mills.

Factory	Total Production (kg)	Total Electricity Use (kWh)	Extracted Groundwater (in Million Liters)	Discharged Wastewater (in Million Liters)
F1	4,330,515.00	15,125,711.00	744.05	624.19
F2	9,088,174.00	20,034,081.00	1455.87	865.66
F3	4,413,619.00	6,673,371.00	687.66	340.77
F4	3,639,554.00	7,566,789.00	533.58	501.68
F5	2,409,076.00	8,099,029.00	477.45	372.53
F6	4,083,918.00	13,265,449.00	703.73	481.55
F7	11,193,569.00	21,806,157.00	1297.86	922.51
F8	4,355,228.00	6,595,024.00	470.36	270.21
F9	4,049,474.00	6,992,793.00	401.49	350.37
F10	10,702,135.00	6,5546454.00	1302.71	851.50
F11	5,223,919.00	9,699,231.00	652.66	358.01
F12	41,530,362.00	57,402,492.00	4526.32	2815.04
F13	2,553,747.00	4,194,361.00	356.67	260.12
F14	2,895,710.00	4,574,060.00	371.86	281.81
F15	3,574,246.00	17,766,474.00	436.65	307.67

Energy Consumption and Carbon Footprint

Carbon footprint (CFP) is a broadly used tool for monitoring global climate change. CFP's impact on the environment is attributed to the emission of greenhouse gases (GHGs) such as CO_2, N_2O, hydrofluorocarbons (HFCs), perfluorocarbons (PFCs), and sulfur hexafluoride (SF6) [4,31]. CFP is measured as grams of CO_2 equivalent to generating per kilowatt-hour of electricity (gCO_2eq/kWh) utilizing hydrocarbon-containing fossil fuel [32,33]. The emission factor (EF) varies from country to country, which depends on the resource utilization of fossil fuels. Table 2 shows country-wise emissions per kWh of electricity depending on carbon heat generation [20]. For example, country-wise emission per kWh in Bangladesh is 0.6371 kg-CO_2, while $KgCH_4$ and KgN_2O's environmental contribution is insignificant. Based on Table 2, India is the highest contributor per kWh equivalent CO_2 emission to the environment, whereas Cambodia and China stood in the second and third positions for CO_2 emission. Additionally, Table 2 can also compare country-wise carbon footprint impact broadly related to textile dyeing production in the global textile supply chain. More elaborately, a fair KPI can be reached based on a particular time of textile dyed fabric amount and electricity consumption using the unit kWh/kg. For instance, 1 kWh of electricity production in Bangladesh contributes 0.6371 kg CO_2 and converts kWh/Kg to 0.6371 kg CO_2/kg. Similarly, this unit can be presented as 0.9746 kg CO_2/kg in China. A list of country-wise emissions factors is adapted from [20]

$$\text{Emission of GHGs} = \text{Energy Consumption (EC)} \times \text{Emission Factor (EF)} \quad (4)$$

Table 2. Country-wise emissions per kWh of electricity generated [20].

Country	kgCO_2/kWh	kgCH_4/kWh	kgN_2O/kWh
Bangladesh	0.6371	0.00001236	0.00000191
China	0.9746	0.00001047	0.00001521
Cambodia	1.1708	0.00004638	0.00000928
India	1.3332	0.00001552	0.00002011
Pakistan	0.4734	0.00001384	0.00000243
Vietnam	0.4668	0.00000705	0.00000420
Sri Lanka	0.4172	0.00001644	0.00000329

Equation (5) shows the emission of GHG based on energy consumption and country-wise emission factor. Equation (5) represents energy consumption based on national grid supply and captive power generation. Calculations use solar energy as a negative emission factor. However, the amount of solar energy is insignificant compared to grid electricity and a captive power source.

$$\text{EC (kWh)} = \sum(\text{Grid electricity} + \text{Captive power generation}) - \sum \text{Solar energy} \quad (5)$$

Equation (6) represents the KPI of energy consumption based on a ratio of total electricity consumption (kWh) and total dyed fabric amount (kg). Equation (7) shows the KPI of CO_2 emission contribution to the environment based on a ratio of total electricity consumption (kWh) times per kWh equivalent emission factor to total dyed fabric amount (kg).

$$\text{KPI}_{\text{Energy}} = \frac{\sum \text{EC (kWh)}}{\sum \text{Dyed fabric amount (kg)}} \quad (6)$$

$$\text{KPI}_{CO_2} = \frac{\sum \text{EC(kWh)} \times \text{Emission Factor(EF)}}{\sum \text{Dyed fabric amount (kg)}} \quad (7)$$

3. Results

3.1. Production vs. Energy Consumption

Figure 4 demonstrates the KPI of 15 factories based on energy consumption and production data for 2019. Based on Table 2 and Equation (6), the annual energy consumption KPI is the total dyed fabric ratio in a year. As a result, researchers calculated the mean KPI of 15 factories to be 2.58 kWh/kg, where the maximum and minimum KPI was 6.12 kWh/kg and 1.38 kWh/kg, respectively.

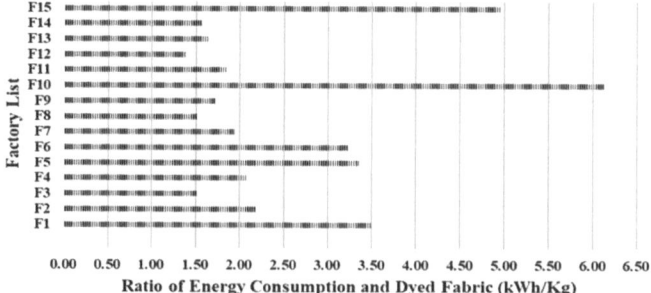

Figure 4. The ratio of energy consumption and dyed fabric (kWh/kg).

3.2. Carbon Footprint Contribution to kWh Electricity and Dyed Fabric

Figure 5 represents carbon footprint contribution equivalent per kWh electricity consumption and dyed fabric amount of 15 factories. Based on Table 1 and Equation (7), the KPI of carbon emission contribution uses a ratio of Bangladesh's yearly energy consumption multiplied by emission factors (Table 2) to the total dyed fabric in a year. As a result, the average carbon emission KPI of 15 factories was 1.64 kg-CO_2/kg, where the maximum and minimum KPI was 3.90 kg-CO_2/kg and 0.88 kg-CO_2/kg, respectively.

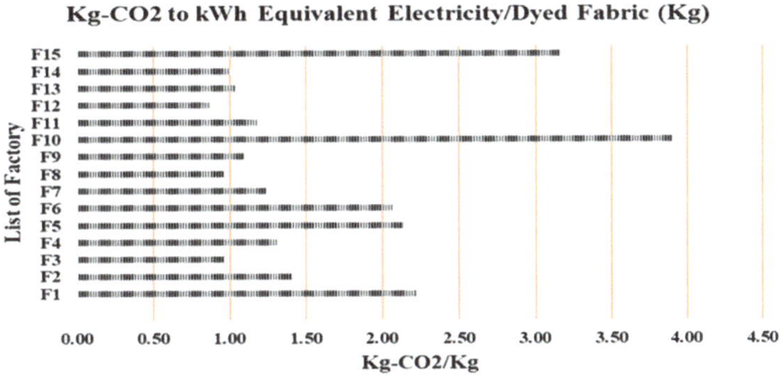

Figure 5. The ratio of energy consumption and dyed fabric (kWh/kg).

3.3. Groundwater vs. Discharged Wastewater Comparison

Figure 6 compares groundwater versus effluent discharged wastewater based on dyed fabric amounts of 15 textile dyeing industries. The maximum, minimum, and average KPI of extracted groundwater were 198.20 L/kg, 99.15 L/kg, and 138.26 L/kg, respectively. Similarly, the maximum, minimum, and average KPI of effluent-treated discharged wastewater was 154.64 L/kg, 62.04 L/kg, and 97.27 L/kg, respectively. Finally, researchers calculated the KPI of water loss in the process by taking the difference between groundwater and effluent discharged wastewater KPI, where maximum, minimum, and average KPI differences were found at 79.0 L/kg, 9.0 L/kg, and 41.0 L/kg, respectively.

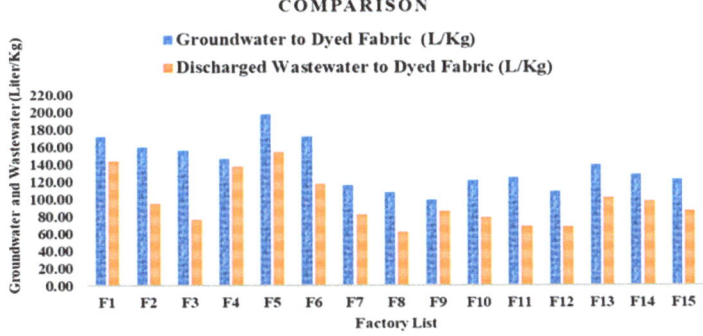

Figure 6. Groundwater vs. effluent discharged wastewater comparison based on dyed fabric amount (kg).

3.4. Heavy Metal Discharge with Treated Wastewater

Figure 7 shows heavy metals released into the environment with effluent-treated wastewater from 15 textile dyeing mills. According to an effluent-treated wastewater analysis report from 15 textile dyeing mills, results detected nine heavy metals (boron, manganese, chromium, zinc, copper, nickel, cobalt, antimony, and lead). Zinc (Zn) was detected in 13 out of 15 factories, while cobalt (Co) and boron (B) were the lowest traced heavy metals found in only two factories. The trace of heavy metals was incorporated from the factory-wise effluent-treated wastewater analysis reports from third-party laboratories.

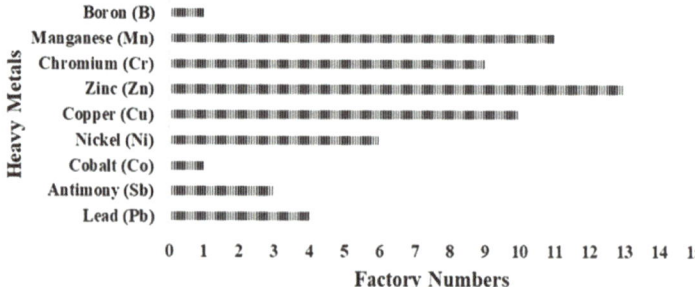

Figure 7. Heavy metal released with effluent-treated wastewater.

3.5. COD and BOD Amount in Effluent Discharged Wastewater

Figure 8 represents the COD, and BOD amount in effluent discharged wastewater according to the factory-provided wastewater analysis report. Using a wastewater analysis report, researchers found maximum, minimum, and mean COD in the wastewater at 216 mg/L, 28 mg/L, and 88 mg/L, respectively. Similarly, the maximum, minimum, and average BOD amounts were traced at 44 mg/L, 4 mg/L, and 21.8 mg/L, respectively.

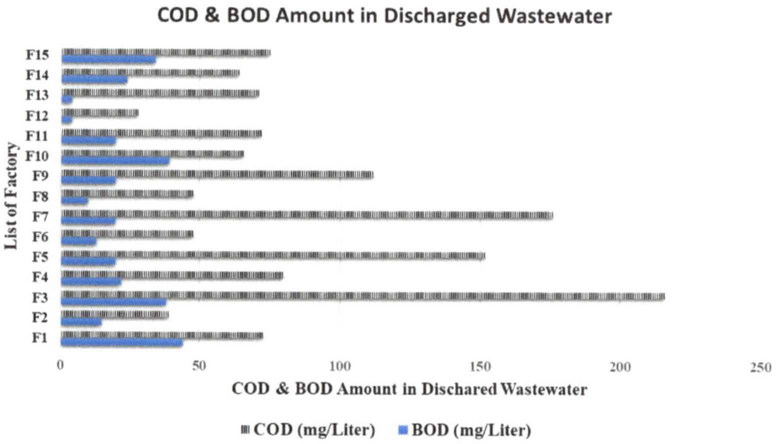

Figure 8. COD & BOD amount in effluent discharged wastewater.

4. Discussion

4.1. Yearly Basis KPI% Reduction Approach and Potential Saving

Figure 9 demonstrates a potential groundwater saving (yearly 5% reduction) approach based on an average groundwater extraction amount of 961.26 million liters per factory. Using 5% reduction strategies of groundwater for each factory can save around 355.43 million liters in 10 years. Figure 10 shows a potential energy-saving approach for a single factory in 10 years. Similarly, a 5% reduction strategy of average energy consumption (17,689.43 MWh) for a single factory can save 6540.68 MWh of electricity in 10 years, equivalent to 4167.08 tons of CO_2 emission reduction to the environment [31].

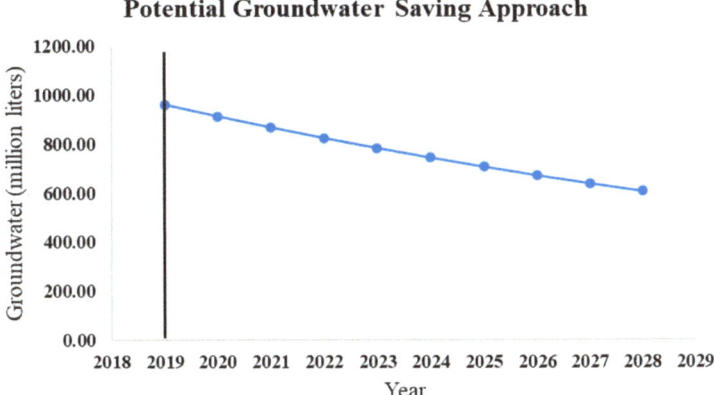

Figure 9. Potential groundwater saving approach in 2019.

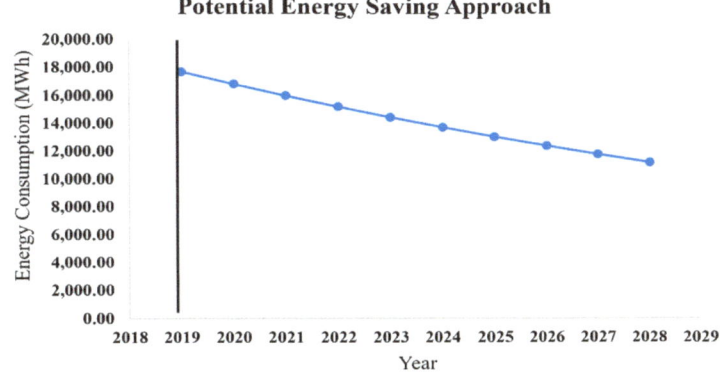

Figure 10. Potential energy saving approach in 2019.

4.2. Recommendations

Factories can adopt different strategies to minimize water and energy losses to save a potential amount of water and energy without significant investment. However, new machinery and equipment setup often require a considerable investment. Some recommendations are highlighted in Sections 4.2.1–4.2.3.

4.2.1. Best Available Techniques for Potential Water-saving Approaches

- Process-wise and machine-wise water consumption should be monitored for individual dyeing machine water consumption and take initiatives where water consumption is comparatively high;
- Use water-efficient machinery and equipment, for instance, substituting a high liquor ratio dyeing machine with a low liquor ratio [13];
- Ensure optimum condensate recovery from all sections by installing steam traps and condensate transfer pump to reuse as boiler feedwater;
- Reuse of effluent-treated wastewater in cleaning the empty chemical drums, printing screens and production floor, car washing and toilet flushing [14];

- Rainwater is much purer than groundwater and can be used in the production process without passing through the WTP (Water Treatment Plant), which is economically suitable and environmentally friendly;
- Prevent all leakages in the waterline and use a trigger nozzle in hose pipes to avoid the excessive flow of water;
- For fabric washing purposes, counter-current rinsing should be followed. Pretreatment washing of the dyed fabric should be conducted according to the requirement to avoid excess washing. Additionally, avoiding the excessive washing of machines;
- Adopting digital printing instead of a standard printing system where chemical wastage is minimal and requires less water;
- Provision of recovering salt from used liquor, which is ultimately drained to ETP, increasing treatment cost and using acid to neutralize the high amount of alkaline effluent.

4.2.2. Best Available Techniques for Potential Energy-Saving Approaches

- Substitute manual blowdown of boilers with an auto blowdown system to save energy;
- The concept of smart lighting involves utilizing natural light from the sun. Smart lighting is also a good initiative that minimizes and saves light by allowing the proper place lighting;
- Proper insulation of all steam valves & flanges to avoid heat loss
- Maintain proper air and fuel ration in boiler through oxygen tuning/oxygen analyzer to reduce excessive natural gas consumption;
- Installation of exhaust gas boiler and heat recovery from flue gas by installing an economizer;
- Performing regular leakage tests and monitoring the leakage level of compressed air lines.

4.2.3. Factory Management Initiatives

- Employee and worker training on water usage also plays a significant role [14]. Conveying the environmental impact and the growing consciousness of illiterate or less-educated workers is very important. The feasibility of waterless dyeing with CO_2 or plasma processing should be investigated as a pilot project basis in Bangladesh as soon as possible. With modern techniques and solid economic background, some countries are introducing absolute recycling of water through the zero liquid discharge (ZLD) plant, which could be the ultimate solution for toxic wastewater. As Dhaka's groundwater level is significantly declining, some researchers have suggested recharging the groundwater artificially [11].
- These approaches could be taken to minimize water and energy without significant investment. However, this study has analyzed the energy and groundwater consumption trend based on 15 textile dyeing mills in Bangladesh in 2019. The article was set up as a critical review of the failure criteria that guide the selection of the most suitable criterion for the chosen case study. Long-term key performance indicator (KPI) reduction is set to a baseline by reducing energy and groundwater consumption in textile dyeing mills. The overall calculation can vary by location of textile dyeing mills worldwide, the number of textile dyeing mills, and the timelines. This case study was limited to energy and groundwater consumption trends in textile dyeing mills in Bangladesh. Future recommendations of this study could be expanded to other textile regions in Bangladesh.

5. Conclusions

Bangladesh is the second-largest exporter of global RMG, followed by China, and this RMG sector has evolved in growing global market share and increasing its export value by approximately 63.40% (from 2009 to 2019). These RMG sectors heavily rely on energy and groundwater consumption during the production process, contributing to carbon footprint and wastewater discharge to the environment. With the shortage of groundwater levels, the energy cost for groundwater extraction will also impact production costs in the RMG sector. However, soon scarcity of sustainable water may hamper the continuous growth of the RMG sector in Bangladesh, mainly relating to use of groundwater. Over the past decades, groundwater decline has been a major threat to Greater Dhaka city and adjacent industrial zones. Meanwhile, the extraction is more than the recharge of aquifers, causing the deterioration of groundwater levels. After comparing dyed fabric amounts of 15 textile dyeing mills and energy consumption, the average KPI of 15 factories was found to be 2.58 kWh/kg. Therefore, on average, COD and BOD in effluent discharged wastewater of 15 factories were 88 mg/L and 21.8 mg/L, respectively. A yearly 5% reduction strategy of groundwater and energy consumption for each factory can save around 355.43 million liters of groundwater and 6540.68 MWh of electricity in the next ten years in Bangladesh (equivalent to 4167.08-ton CO_2 emission). Therefore, without hampering global demands, this saved water and energy could help us survive more sustainably in the future.

Author Contributions: Conceptualization, A.A.M., K.K.B. and M.N.S.R.; methodology, A.A.M., K.K.B. and M.N.S.R.; investigation, A.A.M., K.K.B. and M.N.S.R.; visualization, A.A.M., K.K.B. and M.N.S.R.; data curation A.A.M., K.K.B. and M.N.S.R.; writing—original draft preparation, A.A.M., K.K.B., M.N.S.R. and A.T.; writing—review and editing, A.T., C.F., R.B. and H.C.; supervision, C.F., R.B. and H.C. All authors have read and agreed to the published version of the manuscript.

Funding: This research received no external funding.

Institutional Review Board Statement: Not applicable.

Informed Consent Statement: Not applicable.

Data Availability Statement: The data presented in this study are available on request from the corresponding author.

Acknowledgments: The authors would like to acknowledge the factory personnel who helped and supported collecting relevant data for writing this paper. The authors did not receive any external funding for conducting this assessment.

Conflicts of Interest: The authors declare no conflict of interest.

References

1. Niinimäki, K.; Peters, G.; Dahlbo, H.; Perry, P.; Rissanen, T.; Gwilt, A. The environmental price of fast fashion. *Nat. Rev. Earth Environ.* **2020**, *1*, 189–200. [CrossRef]
2. Choudhury, A.K.R. Environmental impacts of the textile industry and its assessment through life cycle assessment. In *Roadmap to Sustainable Textiles and Clothing*; Muthu, S.S., Ed.; Springer: Singapore, 2014; pp. 1–39. [CrossRef]
3. Palamutcu, S. Energy footprints in the textile industry. In *Handbook of Life Cycle Assessment (LCA) of Textiles and Clothing*; Elsevier: Amsterdam, The Netherlands, 2015; pp. 31–61. [CrossRef]
4. Hasanbeigi, A.; Price, L. A technical review of emerging technologies for energy and water efficiency and pollution reduction in the textile industry. *J. Clean. Prod.* **2015**, *95*, 30–44. [CrossRef]
5. Lehmann, M.; Arici, G.; Martinez-pardo, C. Pulse of the Fashion, Global Fashion Agenda & Boston Consulting Group & Sustainable Apparel Coalition. 2019. Available online: http://media-publications.bcg.com/france/Pulse-of-the-Fashion-Industry2019.pdf (accessed on 28 August 2022).
6. Rather, L.J.; Jameel, S.; Dar, O.A.; Ganie, S.A.; Bhat, K.A.; Mohammad, F. Advances in the sustainable technologies for water conservation in textile industries. In *Water in Textiles and Fashion*; Elsevier: Amsterdam, The Netherlands, 2019; pp. 175–194. [CrossRef]
7. BGMEA. Export Performance. 2022. Available online: https://www.bgmea.com.bd/page/Export_Performance (accessed on 28 August 2022).
8. Agoncillo, J. "Rapid Assessment of Greater Dhaka Groundwater Sustainability", 2030 Water Resources Group, 11 November 2019. Available online: https://2030wrg.org/rapid-assessment-of-greater-dhaka-groundwater-sustainability/ (accessed on 28 August 2022).
9. Mathews, B. Water issues could harm Bangladesh textile sector. *Ecotextile News*, 1 February 2016.
10. Ashraf, A. *Zero Liquid Discharge: A Success Story of Tirupur Textile Cluster*; Partnership for Cleaner Textile (PaCT): Dhaka, Bangladesh, 2015; pp. 41–43.
11. Islam, D.S.; Islam, F.F. Spatial Disparity of Groundwater Depletion in Dhaka City. In Proceedings of the 15th International Conference on Environmental Science and Technology, Rhodes, Greece, 31 August–2 September 2017; Volume 31, p. 7.
12. Hoque, M.A.; Hoque, M.M.; Ahmed, K.M. Declining groundwater level and aquifer dewatering in Dhaka metropolitan area, Bangladesh: Causes and quantification. *Hydrogeol. J.* **2007**, *15*, 1523–1534. [CrossRef]

13. Hussain, T.; Wahab, A. A critical review of the current water conservation practices in textile wet processing. *J. Clean. Prod.* **2018**, *198*, 806–819. [CrossRef]
14. Ozturk, E.; Koseoglu, H.; Karaboyacı, M.; Yigit, N.O.; Yetis, U.; Kitis, M. Minimization of water and chemical use in a cotton/polyester fabric dyeing textile mill. *J. Clean. Prod.* **2016**, *130*, 92–102. [CrossRef]
15. Amar, N.B.; Kechaou, N.; Palmeri, J.; Deratani, A.; Sghaier, A. Comparison of tertiary treatment by nanofiltration and reverse osmosis for water reuse in denim textile industry. *J. Hazard. Mater.* **2009**, *170*, 111–117. [CrossRef] [PubMed]
16. Brik, M.; Schoeberl, P.; Chamam, B.; Braun, R.; Fuchs, W. Advanced treatment of textile wastewater towards reuse using a membrane bioreactor. *Process Biochem.* **2006**, *41*, 1751–1757. [CrossRef]
17. Dasgupta, J.; Sikder, J.; Chakraborty, S.; Curcio, S.; Drioli, E. Remediation of textile effluents by membrane based treatment techniques: A state of the art review. *J. Environ. Manag.* **2015**, *147*, 55–72. [CrossRef]
18. Ozturk, E.; Yetis, U.; Dilek, F.B.; Demirer, G.N. A chemical substitution study for a wet processing textile mill in Turkey. *J. Clean. Prod.* **2009**, *17*, 239–247. [CrossRef]
19. Çay, A. Energy consumption and energy saving potential in clothing industry. *Energy* **2018**, *159*, 74–85. [CrossRef]
20. Brander, M.; Sood, A.; Wylie, C.; Haughton, A.; Lovell, J. Electricity-specific emission factors for grid electricity. *Ecometrica* **2011**, 1–22.
21. Ozturk, E.; Karaboyacı, M.; Yetis, U.; Yigit, N.O.; Kitis, M. Evaluation of integrated pollution prevention control in a textile fiber production and dyeing mill. *J. Clean. Prod.* **2015**, *88*, 116–124. [CrossRef]
22. Samanta, K.K.; Pandit, P.; Samanta, P.; Basak, S. Water consumption in textile processing and sustainable approaches for its conservation. In *Water in Textiles and Fashion*; Elsevier: Amsterdam, The Netherlands, 2019; pp. 41–59. [CrossRef]
23. Asghar, A.; Raman, A.A.A.; Daud, W.M.A.W. Advanced oxidation processes for in-situ production of hydrogen peroxide/hydroxyl radical for textile wastewater treatment: A review. *J. Clean. Prod.* **2015**, *87*, 826–838. [CrossRef]
24. Hossain, L.; Sarker, S.K.; Khan, M.S. Evaluation of present and future wastewater impacts of textile dyeing industries in Bangladesh. *Environ. Dev.* **2018**, *26*, 23–33. [CrossRef]
25. Islam, M.; Chowdhury, M.; Billah, M.; Tusher, T.; Sultana, N. Investigation of effluent quality discharged from the textile industry of purbani group, Gazipur, Bangladesh and It's textile industry of purbani group, Gazipur, Bangladesh and It's management. *Bangladesh J. Environ. Sci.* **2012**, *23*, 123–130.
26. Sultana, M.S.; Islam, M.S.; Saha, R.; Al-Mansur, M. Impact of the effluents of textile dyeing industries on the surface water quality inside D.N.D embankment, Narayanganj. *Bangladesh J. Sci. Ind. Res.* **1970**, *44*, 65–80. [CrossRef]
27. Bhuiyan, M.A.H.; Suruvi, N.I.; Dampare, S.B.; Islam, M.A.; Quraishi, S.B.; Ganyaglo, S.; Suzuki, S. Investigation of the possible sources of heavy metal contamination in lagoon and canal water in the tannery industrial area in Dhaka, Bangladesh. *Environ. Monit. Assess.* **2011**, *175*, 633–649. [CrossRef]
28. Banglapedia. Dhaka District. 2022. Available online: https://en.banglapedia.org/index.php/Dhaka_District (accessed on 28 August 2022).
29. Ali, T. Waste of Water, Way to Disaster, the Daily Star. 2019. Available online: https://www.thedailystar.net/supplements/news/waste-water-way-disaster-1718767 (accessed on 28 August 2022).
30. Khan, M.R.; Koneshloo, M.; Knappett, P.S.K.; Ahmed, K.M.; Bostick, B.; Mailloux, B.J.; Mozumder, R.; Zahid, A.; Harvey, C.; Van Geen, A.; et al. Megacity pumping and preferential flow threaten groundwater quality. *Nat. Commun.* **2016**, *7*, 12833. [CrossRef]
31. Akan, A.E.; Akan, A.P. Potential of reduction in carbon dioxide equivalent emissions via energy efficiency for a textile factory. *J. Energy Syst.* **2018**, *2*, 57–69. [CrossRef]
32. Laurent, A.; Olsen, S.I.; Hauschild, M.Z. Carbon footprint as environmental performance indicator for the manufacturing industry. *CIRP Ann.* **2010**, *59*, 37–40. [CrossRef]
33. Muthu, S.S.; Li, Y.; Hu, J.Y.; Ze, L. Carbon footprint reduction in the textile process chain: Recycling of textile materials. *Fibers Polym.* **2012**, *13*, 1065–1070. [CrossRef]

Article

Numerical and Experimental Investigation on Bending Behavior for High-Performance Fiber Yarns Considering Probability Distribution of Fiber Strength

Yu Wang [1,2,3], Xuejiao Li [4], Junbo Xie [1,2], Ning Wu [1,2], Yanan Jiao [1,2,*] and Peng Wang [3]

1. Ministry of Education Key Laboratory of Advanced Textile Composite Materials, Institute of Composite Materials, Tiangong University, Tianjin 300387, China
2. School of Textile Science and Engineering, Tiangong University, Tianjin 300387, China
3. ENSISA, LPMT, University of Haute-Alsace, F-68000 Mulhouse, France
4. Jiangsu Hengli Chemical Fibre Co., Ltd., Suzhou 215228, China
* Correspondence: jiaoyn@tiangong.edu.cn

Abstract: The performance of fiber-reinforced composite materials is significantly influenced by the mechanical properties of the yarns. Predictive simulations of the mechanical response of yarns are, thus, necessary for fiber-reinforced composite materials. This paper developed a novel experiment equipment and approach to characterize the bending behavior of yarns, which was also analyzed by characterization parameters, bending load, bending stiffness, and realistic contact area. Inspired by the digital element approach, an improved modeling methodology with the probability distribution was employed to establish the geometry model of yarns and simulated bending behavior of yarns by defining the crimp strain of fibers in the yarn and the effective elastic modulus of yarns as random variables. The accuracy of the developed model was confirmed by the experimental approach. More bending behavior of yarns, including the twisted and plied yarns, was predicted by numerical simulation. Additionally, models revealed that twist level and number of plies affect yarn bending properties, which need to be adopted as sufficient conditions for the mechanical analysis of fiber-reinforced composite materials. This efficient experiment and modeling method is meaningful to be developed in further virtual weaving research.

Keywords: fiber/yarn; mechanical properties; finite element analysis (FEA); probability distribution

1. Introduction

Usage of fiber-reinforced composite materials is growing in the aerospace, automotive sectors, and protective clothing, due to excellent mechanical properties, structural designability, and fatigue resistance, where there is an increasing need to produce complex parts in a cost-effective manner [1–3]. Such parts usually consist of two components, including the fiber preforms and the matrix. In the fiber preforms, consisting of straight weft yarns and crimped warp yarns, architecture has an important influence on the final mechanical properties of the composites. Hence, as a vital part of fiber-reinforced composites, the mechanical behavior of the yarns will significantly disperse the mechanical properties of the composite [4–6], such as tensile strength, compression and bending, etc. As a matter of fact, the yarns themselves are formed by many thousands of individual continuous fibers, whose fiber structures are extremely complex and variable under force. Thus, a classification into two scales can be made: meso (yarn level) and micro (fiber level) [7]. The geometrical deformations that occur across these scales are critical in determining the final mechanical properties of the composites.

Depending on that described above, the geometrical deformation and mechanical properties of the two scales need to be explored under force. However, experimentally predicting the geometrical deformation and mechanical properties of yarn, made of large fibers, remains an obstacle to the understanding of these behaviors due to the stochastic effects of micro-scale interactions [8]. Numerical simulation [9–11] is used to analyze the mechanical behavior of fiber-reinforced composite, in which the yarns should be modeled as realistically as possible to make sure that the obtained behaviors are reliable. Some simulation models [12–14] are established based on the simplified meso scale geometries of yarns, which does take into account the geometries and mechanical properties of fibers. The yarn structures can be generated by specialist software such as TexGen [15,16] and WiseTex [17]. To simulate the realistic geometries of yarn, approaches are generally carried out at the micro-meso scale based on previous research [18,19]. Wang et al. [20,21] proposed the digital-element method to model yarns, which were constituted by a bundle of digital fibers, which were represented by a chain of digital elements between every two digital nodes. Based on the digital-element method, the virtual fiber method was employed to model yarns [12,14]. Durville et al. [22] employed beam elements to represent fibers, which considered the geometrical and mechanical aspects. Durville's approach starts from an arbitrary of virtual fiber configuration, in which all yarns have straight trajectories and all fibers constituting these yarns are taken into account in the model. Daelemans et al. [23] proposed a numerical modeling methodology to simulate yarns mechanics and geometries in Abaqus/Explicit. The idealized fibers are produced, which are constituted as a yarn by a Python script. The main advantage of this methodology is that the simulations are able to predict the mechanical response of the fabric by considering the sub-yarn behavior without the requirement of complex constitutive laws. High precision models of yarns can be generated using the above methods. These yarn modeling techniques, however, differ significantly from the realistic weaving process and do not consider the geometrical deformation and mechanical properties of the heterogeneous yarns, namely twisted and plied yarns, independently.

To acquire a more accurate model and character realistic mechanical properties described of heterogeneous yarns, an improved modeling methodology with the probability distribution constitutive laws needs to be established. A few constitutive laws have been utilized to characterize the mechanical properties of heterogeneous yarns [24,25]. Lamon et al. [26] proposed a solution based on a plot of quintile–quintile vs. strength to characterize flaw strengths using tensile tests on various fiber two types including, SiC, carbon, glass, basalt, and alumina. The results indicate that the normal distribution function is used to construct reference empirical distributions of flaw strengths that allow the evaluation of the Weibull plot and Maximum Likelihood Estimation methods as functions of sample size and composition. Wang et al. [27] proposed an analytical model based on the statistic theory to evaluate the random tensile response of jute fiber yarns which considered fiber crimp and properties distribution. The Beta probability distribution function was employed to characterize the stochastic tensile response of yarns. Wang et al. [28] used a constitutive model with stochastic damage properties to simulate axial tensile and transverse compressive behavior of the twisted yarns in explicit solver Abaqus/Explicit. The accuracy of the constitutive model, namely geometry and mechanics, was validated experimentally. Excellent agreement between simulation and experiment is obtained for the right set of input parameters. The above research investigated the mechanical properties of heterogeneous yarns focused on the constitutive laws, including the tensile and compressive behavior. However, the bending behavior of twisted and plied yarns, such as slippage, distortion, and shift of positions, have not been adequately explained. As a result, understanding and predicting the bending behavior of yarns, especially twisted and plied yarns, is crucial for final performance of the fiber-reinforced composite materials.

In the current paper, we use a novel experiment equipment and approach to evaluate the bending behavior of yarns, which is characterized by bending load, bending stiffness EI_{yarn}, and realistic contact area Ar. An improved constitutive model with probability distribution constitutive laws was employed to establish the models and simulate the bending behavior of yarns. The probability distribution in terms of crimp strain is introduced to characterize the upper limit of crimp region. The model is demonstrated on carbon fiber yarns using the explicit solver Abaqus/Explicit. Meanwhile, to explore how the yarns' architecture affects the bending behavior, the geometric deformation and bending behavior of yarns with various twists and plies were predicted on alumina and fiber quartz yarns, respectively. The model also provided the calculation approach for the bending stiffness of yarns at specific deflection.

2. Materials and Methods

2.1. Materials

This study was carried out on high-performance yarns, including carbon fiber yarn, alumina fiber yarn, and quartz fiber yarn, which are directly produced by Weihai Tuozhan Fiber Co., Ltd. (Weihai, China), Shandong University (Jinan, China), and Feilihua Quartz Glass Co., Ltd. (Jingzhou,, China). The detailed parameters of the samples are listed in Table 1. In this study, the carbon fiber yarn, alumina fiber yarn, and quartz fiber yarn are called CF, ALF, and QF, respectively. Initially, the samples were in the form of yarn containing thousands of fibers each.

Table 1. Material properties of investigated samples given by the manufacturers.

Yarn and Fiber Type	Density (g/cm^3)	Tex (g/1000 m)	Twist Level (tpm)	Radius of Fiber (μm)
Carbon fiber yarn	1.91	218	0	5.0
Alumina fiber yarn	2.88	330	170	7.0
Quartz fiber yarn	2.20	190	80	7.5

2.2. Experiment Method and Set Up

According to previous research [29,30], the experiment tests were carried out by a UMT-TRIBO LAB tribometer (Bruker Nano, Inc., Campbell, CA, USA) which were capable of being set at gauge lengths of 350 ± 10 mm. A pretension force of 2–6% of the maximum tensile load was applied to the samples and the deflection is 3 mm. To obtain more accurate test results, the particular fiber carrier was designed to fix the sample under a certain tension, as shown in Figure 1. Two ends of the sample were screwed together under strain during sample fixation. The screwed part of sample was protected by a plastic tube. The whole test was repeated 20 times at a room temperature of 23 ± 2 °C and a humidity of 60 ± 5%.

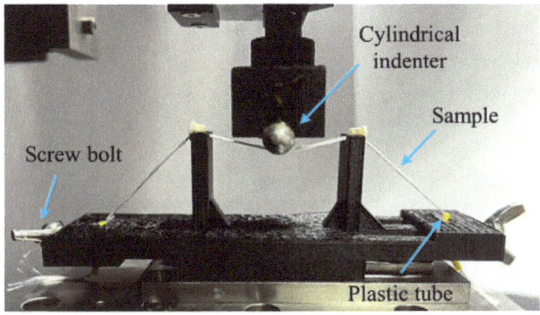

Figure 1. The picture of the bending test setup developed in the laboratory.

Based on the previous research on fiber/yarn contact surface [7,31], the yarn contact surface extraction experiment was conducted as shown in Figure 2. A silicone film mixture was prepared with silicone grease and curing agent in the ratio of 1:5, and 2 mL of the mixture was dropped evenly onto the surface of the cylindrical indenter. After 5–8 min of contact between the sample and the cylindrical indenter, a silicone film containing information on the contact surface for the sample was obtained. Furthermore, the silicone film was analyzed by a 3D contour measuring instrument VR-5200 (KEYENCE Inc., Osaka, Osaka Prefecture, Japan).

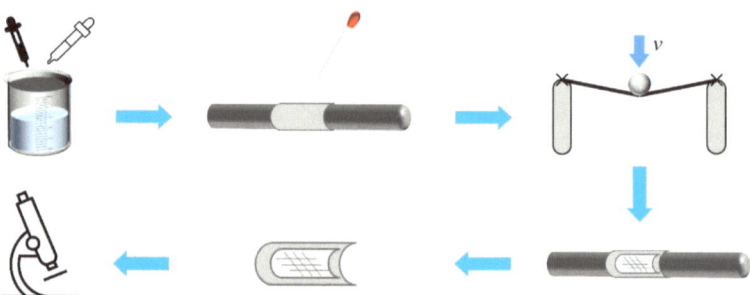

Figure 2. Experimental method for extracting realistic contact surface of the yarn.

3. Simulation Details
3.1. Constitutive Model Description

Each yarn is composed of thousands of fibers. However, modeling a yarn with realistic fiber amount is infeasible. A segmented constitutive model is utilized to describe the bending behavior of yarn. The model proposed in this research is based on the Weibull distribution described in the literature [28,32,33]. The terms apparent cross-section area and realistic cross-section area are used to describe the cross-section area of fiber strands. Images of yarns taken under an optical microscope can be used to determine the apparent yarn cross-section area, which includes the cross-section areas of the fibers inside the yarn and the gap between them. The geometry model of yarn is established in Abaqus by Python, which initially establishes the fiber arrangement of each yarn cross-section according to the twist level, and further connects each fiber node with a beam element, as shown in Figure 3. The geometry of yarn was determined based on the following equations:

$$\theta_0 = \pi - n \times \arcsin(z_0/r_{point}) \quad (1)$$

$$\begin{cases} x' = x_0 + \kappa \times Le \\ y' = r_{point} \times \cos(\theta_0 + \kappa\theta_t) \\ z' = r_{point} \times \sin(\theta_0 + \kappa\theta_t) \end{cases} \quad (2)$$

where x', y', and z' are coordinates along x, y and z directions, separately. κ is the number of the element. $r_{point} = \sqrt{x_0^2 + y_0^2}$ is the distance between the node and the center of the circle on the section. θ_t and θ_0 are the twist level of yarns and the deflection angle of the element, which usually is normalized by n.

The constitutive structure model of the fiber and yarn needs to be clarified. Initially, l_i ($i = 1 \sim n$) is defined as crimp length of the i-th fiber, which is arranged from small to large. Hence, the bending strain outside the neutral axis of the i-th fiber is obtained as:

$$\varepsilon_i^f = \frac{l - l_0 - l_i}{l_0} = \varepsilon - \varepsilon_i^l \quad (3)$$

where l is the stretched yarn length and l_0 is the initial yarn length. ε is the yarn strain. $\varepsilon_i^l = \frac{l_i}{l_0}$ is defined as crimp strain of the i-th fiber and is used to characterize the crimp behavior of fibers within a yarn [21].

$$\sigma_i = \begin{cases} E_f \varepsilon_i^f & \varepsilon_i^l \leq \varepsilon \leq \varepsilon_i^b \\ 0 & otherwise \end{cases} \quad (4)$$

where E_f is Young's modulus of the fiber, ε_i^b and ε_i^l are breaking strain and crimp strains of the i-th fiber.

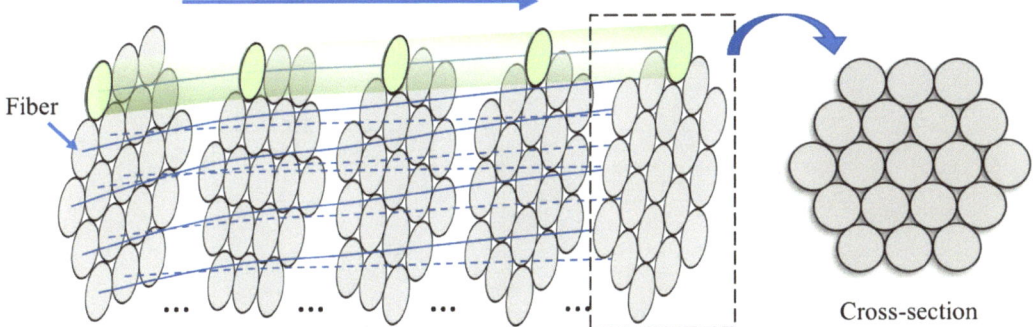

Figure 3. The 3D segmented clockwise twisted yarns. (Note: the gray is a cross-section of yarn, and the green is a cross-section of fiber).

The attributes with Weibull distribution must be assigned to each virtual fiber based on the stochastic distribution algorithm to investigate a more precise mechanical response of yarn. The factors with Weibull distribution of fibers—scale and shape factor—are obtained from the research [26,28], probability of fiber can be obtained by Equation (5).

$$P(\varepsilon) = 1 - \exp\left[-\left(\frac{\varepsilon}{\varepsilon_i^l}\right)^m\right] = \frac{N'}{N} \quad (5)$$

where m is the Weibull modulus or shape factor. Assuming that yarn with N fiber filaments has N' fibers under tension and other fibers not, the strain with n' fibers straightened is ε_i^l. Thus, the yarn force can be calculated as:

$$F = NA_f E_f \left[P(\varepsilon)\varepsilon - \int_0^\varepsilon xp(x)dx\right] \quad (6)$$

where A_f is the apparent cross-section area of the fiber. The probability density function $p(x)$ is the derivative of $P(x)$. The stress of yarn with N fibers can be defined as:

$$\sigma = E_f \left[P(\varepsilon)\varepsilon - \int_0^\varepsilon xp(x)dx\right] \quad (7)$$

Indeed, yarn failure strain and yarn strength are two commonly used yarn failure criteria [28,34,35]. Assuming that a brittle yarn breaks when its stress exceeds its yarn strength σ_s, therefore, the stress–strain relationship in the axial direction of the yarn can be given as:

$$\sigma = \begin{cases} E_f\left[P(\varepsilon)\varepsilon - \int_0^\varepsilon xp(x)dx\right] & \varepsilon < \left|\varepsilon_N^l\right| \\ E_f\left[P(\varepsilon)\varepsilon - \int_0^{\varepsilon^l} xp(x)dx\right] & \varepsilon = \left|\varepsilon_N^l\right| \\ 0 & \text{otherwise} \end{cases} \quad (8)$$

The proposed constitutive model is carried out by User subroutine VUMAT of Abaqus nonlinear finite element codes, which is based on the classic step-by-step iterative method in an incremental form, shown in Figure 4.

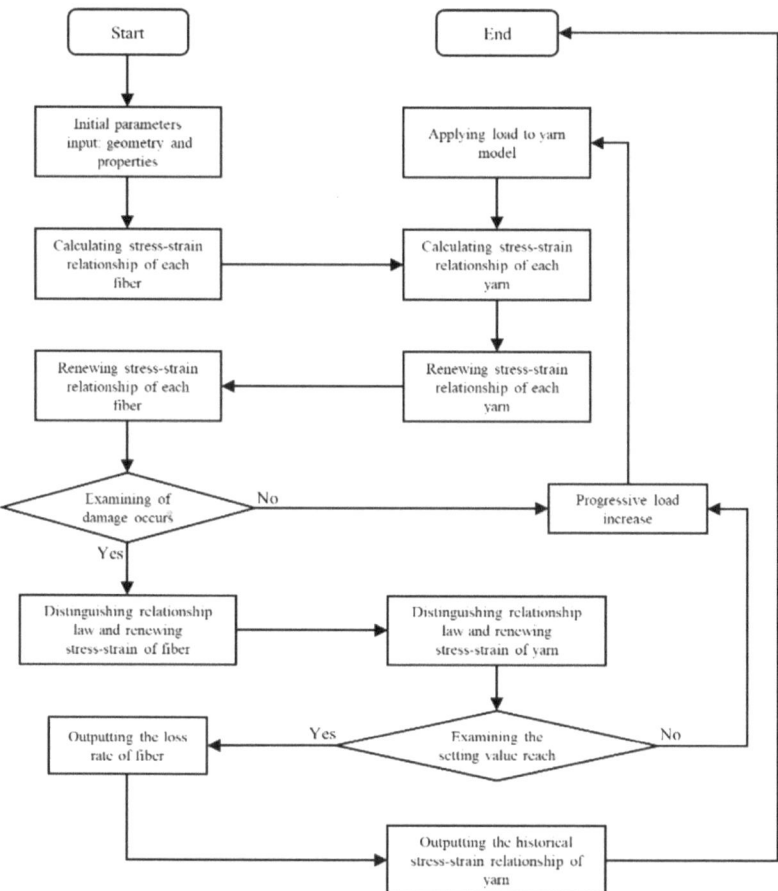

Figure 4. Flow chart for mechanical properties prediction of yarns.

3.2. Simulation of Bending Test

A quasi-fiber scale model of yarn for bending simulation was established in an explicit solver Abaqus/Explicit. The parameters of yarn simulation, COF (coefficient of friction)—0.35, 112 fibers and element length of L_e—0.3, were validated based on previous research [28]. The bending behavior was simulated using two dynamic/explicit steps by three rigid parts (dimensions consistent with experiments, rigid shell elements R3D4), which move (displacement-controlled) towards yarn in between the pre-tension in Figure 5. The dimensional parameters of specimen were identical to experiment in the bending simulation. The two steps, both 0.5 s, were performed to achieve the bending simulation. In step 1 the pre-tension was applied, while in step 2 the cylindrical indenter moved to a set displacement of 3 mm. To achieve a balance between the modeling precision and computational efficiency and ensure the convergence of the simulation, it is necessary to use mass scaling to increase the mass of the model artificially. During the analysis, a fixed mass scaling factor of 100 was introduced to the virtual fiber model. To conveniently apply the boundary conditions, the nodes at the start and end of the geometric model are coupled to points RP-1 and RP-2. During the bending simulation of the yarn, the reaction forces and the distance on the cylindrical indenter are recoded to characterize the bending behavior of yarns.

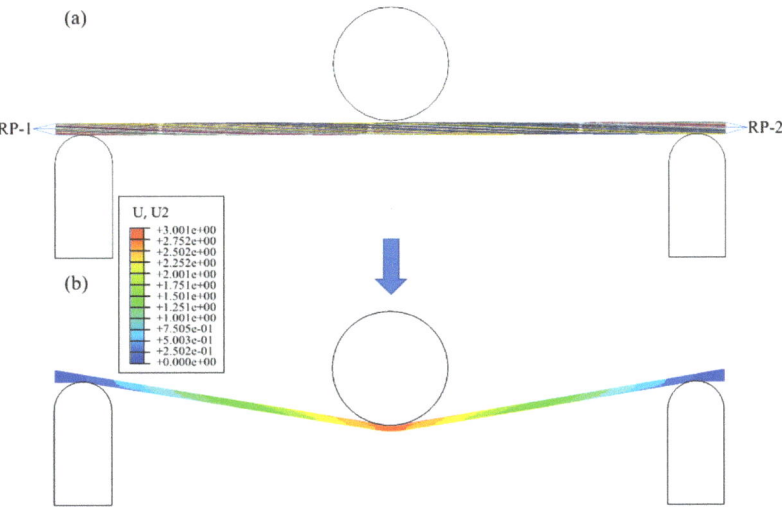

Figure 5. The bending simulation of the yarn: (**a**) boundary condition and (**b**) simulation result.

4. Results and Discussion

In this section, the proposed analytical model was validated against the data obtained from the experiment. The predictions of mechanical properties for twisted and plied yarns were also performed to understand the generality of the present model. Various parameters were employed to characterize the bending properties of the yarn and to demonstrate the universality of the present model.

4.1. Verification of Numerical Simulation Model

Following the experimental and simulation methods described above, taking carbon fiber yarn as an example, the bending properties were characterized and compared by the curve between bending load and deflection, as shown in Figure 6. The relationships between bending load and deflection under pretension are shown in Figure 6a. The bending load increases non-linearly with increasing deflection, whose shape is similar to the load–deflection curve of compression behavior [12,36]. Similarly, the curve can be divided into two phases, the deformation and bending phase, to explain the bending behavior. Both curves have the same trend for relationships between bending load and deflection during the whole bending. Additionally, the characteristic value is located in 95% prediction bounds of the experiment range.

Furthermore, the EI_{yarn} and Ar of yarn were compared to experimental results using the present model. Here, EI_{yarn} was calculated based on Equation (9) [12]:

$$EI_{yarn} = \sum E_f(I + d^2 A) \tag{9}$$

where I is the second moment of area, d is the projected distance to the centerline, and A is the cross-section area of the fiber.

Figure 6b shows the values for the EI_{yarn} and Ar at the deflection of 3 mm. It was found that the error of the experiment is larger than the simulation. This may be due to the fact that fiber rearrangement, and the accompanying cross-sectional change, is a dominant influence mechanism in the bending process of yarns in the experiment. Additionally, it is worth noting that the determined EI_{yarn} by simulation falls within the range determined by the experiment whose changing rate is 1.7%. For contact area morphology, both showed a low center and high surrounding, the height changing rate was about 0.7 mm. The Ar of experiment and simulation were 17.5 and 15.8 mm^2, respectively, whose changing rate for Ar was 12.6%.

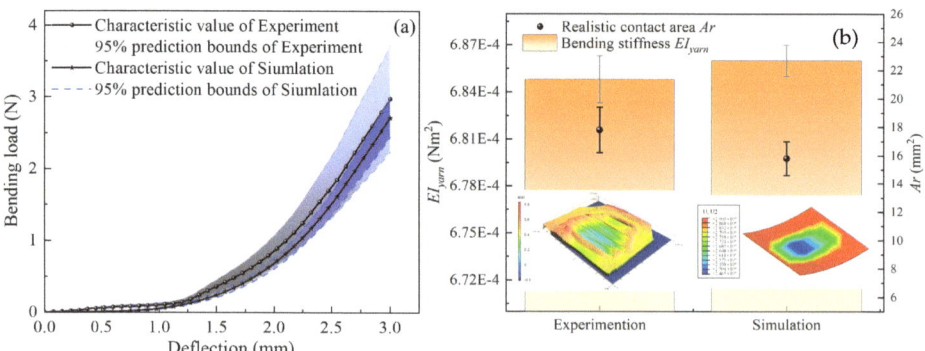

Figure 6. Comparison of experimental and simulation results of yarn bending behavior: (**a**) bending load–deflection curve, (**b**) resulting values of EI_{yarn} and Ar (taking carbon fiber yarn as an example).

The bending response showed better agreement with the experimentally determined curve, both in terms of its macroscopic (bending load–deflection) and microscopic (EI_{yarn} and Ar) parameters, which validates the universality of the present model. Hence, it shows that the simulation model can be employed to explore the effect of more details on yarn bending properties.

4.2. Influence of Twist Level on Bending Behavior

Based on the validated simulation model, yarn models with different twist levels were established and the bending behavior was predicted for 50, 80, 100, 150, and 200 tpm, respectively. Geometry models of various twisted yarns are generated using 122 beam elements with different properties. Figure 7 shows the local enlargement of the yarn model, from which the variation of yarn shape and twist level can be clearly observed. In each twisted yarn, the twist angles α were obtained from the geometric model of yarns, which indicates an increasing trend with the increasing twist level. Simultaneously, as the twist level increased, the fibers within the yarn became entangled and in contact with each other, gradually squeezing toward the central axis of the central yarn. At the same time, the curvature of the fiber spatial path gradually increased as the twist level increased, which led to fiber volume content with increasing twist level.

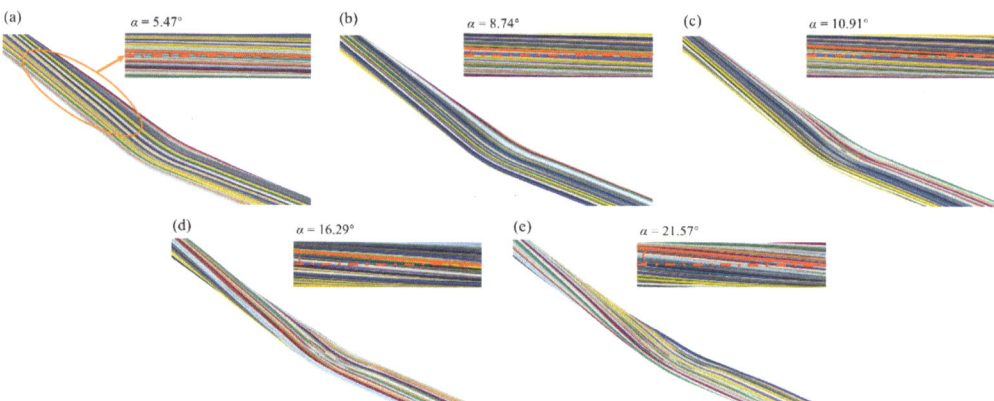

Figure 7. Prediction of the geometric model with different twist levels: (**a**) 50 tpm; (**b**) 80 tpm; (**c**) 100 tpm; (**d**) 150 tpm; and (**e**) 200 tpm.

The simulation results are shown in Figure 8a. It can be seen that when the deflection was in the range of 0 and 0.1 mm, the effect of twist level on the bending load was insignificant, while the effect of twist level on the bending load was significant for deflections between 1.0 and 3.0 mm, that is, the bending load decreased with increasing twist level. This may be due to the increase in twist level, therefore, migration of external fibers to the center of the yarn as the arrangement was not yet steady and the increase in fiber cohesion was achieved. Netherlands, nonlinear processes are observed in Figure 8a, that is, the rates of change in bending load of yarn between 50, 80 and 100 tpm were not significant. The rate of change between 50 and 100 tpm with a deflection of 3.0 mm was just 2.9%, while the rate of change with 200 tpm was 6.9%. Theoretically, the rates of change will probably continue to increase as the twist level continues to increase.

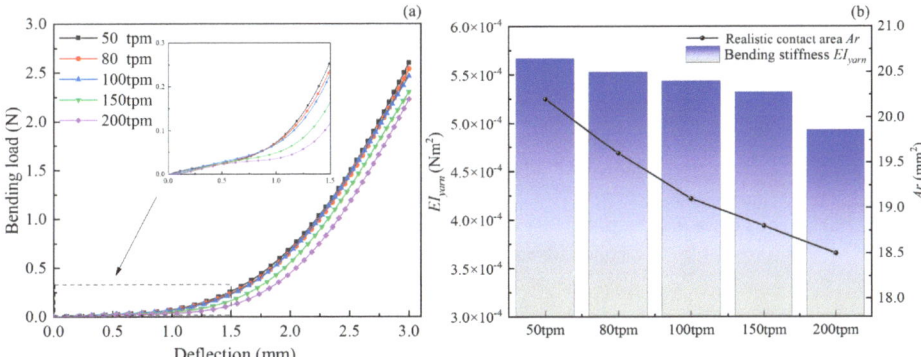

Figure 8. Prediction of bending behavior of twisted yarn: (**a**) curve of deflection and bending load, (**b**) realistic contact area and bending stiffness for different twisted yarns at a deflection of 3 mm.

Figure 8b showed that the values for the EI_{yarn} and Ar that are determined according to the methods described in the previous section for different twisted yarns at a deflection of 3 mm. Firstly, one can see that EI_{yarn} is inversely proportional to twist level, especially 150 tpm to 200 tpm, which can be explained using the deformation mechanisms described. Yarns with a small twist result in weak cohesion and a great degree of freedom of fibers. Therefore, the EI_{yarn} can be considered as the sum of each fiber. Conversely, yarns with a greater twist are susceptible to deformation. In addition, the results of bending simulation show that the effect of twist level on EI_{yarn} is mainly concentrated in a finite range, whose conclusions are similar to the ones in the literature. Furthermore, the relationship between the realistic contact area and twist level was recorded at a deflection of 3 mm by the numerical simulation method in Figure 8b. It is shown that the Ar decreased non-linearly as the twist level increased, which is the same trend as the EI_{yarn}'s. This can be explained by fiber rearrangement theory [7,37,38], that is, mainly due to the great degree of freedom between fibers within small twist yarns, a new contact surface of yarn is generated by rearrangement of inner layer fibers to the outer layer. As the twist level continues to increase, it is more difficult for the fibers to move with each other and the rate of change of Ar gradually decreases. It can be obtained that the rates of change of the realistic contact areas with different twists are 2.9, 2.5, 1.5, and 1.4%, respectively, further validating the above explanation from Figure 8b.

4.3. Influence of Ply on Bending Behavior

To prediction of the effect of the number of plies on the bending behavior of yarns, similar simulation approach and characterization parameters were employed for analysis using the quartz fiber yarn. Figure 9 illustrates the five kinds of plied yarns consisting of 1, 2, 3, 6, and 10 plies all with 80 tpm, which clearly indicates single ply yarns by different colors. It can be seen that the yarn of 10 plies model was relatively "strong", that is, larger diameter and more fibers. In addition, the cross-sections of plied yarns are also shown. With the increase of fiber amount, the resolution of the yarn's cross-sectional shape improves. The calculation time, however, increased from 1.5 h to 8.3 h.

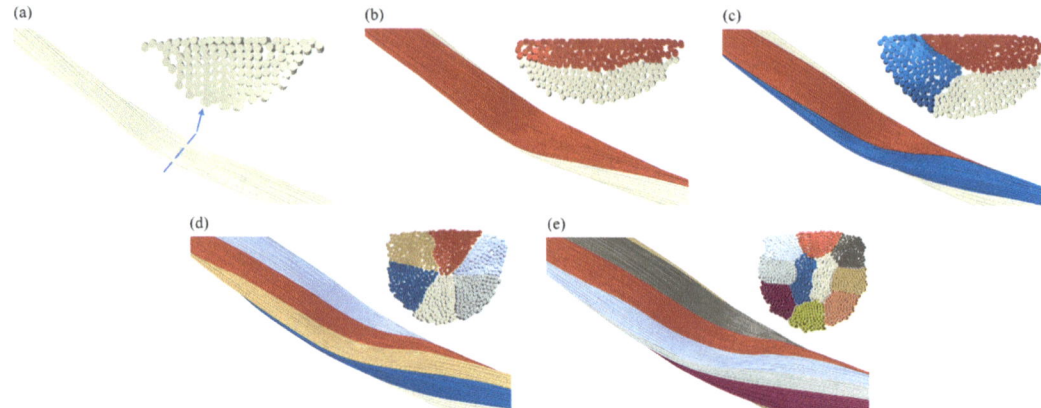

Figure 9. The prediction of the geometric model with the different number of plies: (**a**) 1 ply; (**b**) 2 plies; (**c**) 3 plies; (**d**) 6 plies; and (**e**) 10 plies.

Similar to the effect of twist level, the curve between the deflection and bending load, EI_{yarn} and Ar were employed to characterize the bending behavior of plied yarn. Figure 10 shows the values for the EI_{yarn} and Ar after the bending of plied yarns, which illustrates that the effect of ply number on the bending behavior is significant. The bending load of plied yarn showed a non-linear increasing trend with increasing deflection during the bending, which can be divided into two phases, that is, the deformation and bending phase in Figure 10a. The bending load of the defection of 3 mm was proportional to the number of plies. However, it is seen that as the number of plies increased, the non-linear variation trend of the bending load at 3 mm gradually became obvious, especially in 3-ply and 6-ply. This means that the effect of inter-fiber friction force on the bending load of multi-plies yarns is significant. Figure 10b represents the relationships of EI_{yarn} and Ar to the number of plies. There was an increase in EI_{yarn} of yarns with the increasing number of plies at the deflection of 3 mm. This is a result of the increase in the number of fibers in the cross-section of yarn with the increase in the number of plies. Furthermore, the trend described is explained by the following equation:

$$EI_{yarn} = \frac{N^2 EI_{fiber}}{\varphi} \tag{10}$$

where φ is the filling coefficient of fiber, N is the number of fibers within the cross-section, and EI_{yarn} is the bending stiffness of fiber.

Figure 10b shows that the Ar increases with the increasing number of plies at the deflection of 3 mm. Apparently, the Ar of 10-ply was the largest within the current research, which was also influenced by inter-fiber friction force. During the bending process, the migration of the inter fibers to the outer layer was obstructed, which resulted in a non-linear change for Ar. Here, Ar varied from 16.5 mm^2 of 1-ply to 96.2 mm^2 of 10-ply. The overall behavior was similar to Ar in Figure 8b, but with a few notable differences: somewhat greater range of variation in the results, and finally, a more marked transition was evident at around 3-ply where the rate of change of Ar changed. The rates of change were 31%, 30%, 47%, and 31%, respectively, within the scope of the current research, for which the arrangement of fibers may be responsible.

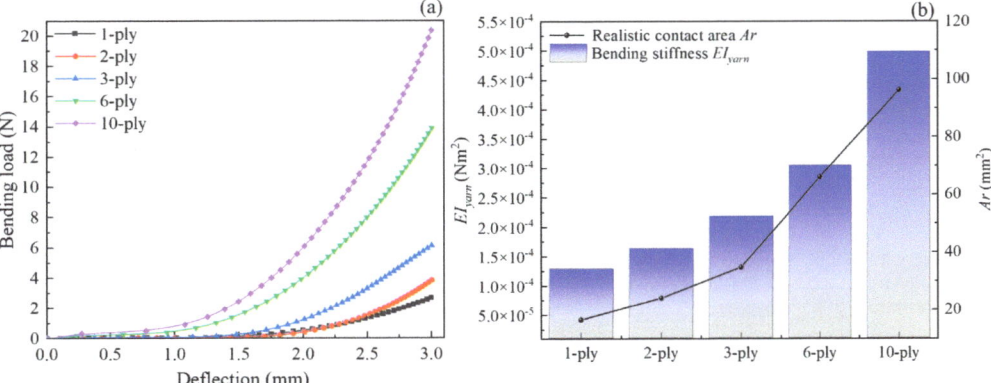

Figure 10. Prediction of bending behavior of plied yarn: (**a**) curve of deflection and bending load, (**b**) realistic contact area and bending stiffness at a deflection of 3 mm.

5. Conclusions

Predictive simulations of the mechanical response of yarns are necessary for the structural design of reinforcement, as well as for the development of fiber-reinforced composite materials. In the present research, experimental equipment and approach were utilized to characterize the bending behavior of yarns. The modeling approach was used to simulate the bending behavior of yarns based on our previous research. There are excellent agreements in characterization parameters including bending load, bending stiffness EI_{yarn}, and realistic contact area Ar after the comparison between the experimental measurements and the simulation. Furthermore, the predictions of bending behavior for twisted and plied yarns were carried out by the described model. It is shown that the bending load decreases gradually as the twist level increases and increases as the number of plies increases at the deflection of 3 mm. The EI_{yarn} and Ar are inversely proportional to twist level of yarns though, which varies on a minor interval. However, mainly influenced by the number of fibers, the EI_{yarn} and Ar increase with the increasing number of strands, which varies on a large interval. In the future, more detailed characterization parameters that cannot be experimentally obtained need to be explored to analyze the mechanical behavior of yarns through simulation models.

Author Contributions: Y.W.: investigation, methodology, software, validation, visualization, and writing—original draft. X.L.: investigation, methodology, visualization, and writing. J.X.: investigation, software, and validation. N.W.: investigation, methodology, and validation. Y.J.: conceptualization, methodology, and supervision. P.W.: conceptualization, investigation, and supervision. All authors have read and agreed to the published version of the manuscript.

Funding: The authors gratefully appreciate the financial support from the China Scholarship Council (Project no. 202108120054).

Institutional Review Board Statement: Not applicable.

Informed Consent Statement: Not applicable.

Data Availability Statement: The data presented in this study are available on request from the corresponding author.

Conflicts of Interest: The authors declare no conflict of interest.

References

1. Boisse, P. *Advances in Composites Manufacturing and Process Design*; Woodhead Publishing: Cambridge, UK, 2015; ISBN 978-1-78242-320-1.
2. Emonts, C.; Grigat, N.; Merkord, F.; Vollbrecht, B.; Idrissi, A.; Sackmann, J.; Gries, T. Innovation in 3D Braiding Technology and Its Applications. *Textiles* **2021**, *1*, 185–205. [CrossRef]
3. Xiang, H.; Jiang, Y.; Zhou, Y.; Malengier, B.; Van Langenhove, L. Binocular Vision-Based Yarn Orientation Measurement of Biaxial Weft-Knitted Composites. *Polymers* **2022**, *14*, 1742. [CrossRef] [PubMed]
4. Janicki, J.C.; Bajwa, D.S.; Cairns, D.; Amendola, R.; Ryan, C.; Dynkin, A. Gauge Length and Temperature Influence on the Tensile Properties of Stretch Broken Carbon Fiber Tows. *Compos. Part A Appl. Sci. Manuf.* **2021**, *146*, 106426. [CrossRef]
5. Xie, J.; Guo, Z.; Shao, M.; Zhu, W.; Yang, Z.; Chen, L. Mechanics of Textiles Used as Composite Preforms: A Review. *Compos. Struct.* **2023**, *304*, 116401. [CrossRef]
6. Li, M.; Wang, P.; Boussu, F.; Soulat, D. Effect of Fabric Architecture on Tensile Behaviour of the High-Molecular-Weight Polyethylene 3-Dimensional Interlock Composite Reinforcements. *Polymers* **2020**, *12*, 1045. [CrossRef]
7. Mulvihill, D.M.; Smerdova, O.; Sutcliffe, M.P.F. Friction of Carbon Fibre Tows. *Compos. Part A Appl. Sci. Manuf.* **2017**, *93*, 185–198. [CrossRef]
8. Chakladar, N.D.; Mandal, P.; Potluri, P. Effects of Inter-Tow Angle and Tow Size on Carbon Fibre Friction. *Compos. Part A Appl. Sci. Manuf.* **2014**, *65*, 115–124. [CrossRef]
9. Wang, J.; Wang, P.; Hamila, N.; Boisse, P. Meso-Macro Simulations of the Forming of 3D Non-Crimp Woven Fabrics. *Textiles* **2022**, *2*, 112–123. [CrossRef]
10. Orlik, J.; Krier, M.; Neusius, D.; Pietsch, K.; Sivak, O.; Steiner, K. Recent Efforts in Modeling and Simulation of Textiles. *Textiles* **2021**, *1*, 322–336. [CrossRef]
11. Gao, Z.; Chen, L. A Review of Multi-Scale Numerical Modeling of Three-Dimensional Woven Fabric. *Compos. Struct.* **2021**, *263*, 113685. [CrossRef]
12. Daelemans, L.; Tomme, B.; Caglar, B.; Michaud, V.; Van Stappen, J.; Cnudde, V.; Boone, M.; Van Paepegem, W. Kinematic and Mechanical Response of Dry Woven Fabrics in Through-Thickness Compression: Virtual Fiber Modeling with Mesh Overlay Technique and Experimental Validation. *Compos. Sci. Technol.* **2021**, *207*, 108706. [CrossRef]
13. Pham, Q.H.; Ha-Minh, C.; Chu, T.L.; Kanit, T.; Imad, A. On Microscopic and Homogenized Macroscopic Analysis of One Kevlar®KM2 Yarn under Transverse Compressive Loading. *Mech. Res. Commun.* **2020**, *104*, 103496. [CrossRef]
14. Yang, Z.; Jiao, Y.; Xie, J.; Chen, L.; Jiao, W.; Li, X.; Zhu, M. Modeling of 3D Woven Fibre Structures by Numerical Simulation of the Weaving Process. *Compos. Sci. Technol.* **2021**, *206*, 108679. [CrossRef]
15. Buchanan, S.; Grigorash, A.; Archer, E.; McIlhagger, A.; Quinn, J.; Stewart, G. Analytical Elastic Stiffness Model for 3D Woven Orthogonal Interlock Composites. *Compos. Sci. Technol.* **2010**, *70*, 1597–1604. [CrossRef]
16. Aziz, A.R.; Ali, M.A.; Zeng, X.; Umer, R.; Schubel, P.; Cantwell, W.J. Transverse Permeability of Dry Fiber Preforms Manufactured by Automated Fiber Placement. *Compos. Sci. Technol.* **2017**, *152*, 57–67. [CrossRef]
17. Vanaerschot, A.; Cox, B.N.; Lomov, S.V.; Vandepitte, D. Experimentally Validated Stochastic Geometry Description for Textile Composite Reinforcements. *Compos. Sci. Technol.* **2016**, *122*, 122–129. [CrossRef]
18. Hoan-Pham, Q.; Ha-Minh, C.; Long-Chu, T.; Kanit, T.; Imad, A. Analysis of the Transverse Compressive Behavior of Kevlar Fibers Using Microscopic Scale Approach. *Int. J. Mech. Sci.* **2019**, *164*, 105149. [CrossRef]
19. Döbrich, O.; Gereke, T.; Hengstermann, M.; Cherif, C. Microscale Finite Element Model of Brittle Multifilament Yarn Failure Behavior. *J. Ind. Text.* **2018**, *47*, 870–882. [CrossRef]
20. Wang, Y.; Miao, Y.; Swenson, D.; Cheeseman, B.A.; Yen, C.-F.; LaMattina, B. Digital Element Approach for Simulating Impact and Penetration of Textiles. *Int. J. Impact Eng.* **2010**, *37*, 552–560. [CrossRef]
21. Wang, Y.; Sun, X. Digital-Element Simulation of Textile Processes. *Compos. Sci. Technol.* **2001**, *61*, 311–319. [CrossRef]
22. Durville, D.; Baydoun, I.; Moustacas, H.; Périé, G.; Wielhorski, Y. Determining the Initial Configuration and Characterizing the Mechanical Properties of 3D Angle-Interlock Fabrics Using Finite Element Simulation. *Int. J. Solids Struct.* **2018**, *154*, 97–103. [CrossRef]
23. Daelemans, L.; Faes, J.; Allaoui, S.; Hivet, G.; Dierick, M.; Van Hoorebeke, L.; Van Paepegem, W. Finite Element Simulation of the Woven Geometry and Mechanical Behaviour of a 3D Woven Dry Fabric under Tensile and Shear Loading Using the Digital Element Method. *Compos. Sci. Technol.* **2016**, *137*, 177–187. [CrossRef]
24. Hu, Y.; Zhao, Y.; Liang, H. Refined Beam Theory for Geometrically Nonlinear Pre-Twisted Structures. *Aerospace* **2022**, *9*, 360. [CrossRef]
25. Li, Z.; Liu, Z.; Lei, Z.; Zhu, P. An Innovative Computational Framework for the Analysis of Complex Mechanical Behaviors of Short Fiber Reinforced Polymer Composites. *Compos. Struct.* **2021**, *277*, 114594. [CrossRef]
26. Lamon, J.; R'Mili, M. Investigation of Flaw Strength Distributions from Tensile Force-Strain Curves of Fiber Tows. *Compos. Part A Appl. Sci. Manuf.* **2021**, *145*, 106262. [CrossRef]
27. Wang, J.; Zhou, H.; Liu, Z.; Peng, X.; Zhou, H. Statistical Modelling of Tensile Properties of Natural Fiber Yarns Considering Probability Distributions of Fiber Crimping and Effective Yarn Elastic Modulus. *Compos. Sci. Technol.* **2022**, *218*, 109142. [CrossRef]
28. Wang, Y.; Jiao, Y.; Wu, N.; Xie, J.; Chen, L.; Wang, P. An Efficient Virtual Modeling Regard to the Axial Tensile and Transverse Compressive Behaviors of the Twisted Yarns. *J. Ind. Text.* **2022**, *52*, 15280837221137352. [CrossRef]

29. Wu, N.; Xie, X.; Yang, J.; Feng, Y.; Jiao, Y.; Chen, L.; Xu, J.; Jian, X. Effect of Normal Load on the Frictional and Wear Behaviour of Carbon Fiber in Tow-on-Tool Contact during Three-Dimensional Weaving Process. *J. Ind. Text.* **2020**, *51*, 152808372094461. [CrossRef]
30. Wu, N.; Li, S.; Han, M.; Zhu, C.; Jiao, Y.; Chen, L. Experimental Simulation of Bending Damage of Silicon Nitride Yarn during 3D Orthogonal Fabric Forming Process. *J. Ind. Text.* **2021**, *51*, 152808372110106. [CrossRef]
31. Ismail, N.; Vries, E.G.D.; Rooij, M.B.D.; Zini, N.H.M.; Schipper, D.J. An Experimental Study of Friction in Fibre-Fibre Contacts. *IJMPT* **2016**, *53*, 240. [CrossRef]
32. Weibull, W. A Statistical Theory of the Strength of Materials. *J. Appl. Mech.* **1939**, *9*, 293–297.
33. Zhou, W.; Wang, H.; Chen, Y.; Wang, Y. A Methodology to Obtain the Accurate RVEs by a Multiscale Numerical Simulation of the 3D Braiding Process. *Polymers* **2022**, *14*, 4210. [CrossRef] [PubMed]
34. Wang, J.; Zhou, H.; Ouyang, Z.; Peng, X.; Zhou, H. A Mesoscale Tensile Model for Woven Fabrics Based on Timoshenko Beam Theory. *Text. Res. J.* **2022**, *93*, 00405175221117616. [CrossRef]
35. Chen, Z.; Wang, B.; Pan, S.; Fang, G.; Meng, S.; Zhou, Z.; Zhu, J. Damage Analysis of Shear Pre-Deformed 3D Angle-Interlock Woven Composites Using Experiment and Non-Orthogonal Finite Element Model. *Compos. Commun.* **2021**, *28*, 100978. [CrossRef]
36. Hemmer, J.; Lectez, A.-S.; Verron, E.; Lebrun, J.-M.; Binetruy, C.; Comas-Cardona, S. Influence of the Lateral Confinement on the Transverse Mechanical Behavior of Tows and Quasi-Unidirectional Fabrics: Experimental and Modeling Investigations of Dry through-Thickness Compaction. *J. Compos. Mater.* **2020**, *54*, 3261–3274. [CrossRef]
37. Tourlonias, M.; Bueno, M.-A.; Poquillon, D. Friction of Carbon Tows and Fine Single Fibres. *Compos. Part A Appl. Sci. Manuf.* **2017**, *98*, 116–123. [CrossRef]
38. Mulvihill, D.M.; Sutcliffe, M.P.F. Effect of Tool Surface Topography on Friction with Carbon Fibre Tows for Composite Fabric Forming. *Compos. Part A Appl. Sci. Manuf.* **2017**, *93*, 199–206. [CrossRef]

Disclaimer/Publisher's Note: The statements, opinions and data contained in all publications are solely those of the individual author(s) and contributor(s) and not of MDPI and/or the editor(s). MDPI and/or the editor(s) disclaim responsibility for any injury to people or property resulting from any ideas, methods, instructions or products referred to in the content.

Article

Tactile Perception of Woven Fabrics by a Sliding Index Finger with Emphasis on Individual Differences

Raphael Romao Santos, Masumi Nakanishi and Sachiko Sukigara *

Department of Advanced Fibro Science, Graduate School of Science and Technology, Kyoto Institute of Technology, Matsugasaki, Sakyo-ku, Kyoto 606-8585, Japan
* Correspondence: sukigara@kit.ac.jp

Abstract: Haptic sensing by sliding fingers over a fabric is a common behavior in consumers when wearing garments. Prior studies have found important characteristics that shape the evaluation criteria and influence the preference of consumers regarding fabrics. This study analyzed the tactile perception of selected woven fabrics, with an emphasis on the participants' individual differences. Individual differences generally are discarded in sensory experiments by averaging them. Small differences among consumers can be important for understanding the factors driving consumer preferences. For this study, 28 participants assessed fabrics with very distinct surface, compression, and heat transferring properties by sliding their index fingers along the surface of the fabric. The participants also engaged in a descriptive sensory analysis. The physical properties of the fabric were measured using the Kawabata Evaluation System for Fabrics (KES-F) system. Moreover, parameters at the finger–fabric interface, such as the contact force, finger speed, and skin vibration, were measured during the assessment. This study used analysis of variance to eliminate nonsignificant attributes. Consonance analysis was performed using principal component analysis (PCA) on the unfolded sensory and interface data matrices. Finally, the physical and interface data were regressed onto sensory data. The results showed that the contact force and finger speed were nonsignificant, while skin vibration was a possible replacement for surface physical properties measured by the Kawabata Evaluation System for Fabrics (KES-F) system with an equal or slightly improved explainability.

Keywords: sensory profiling; descriptive analysis; textiles; individual differences; skin vibrations

Citation: Romao Santos, R.; Nakanishi, M.; Sukigara, S. Tactile Perception of Woven Fabrics by a Sliding Index Finger with Emphasis on Individual Differences. *Textiles* **2023**, *3*, 115–128. https://doi.org/10.3390/textiles3010009

Academic Editor: Laurent Dufossé

Received: 22 December 2022
Revised: 8 February 2023
Accepted: 13 February 2023
Published: 16 February 2023

Copyright: © 2023 by the authors. Licensee MDPI, Basel, Switzerland. This article is an open access article distributed under the terms and conditions of the Creative Commons Attribution (CC BY) license (https://creativecommons.org/licenses/by/4.0/).

1. Introduction

Running their fingers over a fabric is a common consumer behavior when choosing and buying garments. In this tactile interaction, consumers may extract information on the special characteristics of the fabrics. Over time, this process can lead to the accumulation of consumer fabric evaluation criteria and can be the reason for their preference for certain fabrics.

Textile industry experts have used sensory analysis to understand consumer preferences for quality. In the case of descriptive sensory analysis (DA), trained panels have been used in product development to obtain consistent data and to find small differences among products. Recently, there has been an increasing interest in engaging consumers in DA. Related studies have also discussed the pros and cons, accuracy, and reliability issues between trained and consumer panels [1,2].

Both subjective and objective evaluations of fabrics are necessary to understand consumer preferences. One approach is to measure a series of physical values, such as tensile strength, shear stress, bending, and compression, using the Kawabata Evaluation System for Fabrics (KES-F) [3]. Prior studies have interpreted the values obtained in the DA in terms of physical values by correlation or regression [4,5]. Another reported approach used a user-oriented design to analyze the depth impression of natural and artificial materials using concept networks [6]. To understand individual perceptions and preferences, new characteristic values must be linked to individual senses to produce customized fabrics.

Recently, studies have analyzed the tactile sensation of materials using sensors that capture the movement of fingers running across fabric. The velocity of the moving finger [7], the measurement of the force applied by the finger, the velocity–vibration system [8], and the vibration [9] have been assessed. These attributes are measured very closely to the fabric and are expected to be closely related to the physical properties of the fabric. Given the two-way mechanisms of haptics, some information related to abstract preference layers may also be collected in these new attributes. These abstract layers are closely related to consumer individuality. We expect this information to support the understanding of the individual tactile nature of consumers [10].

In this study, a fundamental case in which a person slides their index finger over a woven fabric surface was considered. Three parameters were measured: the contact force, sliding speed, and skin vibration (hereinafter referred to as interface attributes). To better mimic a familiar environment, little constraint was imposed on participants when moving their finger from left to right along the length of the sample. Furthermore, a DA and measurements of the physical properties of the fabrics were also conducted.

The study mainly focused on the differences among the participants and the interplay between the DA, interface attributes, and physical attributes. Understanding participants' individuality was prioritized in this study and not sample differences. Samples were selected for easy discrimination, especially regarding surface properties. This had the benefits of reducing the mental burden and speeding up the experiment. Unless explicitly mentioned, the individual averages were not recorded during the experiment.

The analysis strategy comprised (a) elimination of non-significant sensory and interface attributes following a qualitative investigation of (b) the level of agreement among the participants for each attribute (consonance analysis). Finally, the relationship between the sensory attributes, interface attributes, and physical properties was investigated using (c) linear regression. (a) ANOVA; (b) PCA; and (c) principal component regression (PCR).

2. Materials and Methods

2.1. Samples

Six woven fabric samples with different weave structures and surface roughness values were used in this study. Photographs of the sample surfaces are shown in Figure 1. The specifications of each sample are listed in Table 1. The surfaces of samples D and F were hairy; however, the surface of sample D was smooth and composed of fine wool fibers. The color of the fabrics was assessed using a colorimeter (CM-3600d; Konica Minolta, Inc., Tokyo, Japan) under illuminant D65 conditions with a 10° field of vision. The values of L^*, a^*, and b^* are listed in Table 1. In sample B (3×1 twill), a large weave rib is clearly observed compared to the other samples, as shown in Figure 1.

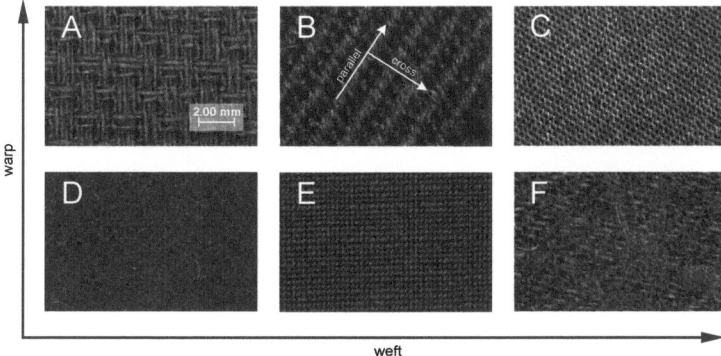

Figure 1. Photographs of the surface structures of the samples with weave directions. (**A–F**) Six woven fabric samples with different weave structures and surface roughness values were used in this study.

Table 1. Sample specifications.

Sample	Fiber Ratio (%)	Weave Structure	Density (cm) Ends	Density (cm) Picks	Thickness mm	Weight g/m²	L* (D65), a*, b*, c* (*1)
A	Polyester 99/polyurethane 1	Wedge slab	28	20	0.90	214	15.28, 0.21, 0.08, 0.23
B	Wool 100	3 × 1 twill	22	12	1.80	406	13.81, 0.03, −0.84, 0.80
C	Polyester 100	Satin	95	39	0.24	9	18.78, 0.47, −0.56, 0.73
D	Wool 100	2 × 1 twill	38	30	1.13	243	13.22, 0.17, −1.37, 1.38
E	Mohair 56/wool 35/water soluble vinylon 9	Plain	35	30	0.38	155	15.98, −0.10, −1.38, 1.34
F	Wool 56/paper 40/cotton 4	2 × 2 twill	60	64	3.40	415	16.91, −0.18, −1.39, 1.40

(*1) Calculated by the SCI method at 10° viewing angle.

2.2. Participants

Twenty-eight participants (14 men and 14 women) took part in the finger-sliding experiment for a total of 14 specimens, as listed under the 'Code No.' column in Table 2. The participants evaluated the samples by sliding the fingers along the warp and weft directions of all samples, and additionally for two extra directions for sample B (parallel and cross). All participants were university students (aged 22–25 years) with normal color vision and normal or corrected-to-normal visual acuity. The order of evaluation of the samples was randomized.

Table 2. Sample physical properties.

Code No	Direction	Surface			Heat Flow	Compression		
		SMD mm	MIU —	MMD —	q_{max} W/cm^2	WC J/m^2	RC %	LC —
A-1	Weft	5.91	0.160	0.016	0.124	0.29	40.0	0.32
A-2	Warp	7.92	0.177	0.012				
B-3	Weft	8.16	0.286	0.018	0.057	0.59	50.6	0.34
B-4	Warp	2.90	0.170	0.009				
B-5	Cross	32.0	0.213	0.019				
B-6	Parallel	4.85	0.197	0.011				
C-7	Weft	1.54	0.177	0.003	0.208	0.07	37.6	0.38
C-8	Warp	0.68	0.140	0.002				
D-9	Weft	2.53	0.134	0.006	0.086	0.37	56.5	0.30
D-10	Warp	2.56	0.124	0.006				
E-11	Weft	3.52	0.131	0.015	0.188	0.11	52.7	0.30
E-12	Warp	7.52	0.158	0.020				
F-13	Weft	3.81	0.212	0.008	0.050	1.82	49.0	0.45
F-14	Warp	3.27	0.215	0.007				

2.3. Measurements

2.3.1. Physical Properties: Surface, Compression, and Heat Flow Properties

Surface properties of the fabrics, such as surface roughness (SMD), coefficient of friction (MIU), and mean deviation of MIU (MMD), were measured under standard measurement conditions along the warp and weft directions using KES-SE-SR and KES-SE surface testers (Kato Tech Co., Ltd., Kyoto, Japan), respectively. The shapes of the surface contact sensors used to measure MIU, MMD, and SMD are shown in Figure 2a,b, respectively. For sample B, the surface properties were measured along four directions, as shown in Figure 1, because a difference in the SMD values along different directions was expected. The compression properties were measured using KES-G5 (Kato Tech Co., Ltd., Japan) under standard measurement conditions (Table A1, Appendix A). The parameters LC (linearity of compression–thickness curve), WC (compression energy, J/m), and RC (compression resilience, %) were obtained. All measurements were performed on three fabric specimens from each sample (20 cm × 20 cm) at room temperature (23 ± 2 °C) and (60 ± 5) % relative humidity. The maximum value of the heat flow (q_{max}, W/cm^2), which is related to the warm/cool feeling, was also measured using a KES Thermo Labo II (Kato Tech Co., Ltd., Kyoto, Japan) in accordance with the JIS L1927 standard.

2.3.2. Descriptive Sensory Evaluation

Eleven sensory attribute pairs divided into two groups were evaluated in this study. The first group was related to the fundamental physical properties of the fabric: warm/cool, hard/soft, flat/bumpy, rough/smooth, thin/thick, slippery/sticky, and weak/strong. The other group was related to personal preferences and experiences: new/familiar, expensive/cheap, uncomfortable/comfortable, and like/dislike. Participants rated the fabric on a scale ranging from 1 to 7 in accordance with the semantic differential method [11]. Each participant was presented with samples in random order. All four edges of the fabric were attached to a 10 cm × 60 cm cardboard of 1 mm thickness to prevent movement of the sample.

Figure 2. Measurement of surface properties. (**a**) Measurement of surface friction using 20 piano wires ($\phi = 0.5$ mm, probe size of 1 cm × 1 cm) (MIU and MMD). (**b**) Surface geometry measurement using a U-type piano wire ($\phi = 0.5$ mm) (SMD).

2.3.3. Interface Parameters: Index Finger Skin Vibration, Contact Force, and Translation Speed

Participants assessed the sample by sliding their right index finger over the length of the sample four times, from left to right. A force plate (3-axis force plate; 20 cm × 20 cm, TF-2020, Tec Gihan Co., Ltd., Kyoto, Japan) was placed under the sample to measure the contact force. A skin vibration sensor (Yubi recorder, Tec Gihan Co., Ltd., Kyoto, Japan) was wrapped around the distal interphalangeal joint of each participant's index finger to measure the vibrations of the skin triggered by the sliding movement of the finger over the fabric surface. A motion capture system (Motive, Nobby Tech, Ltd., Tokyo, Japan) comprising six cameras and three markers (4 mm diameter) was used in this study. One marker was used on the finger to track its location and calculate its speed. Two markers were used to indicate the location of the force plate in the motion capture space. Data logging was synchronized over the period of each assessment. The ambient conditions were (23 ± 2) °C with a relative humidity of 60.5%. The luminance over the fabric was 150 lx, and the effect of the color difference on visual perception was marginal. The details of the setup are shown in Figure 3. In this experiment, the following characteristic parameters were obtained: area under the power spectral density (PSD) of the vibration signal between 60 and 1000 Hz, delta power (DP, mV); contact force (F_z, N), and finger speed along the sample (v, m/s). The contact force and speed were averaged over the finger slides.

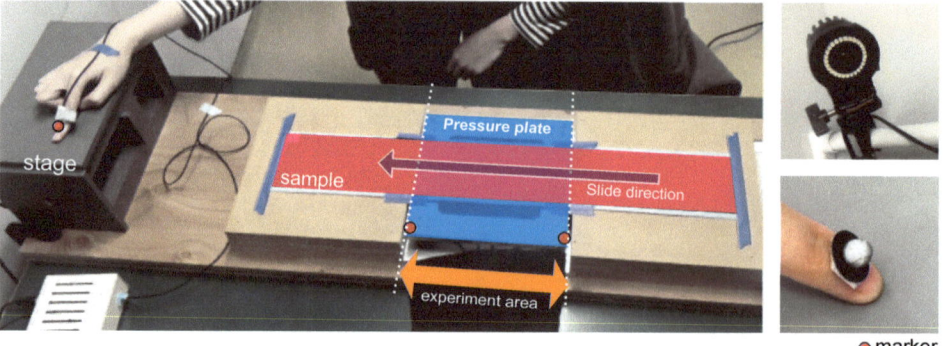

Figure 3. Experimental setup. Starting from the stage, the participants slid the index finger over the sample from left to right (slide direction) in a smooth and continuous motion. On the index finger, a motion capture marker and a skin vibration sensor were attached to measure the speed and the vibration signal, respectively. Beneath the sample, a pressure plate was used to measure contact force. Two markers on the pressure were defined as the effective experimental areas where all the parameters were calculated.

2.4. Analysis

It is important to recognize and consider the various differences that may exist among participants in sensory description experiments. This recognition and consideration play a crucial role in deciding n an appropriate model for data analysis, as well as in the interpretation of the results and determination of the most suitable analysis strategy. The following is a non-exhaustive list of differences commonly observed among participants:

- Differences in the use of attributes and measurement scale;
- Discrimination ability;
- Differences in sensitivity (perception and recognition);
- Misunderstanding of the meaning of attributes;
- Confusion of similar attributes;
- Repeatability.

In contrast to a trained panel, where the above differences are considered undesirable, it is expected that such differences will manifest randomly among consumers (represented by the participants). Some of these differences may be mitigated. For example, centering of sensory attributes reduces the differences in the use of the measuring scale. Other differences may require more specialized experimental designs. However, it is important to acknowledge that, from the participant's viewpoint in a decision-making scenario, these differences are of secondary importance as they collectively contribute to the participant's unique individuality. This raises the question of the extent to which these differences should be taken into account. By focusing on the most general case, we have chosen to examine the degree of agreement among participants, also known as consonance analysis.

The overall analysis procedure included the following steps: (1) initial health check-up of the data using descriptive statistical analysis; (2) elimination of attributes with low significance through the use of ANOVA; (3) evaluation of the agreement among participants through the application of PCA and Tucker-1 modeling as part of a consonance analysis; and (4) analysis of the correlation between the sensory data, the physical properties of the samples, and the interface parameters through Principal Component Regression (PCR).

2.4.1. Datasets

The three datasets used in the study consisted of sensory descriptive analysis, physical properties, and interface data. The sensory dataset served as the master dataset. All datasets used in the analysis were centered. The interface data and physical properties were standardized. Unless otherwise specified, the data were not averaged to maintain the individuality of the participants. The structures of the three datasets are listed in Table 3.

Table 3. Structure of the datasets.

Dataset	Samples	Attributes	Participants
Sensory		11	28
Interface	14	3	28
Physical		7	—

2.4.2. Structure of the Sensory Dataset

In the sensory analysis experiment, m participants evaluated n samples for p attributes. The data formed a three-way table x_{ijk}, with $i = 1, \ldots, m$; $j = 1, \ldots, n$, and $k = 1, \ldots, p$. The variance model, represented by Equation (1), comprised of participant effects, α_{ik}, sample effects, β_{jk}, and random error, ϵ_{ijk}:

$$x_{ijk} = \mu_k + \alpha_{ik} + \beta_{ij} + \epsilon_{ijk}, \quad (1)$$

The participant effects represented the differences between the average score of participant i (and attribute k) and the overall average for that attribute. Similarly, the sample effects described the differences between the average score of the jth sample for a particular attribute k and the overall average value for that sample and attribute. Since there were no replicate samples, the interaction term was confounded with the error. The interface dataset followed a similar structure.

2.4.3. Descriptive Statistical Analysis

The overall impression of the sensory and interface datasets was assessed by distribution plots of all participants and attribute scores for each sample. The presence of extreme outliers was investigated.

2.4.4. ANOVA

A two-way analysis of variance based on the variance model in Equation (1) was carried out. Attributes with a p-value < 0.05 were eliminated. Moreover, attributes with low agreement were also eliminated in the subsequent consonance analysis.

2.4.5. PCA and Tucker-1

Given the three-way structure of the data, it was deemed appropriate to employ three-way factor analysis (TWFA) in the study, utilizing the Tucker-1 modeling approach. TWFA represents a generalization of the principal component analysis (PCA) model for matrices of higher dimensions. The standard PCA of an $(n \times p)$ matrix $\mathbf{X}$ conforms to the following model:

$$\mathbf{X} = \mathbf{T}\mathbf{P}^\mathbf{T} + \mathbf{E} \qquad (2)$$

where $\mathbf{T}(n \times a)$ is the sample score, $\mathbf{P}^\mathbf{T}(a \times p)$ is the attribute loading, and $\mathbf{E}(n \times p)$ is the matrix of residuals. The scores are defined to have orthogonal columns and the loadings orthogonal rows. The loadings P are defined to describe as much variation in $\mathbf{X}$ as possible given a dimension a. Normally $\mathbf{P}^\mathbf{T}\mathbf{P} = \mathbf{I}$ and $\mathbf{T}$ are the projection of $\mathbf{X}$ on $\mathbf{P}$. Alternatively, the problem can be stated as finding the $\mathbf{T}$ and $\mathbf{P}$ matrices that minimize the residuals in

$$||\mathbf{X} - \mathbf{T}\mathbf{P}^\mathbf{T}||^2. \qquad (3)$$

In the context of Tucker-1 modeling, if we consider a slice $n \times p$ of the three-way sensory data representing participant i's individual sample-by-attributes matrix $\mathbf{X_i}$, where $i = 1, \ldots, m$, $\mathbf{X_i}$ can be defined as

$$\mathbf{X_i} = \mathbf{T_i}\mathbf{P}^\mathbf{T} + \mathbf{E_i}, \qquad (4)$$

where $\mathbf{P}$ has dimension $p \times a$. $\mathbf{T_i}$ and $\mathbf{P}$ are found for any a by minimization of the residuals in

$$\sum_{i=1}^{m} ||\mathbf{X_i} - \mathbf{T_i}\mathbf{P}^\mathbf{T}||^2. \qquad (5)$$

$\mathbf{P}$ is constrained to have orthogonal rows ($\mathbf{P}^\mathbf{T}\mathbf{P} = \mathbf{I_a}$) and the $\mathbf{T_i}$s are unconstrained. If the matrix $\mathbf{X}$ is unfolded to give an $mn \times p$ matrix, it can be shown that the minimization can be achieved by applying the standard PCA to the unfolded matrix. This corresponds to the vertical unfolding in Figure 4. Similarly, if the matrix $\mathbf{X}$ is unfolded as an $n \times mp$ matrix, it can be shown that the minimization of the criterion can be solved by standard PCA. The criterion is

$$\sum_{i=1}^{m} ||\mathbf{X_i} - \mathbf{T}\mathbf{P_i^T}||^2, \qquad (6)$$

where $\mathbf{T}$ has dimensions $(n \times b)$ and b stands for the reduced dimension. $\mathbf{T}$ is constrained to have orthogonal columns ($\mathbf{T}^\mathbf{T}\mathbf{T} = \mathbf{I_b}$) and $\mathbf{P_i}$s are unconstrained. For a more detailed description of the Tucker-1 model see [12,13].

2.4.6. Consonance Analysis—Visualization of Agreement among Participants

The level of agreement among participants was assessed by applying PCA to the unfolded sensory and interface data. The distinct unfolding direction results in a different perspective of the same underlying information. Vertical unfolding made it easier to see the variability in the sample's score, whereas horizontal unfolding emphasized the variability in the attribute loadings. Cross-validation was performed by leaving one participant out in the case of vertical unfolding.

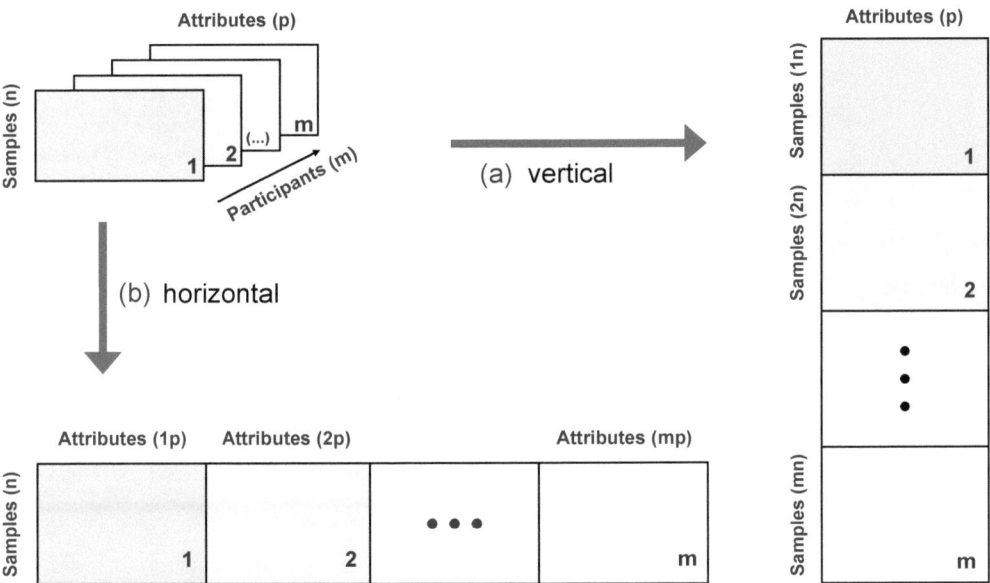

Figure 4. Three-way matrix unfolding. The figure depicts the two unfolding strategies used in Tucker-1 analysis. The three-way dataset is arranged either (a) vertically as an $(mn \times p)$ matrix or (b) horizontally as an $(n \times mp)$ matrix, resulting in unique sample–participant or participant–attribute pairs. This forms the foundation for the agreement (consonance) analysis, where participants with similar perceptions of samples (a) cluster together in the principal component space and participants with similar perceptions of attributes or (b) cluster together similarly.

2.4.7. Principal Component Regression (PCR)

In sensory descriptive experiments, high semantic similarity among sensory attributes often leads to high correlation. Additionally, removing attributes is not always appropriate as the goal is often to comprehend subtle differences in meaning. Performing PCA before conducting a linear regression, also know as PCR, can address the issues caused by collinearity. Principal component regression analysis was performed to assess the influence of the physical properties and interface parameters on the sensory perception of the participants. The physical and interface datasets were joined into a single matrix for attributes not yet eliminated. The two steps of (1) PCA and (2) linear regression for the PCR model were

$$\mathbf{X} = \mathbf{TP}^T + \mathbf{E} \qquad (7)$$

where $\mathbf{X}$ represents the sensory dataset, $\mathbf{T}$ represents the PCA scores, $\mathbf{P}$ represents the loadings, and $\mathbf{E}$ represents the residuals. PCA was conducted on each attribute separately, with rows as samples and columns as participants.

$$\mathbf{y} = \mathbf{Tq} + \mathbf{f} \qquad (8)$$

where **y** represents a physical or interface attribute, **T** is the PCA score, **q** is the regression coefficients for **y** on **T**, and **f** is the residuals. The regression coefficients are also known as regression loadings because they assume the same role as $\mathbf{P}^T$ in Equation (8). The regression loading plots provided a direct relationship between the predictors (sensory data) and the response (physical attributes and interface attributes).

3. Results and Discussion

3.1. Descriptive Analysis

The median values of the distributions indicated discrimination tendencies among the participants. However, the high variance in the distributions makes it unclear whether each participant's score follows the trend of the median.

3.2. ANOVA

The attributes of weak/strong and new/familiar were found to be not significant in the sensory dataset. The speed and force were not significant in the interface dataset (Figure 5). Although the contact force and speed attributes were eliminated, there may still be important correlations not investigated in this study. For example, the time between evaluations of variations in speed may be correlated with the level of confusion over the attributes.

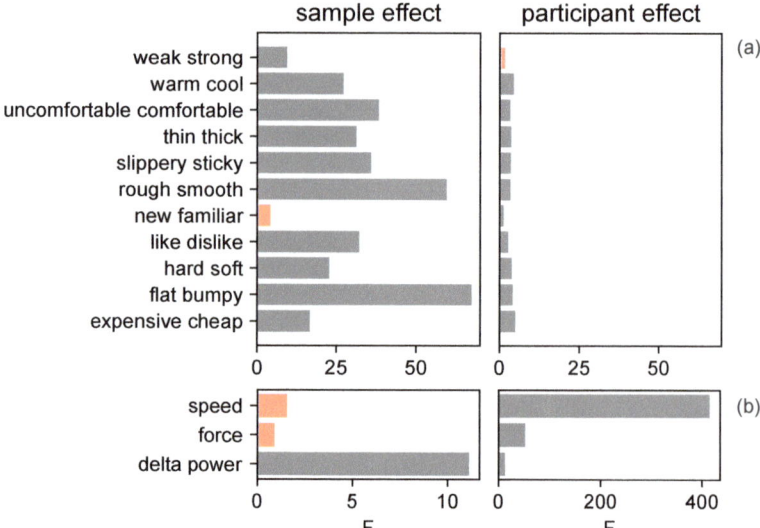

Figure 5. Two-way ANOVA (sensory and interface data). Model with effects for the sample and the participants. (**a**) Sensory attributes and (**b**) fabric–finger interface attributes. The bar length represents the F-value. The attributes in red have a p-value > 0.05 and were regarded as non-significant.

3.3. Consonance Analysis

3.3.1. Vertical Unfolding—Sensory Dataset

Figure 6 shows that the robustness of the PCA model is supported by the invariance of explainability with and without cross-validation. The loading plot shows that the first component spans surface property variations, whereas the second component spans a compression-heat transport variation. The correlation was found to be strong between slippery/sticky and flat/bumpy. As the sample surface properties were quite different, we assumed that the participants interpreted the attributes similarly. The correlation plot suggests that slippery and flat were interpreted as smooth, whereas sticky and bumpy were interpreted as rough. Similarly, but to a lesser degree, we could observe that the participants like samples that appear expensive and comfortable while disliking cheap and uncomfortable. Finally, the soft samples were interpreted as warm and the hard samples as cool. Given the correlation, the attributes expensive/cheap and slippery/sticky were dropped at this point.

3.3.2. Horizontal Unfolding—Sensory Dataset and Delta Power

Scores from the sensory data and delta power (Figure 7) showed similar relationships among the samples. This result highlights the fact that skin vibrations were the main factor affecting the surface perception of the samples. In Figure 8, the loading plots show the level of agreement among the participants for each attribute. A strong agreement in the surface-related attributes was noticed.

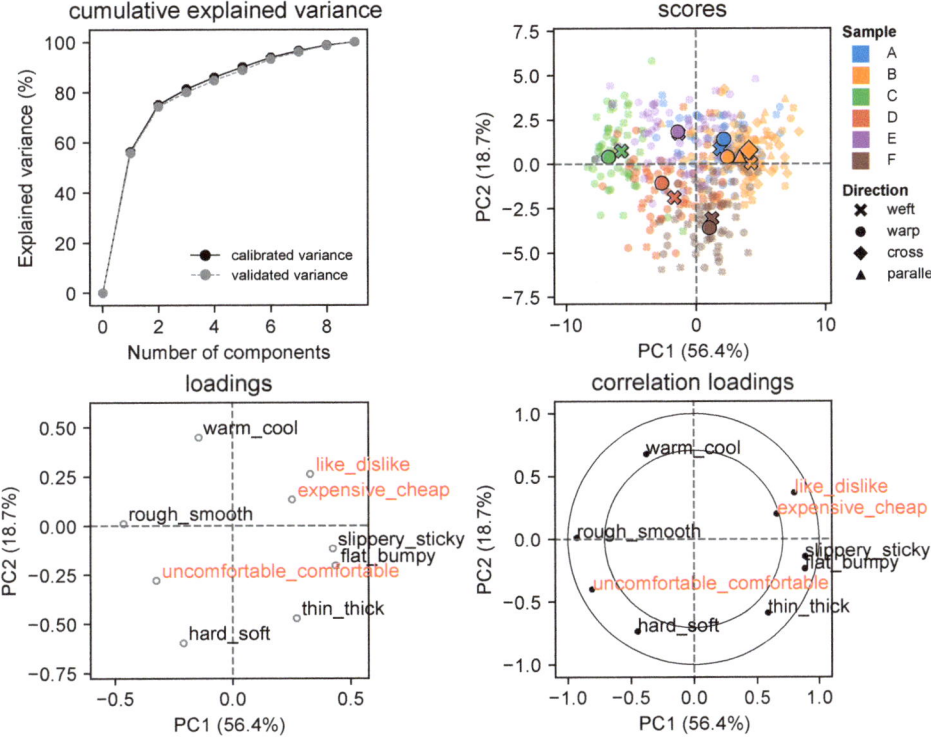

Figure 6. Consonance analysis—sensory data (vertical unfolding). The agreement in the samples can be inspected. Attributes flagged by the ANOVA have been dropped. Cross-validation was performed, leaving one participant out. "Scores" represents the distribution of the samples, as seen by the participants compressed in the first two PCs. Bold colors represent the sample median. "Loadings" represents the attributes of the PCA space. Subjective preference attributes are highlighted in red. "Correlation loadings" is an alternative scaling of the loadings plot, where each original attribute is correlated with the score components. The outer ellipse corresponds to a 100% correlation and the inner to a 50% correlation.

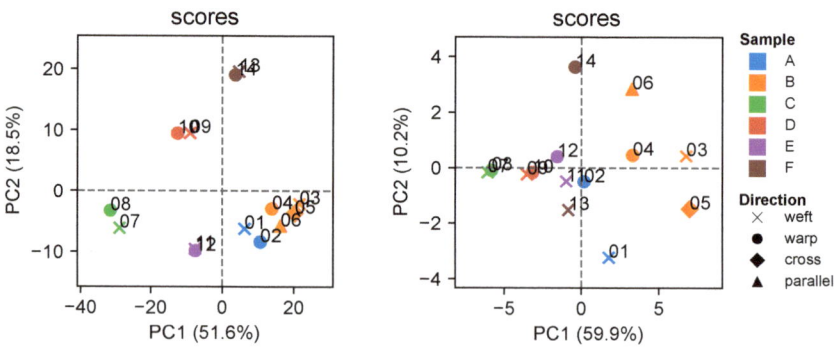

Figure 7. Consonance analysis (scores)—sensory and delta power (horizontal unfolding). The left scores plot corresponds to the sensory data and the right corresponds to delta power.

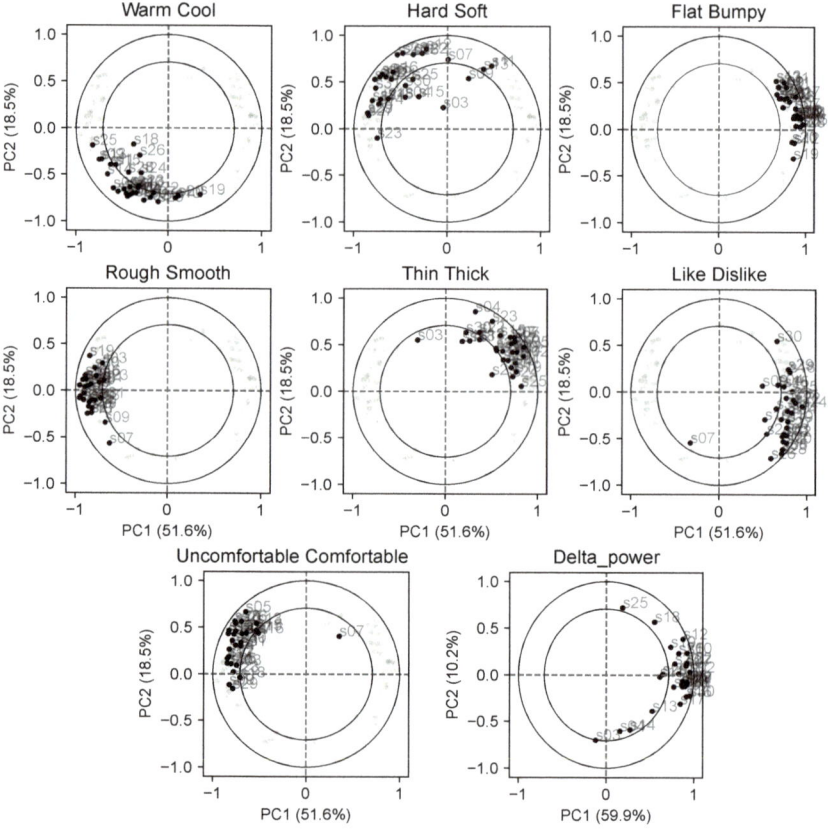

Figure 8. Consonance analysis (loadings)—sensory and delta power (horizontal unfolding). Correlation loadings plot for each sensory attribute. Each graph represents the same data. The same graph is repeated highlighting each sensory attribute at a time. The level of agreement is given by the level of the clustering of the participants.

4. PCR

Regression analysis highlights the relationships among the sensory attributes, physical properties, and delta power. The correlation plots (Figure 9) with 100% and 50% correlation ellipses show which attributes are better at describing the variance in the sensory data. The explained variance values are provided for PCA and regression. The regression-explained variance is the average value for all participants. The relationship between compression, heat transfer, and surface properties depends on the investigated sensory attributes. More importantly, even where agreement exists, the factors driving the variability were slightly different. This is a manifestation of the individuality of participants. The attributes warm/cool and thin/thick were independent of finger vibration. In the case of sensory attributes related to surface properties, the results show that delta power can be a drop-in replacement for physical surface property parameters showing equal or increased explainability.

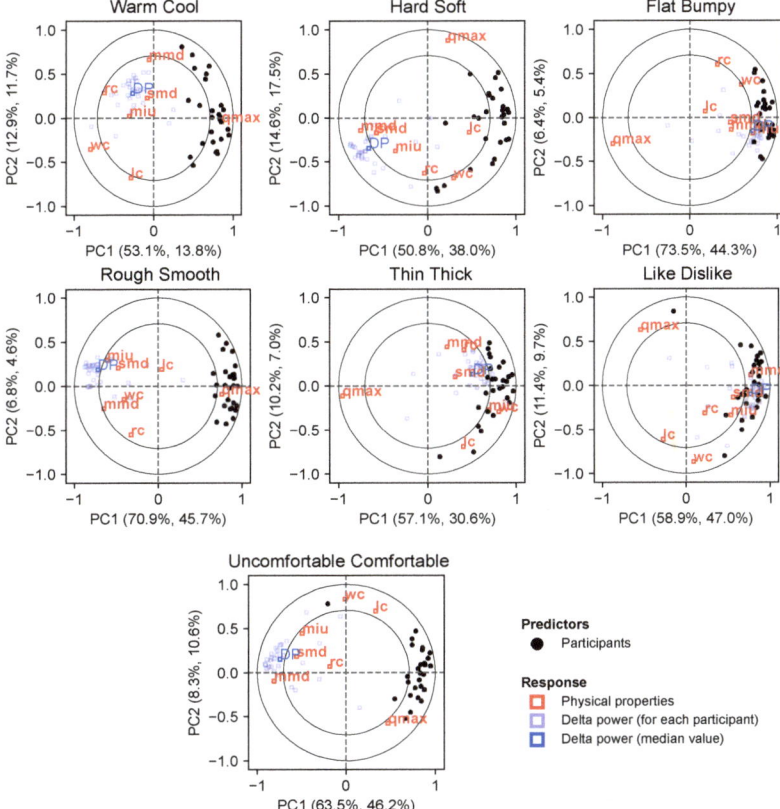

Figure 9. Principal component regression—correlation loadings plots. Physical properties and delta power were regressed onto the principal components of the sensory dataset. The regression coefficients (regression loadings) are plotted in red. The percentages shown on the left indicate the explained variance in the component. On the right, the mean R^2 values for all the response variables in the component are shown. Delta power was regressed without averaging and the medium value (blue) is highlighted. Roughly, values inside the inner ellipse are not significant (p-value > 0.05).

5. Conclusions

In this study, individual tactile perception was analyzed by sliding a finger over a fabric. During this simple finger movement, the relationships between the individual perception, fabric weave structure, physical properties, and hybrid attributes measured at the finger–fabric interface (contact force, speed, and vibration) were investigated. Surface tactile perceptions, such as flat/bumpy, rough/smooth, and slippery/sticky, were influenced by the vibration occurring at the finger–fabric interface. However, warm/cool and thin/thick perceptions were independent of finger vibrations. Skin vibration was shown to be a possible replacement for the surface physical properties measured by the KES-F with equal or slightly improved explainability. Although not significant in this experiment, the interface parameters of contact force and speed could be useful in describing the decision-making characteristics not covered here, for example, the degree of confusion toward a sensory attribute. Finally, individuality, depicted using PCR correlation plots, was shown to be a useful tool to understand sensory data while accounting for participant individuality.

Author Contributions: R.R.S. designed and supervised the experiments, analyzed the data, and drafted the manuscript. M.N. performed the experiments. S.S. guided and supplied feedback and commented on the final format of the paper. All authors have read and agreed to the published version of the manuscript.

Funding: This work was supported by JSPS KAKENHI (grant number 20K02365).

Institutional Review Board Statement: The experiment was approved by the Ethics Committee of the Kyoto Institute of Technology (approval no. 202039).

Informed Consent Statement: Informed consent was obtained from all subjects involved in the study.

Data Availability Statement: Datasets are available upon request.

Conflicts of Interest: The authors declare no conflict of interest.

Abbreviations

The following abbreviations are used in this manuscript:

KES-F	Kawabata Evaluation System for Fabrics
PCA	Principal Component Analysis
DA	Sensory Descriptive Analysis
MIU	Coefficient of Friction
MMD	Mean Deviation of MIU
SMD	Surface Roughness
LC	Linearity of Compression
WC	Compression Energy
RC	Compression Resilience
PSD	Power Spectral Density
PCR	Principal Component Regression
TWFA	Three-way Factor Analysis

Appendix A

Table A1. Characteristic values and standard conditions of measurement of the physical properties.

Property	Symbol	Characteristic Value	Unit	Measurement Condition
Compression	LC	Linearity of compression displacement curve	—	Maximum pressure, Pm, = 5 kPa
	WC	Compression energy	J/m^2	Rate of compression = 20 μm/s
	RC	Compression resilience	%	
Surface	MIU	Coefficient of friction	—	20 steel piano wires with 0.5 mm diameter and 10 mm length.
	MMD	Mean deviation of MIU	—	Contact force = 0.49 N
	SMD	Geometrical roughness	μm	Steel piano wire with 0.5 mm diameter and 5 mm length. Contact force = 0.1 N
Thickness	T_0	Thickness at pressure of 49.0 Pa	mm	
Weight	W	Fabric weight per unit area	g/m^2	
q_{max}	q_{max}	Maximum value of heat flux	W/cm^2	$\Delta T = 10\ °C$

References

1. Ares, G.; Varela, P. Trained vs. consumer panels for analytical testing: Fueling a long lasting debate in the field. *Food Qual. Prefer.* **2017**, *61*, 79–86. [CrossRef]
2. Flora, P.; Schacher, L.; Adolphe, D.C.; Dacremont, C. Tactile feeling: Sensory analysis applied to textile goods. *Text. Res. J.* **2004**, *74*, 1066.
3. Kawabata, S. *The Standardization and Analysis of Hand Evaluation*, 2nd ed.; Textile Machinery Society of Japan: Osaka, Japan, 1980; pp. 23–34.
4. Isami, C.; Kondo, A.; Goto, A.; Sukigara, S. Effects of Viewing Distance on Visual and Visual-Tactile Evaluation of Black Fabric. *J. Fiber Sci. Technol.* **2021**, *77*, 56–65. [CrossRef]
5. Phoophat, P.; Yamamoto, H.; Sukigara, S. Visual aesthetic perception of handwoven cotton fabrics. *J. Text. Inst.* **2019**, *110*, 412–425. [CrossRef]
6. Nagai, Y.; Georgiev, G.V. The role of impressions on users' tactile interaction with product materials: An analysis of associative concept networks. *Mater. Des.* **2011**, *32*, 291–302. [CrossRef]
7. Kimura, K.; Natsume, M.; Tanaka, Y. Influence of Scanning Velocity on Skin Vibration for Coarse Texture. In Proceedings of the International Conference on Human Haptic Sensing and Touch Enabled Computer Applications, Pisa, Italy, 13–16 June 2018; Springer: Berlin/Heidelberg, Germany, 2018; pp. 246–257.
8. Asaga, E.; Takemura, K.; Maeno, T.; Ban, A.; Toriumi, M. Tactile evaluation based on human tactile perception mechanism. *Sens. Actuators A Phys.* **2013**, *203*, 69–75. [CrossRef]
9. Tanaka, Y.; Nguyen, D.P.; Fukuda, T.; Sano, A. Wearable skin vibration sensor using a PVDF film. In Proceedings of the 2015 IEEE World Haptics Conference (WHC), Evanston, IL, USA, 22–26 June 2015; IEEE: Piscataway, NJ, USA, 2015; pp. 146–151.
10. Kanai, R.; Rees, G. The structural basis of inter-individual differences in human behaviour and cognition. *Nat. Rev. Neurosci.* **2011**, *12*, 231–242. [CrossRef] [PubMed]
11. Union of Japanese Scientists and Engineers. *Sensory Evaluation Hand Book*; JUSE Press Ltd.: Tokyo, Japan, 1999.
12. Tucker, L.R. The extension of factor analysis to three-dimensional matrices. *Math. Psychol.* **1964**, *110119*.
13. Næs, T.; Risvik, E. *Multivariate Analysis of Data in Sensory Science*; Elsevier: Amsterdam, The Netherlands, 1996.

Disclaimer/Publisher's Note: The statements, opinions and data contained in all publications are solely those of the individual author(s) and contributor(s) and not of MDPI and/or the editor(s). MDPI and/or the editor(s) disclaim responsibility for any injury to people or property resulting from any ideas, methods, instructions or products referred to in the content.

Article

Prediction of Shrinkage Behavior of Stretch Fabrics Using Machine-Learning Based Artificial Neural Network

Meenakshi Ahirwar * and B. K. Behera

Department of Textile and Fiber Engineering, Indian Institute of Technology Delhi, New Delhi 110016, India
* Correspondence: ttz188455@textile.iitd.ac.in

Abstract: Stretch fabric provides good formability and does not restrict the movement of the body for increased tension levels. The major expectations of a wearer in an apparel fabric are a high level of mechanical comfort and good aesthetics. The prediction of shrinkage in stretch fabric is a very complex and unexplored topic. There are no existing formulas that can effectively predict the shrinkage of stretch fabrics. The purpose of this paper is to develop a novel model based on an artificial neural network to predict the shrinkage of stretch fabrics. Different stretch fabrics (core-spun lycra yarn) with stretch in the weft direction were manufactured in the industry using a miniature weaving machine. A model was built using an artificial neural network method, including training of the data set, followed by testing of the model on the test data set. The correlation of factors, such as warp count, weft count, greige PPI, greige EPI, and greige width, was established with respect to boil-off width.

Keywords: shrinkage behavior; stretch fabrics; weave structure; correlation; boil-off width

Citation: Ahirwar, M.; Behera, B.K. Prediction of Shrinkage Behavior of Stretch Fabrics Using Machine-Learning Based Artificial Neural Network. *Textiles* **2023**, *3*, 88–97. https://doi.org/10.3390/textiles3010007

Academic Editor: Laurent Dufossé

Received: 2 November 2022
Revised: 29 December 2022
Accepted: 3 January 2023
Published: 2 February 2023

Copyright: © 2023 by the authors. Licensee MDPI, Basel, Switzerland. This article is an open access article distributed under the terms and conditions of the Creative Commons Attribution (CC BY) license (https:// creativecommons.org/licenses/by/ 4.0/).

1. Introduction

Stretch textiles, which range from infant clothes to upholstery, provide a better fit for a wide range of products. Some fabrics offer tremendous control, while others offer ease of movement or a comfortable fit. These stretch fabrics can be divided into two types. One type is power stretch, which can be found in support garments, swimsuits, and other items. The other is comfort, sometimes known as action stretch. It can be found in athletic and casual wear, and other products with a lot of mobility and a smooth fit [1,2]. Stretch fabric is a fabric that stretches when force is applied in a single or several directions during normal wear. Power stretching is well known, while comfort stretching is a relatively recent notion [3]. Stretch fabric is made from materials that have specialized extension qualities, with at least 20% elongation in the warp direction [4]. Cotton fabrics are used along with spandex to provide greater stretch and recovery than cotton alone can provide. Cotton/spandex blends are most common in women's clothing, but are now spreading into other product categories, such as knit products, skirts, leggings, and tops, as well as almost all sorts of woven fabrics, such as stretch jeans. Spandex is typically found in woven textiles as a core spun yarn; however, the form in which the spandex is used relies on the fabric construction, performance needs, and designer's knowledge. Cotton knit fabrics with spandex often have stretch qualities ranging from 50 to 100%; this figure is typically 15 to 50% for woven structures.

In general, a 'stretch fabric' is defined as a material with a minimum stretch of 20% in the warp direction. Natural rubber is the main choice for adding elasticity to fibers, and mercerization and texturization are used to process normal woven fabrics to develop elasticity. Since conventional rubber is used, there is a need for other synthetic materials to take on increasing demand, and spandex has emerged as a solution to this problem. A popular way to produce stretch fabrics is to use elastane yarns, because the desired properties can be achieved by using 2–3% of elastane yarns. Spandex-containing stretch fabrics are less stiff, more easily stretchable, and more recoverable than non-spandex fabrics. Value may be added to garments due to spandex by emerging properties such as wash and wear, which add comfort during wear [5].

Stretch fabrics have lower drapability and wrinkle recoverability. Stretch textiles can be compressed more than others and have larger surface volumes when it comes to compression. Furthermore, their higher compression recovery reflects the inherent property of stretch materials as opposed to non-stretch fabrics in general [6]. Stretch fabric is either a two-way stretch or a four-way stretch. According to researchers, a stretch range of 20 to 45% from the wearer's comfort viewpoint depends on the end use [7]. Stretch fabrics can be woven as well as knitted fabrics. The various methods used in the production of stretch fabric can be listed as slack, the mercerization of fabrics, weaving fabric with high twist yarn, using textured synthetic yarn, using covered rubber yarns, using spandex yarns, and so on [8]. In one study, the linear density of weft threads and the preliminary tension of the ground yarn, which produced a pillar stitch, were found to affect the stretch qualities of elastic warp knitted fabrics, the full deformation, and the constituent parts [9].

As spandex is generally used in the weft direction in stretch fabrics, the properties based on weft input become crucial in testing the performance of stretch fabrics. These parameters include fabric growth, fabric stretch, and elastic recovery of the stretch fabric. Stretch fabric is popular among sportswear companies. Stretch fabrics have a wide range of applications in sports apparel, including yoga suits, bicycle shorts, skiwear, swimsuits, leotards, sports bras, and all other types of apparel designed for intense stretching of body parts.

Cotton and spandex fabrics will always default to a lower energy level. Relaxing the cloth from intrinsic construction and processing tensions improves dimensional stability. The boil-off bath should contain a good textile detergent or the scouring agents to be used in processing. A useful way to quickly determine the relaxed state is to "boil-off" a two yard long, full-width sample. Following the boil-off, the samples can be dried in a sample oven and the fabric width measured. This is the absolute smallest width to which the fabric will shrink. This understanding will prohibit the finisher from attempting to complete the fabric at a width less than the relaxed width.

Stretch fabric, when subjected to heat and moisture during processing, tends to release its internal stresses and shrink to a certain extent. This shrinkage is dependent on the following parameters: yarn count, fabric construction (EPI, PPI), weave structure, the percentage content of spandex in the fabric, loom width, and so on. Shrinkage occurs when moisture, heat, and mechanical action (movement during fabric formation, drying, and washing) are mixed [3,10–12]. The combination of these factors causes the fibers to release the stresses created during the fabrication of the stretch fabric. Higher moisture content also increases the rate of shrinkage. There are two methods to control fabric shrinkage. First, shrinkage is appropriately controlled at feeding and is reduced with length and width to apply clothing with lower or greater feed speed. The maximum is controlled using 70–75% shrinkage. Second, shrinkage fiber is raised to its original position in the thermoplastic and cooled after the reaction of a hot steam process [13–15]. This process will mold the fiber to its original position. Shrinkage percentage plays a very important role in stretch fabric manufacturing as the stretch of the fabric is indirectly dependent on it, and stretch is very important in specific end-use fabrics, such as sportswear and infant wears. Therefore, it is very essential to predict the shrinkage percentage of stretch fabrics in the manufacturing textile industries.

In one study based on the basic fabric constructional variables, it was discovered that an artificial neural network can accurately assess the tear strength of a bed sheet fabric. The weave structure of the material was discovered to be significantly related to the tearing strength properties. Including the weave factor parameter as a training input can improve prediction performance even further. Therefore, artificial neural networks have great importance in the textile industry for advanced fabric manufacturing, product development, etc., wherein developing new specialty fabrics is vital (Ahirwar and Behera, 2022).

Artificial intelligence (AI) is a machine's abiltiy to carry out tasks in a way that would be considered smart. The AI approach is founded on the assumption that data are fed into machine learning models which then gain intuition on their own, with no manual intervention [16–19]. ANNs, or artificial neural networks, are forms of neural networks based on a collection of units called artificial neurons which are connected to the next layer's units, as shown in Figure 1. Arrows represent the connection or interchange of information through an arrow from one layer to another, and on the other side, we obtain the output of our neural network [20–23]. This design is based on our brains. Numerical information from these neurons passes on through deeper layer neurons.

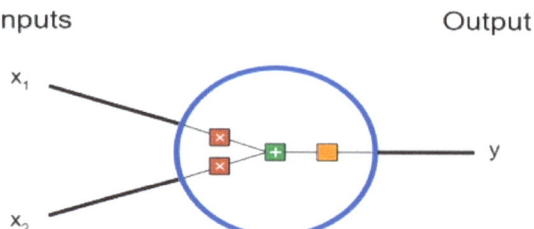

Figure 1. Two input, one output model.

Algorithms in machine learning are primarily based on methods to easily process a wide variety of data that are difficult to process manually. In the production of stretch fabric, stretch yarns were used. These yarns are very sensitive to low-standard spinning methods compared with core-spun yarns. Due to improper heat settings or machine parameters, some problems may occur, such as variation of dyeability or shrinkage throughout the fabric or color fading or stripe effect. Since a two-dimensional fabric is used to mimic the surface of a three-dimensional body, the result can be problematic, and the main important thing is the shape and the frequent posture of the person and how our fabric performs in that condition. The shrinkage in the stretch fabric is dependent on every factor that comes into account in its production. A lot of these variables are unable to derive a formula for shrinkage manually. Additionally, other studies predict fabric properties that are hard to predict manually using machine learning models. This gives us a direction of research for predicting shrinkage using machine learning. In light of the above discussion, the present study focuses on predicting the shrinkage of stretch fabrics by using machine-learning-based artificial neural networks.

2. Materials and Methods

2.1. Materials

Stretch fabrics (core-spun lycra yarn) with stretch in the weft direction were manufactured by Vardhman Industries, India. The fiber blends used were 96–98% cotton and 2–4% elastane. Different samples were manufactured in an industry using a miniature weaving machine. The parameters and their numerical range are given in Table 1 and were used as the data points in the artificial neural network.

2.2. Methodology

2.2.1. Process Sequence

The grey fabrics were processed using the standard process settings used in commercial production, as shown below.

Grey fabric → Grey wash → Heat Set→ Pre-treatment → Mercerization → Peach → Wash → Dry pad → Pad steam → Finish→ Dry

Table 1. Particulars of fabric samples.

Parameters	Minimum	Maximum	Average
Warp count (Ne)	7	50	22
Weft count (Ne)	6	50	18
Spandex in weft core (Ne)	7	140	68
Ends/inch (EPI)	50	186	101
Picks/inch (PPI)	38	108	65
Greige width (cm)	125	210	180
Boil-off width (cm)	38	177	130
Areal density (g/m^2)	96	327	199

2.2.2. Artificial Neural Network to Predict Shrinkage

Machine learning-based artificial neural networks, are now widely being used to forecast the performance of any parameter. The artificial intelligence (AI) concept is based on the idea that data are supplied to machine learning models which then learn intuitively, on their own, with no manual intervention [16–19]. Artificial neural networks are varieties of neural networks based on a collection of units called artificial neurons that are connected to the next layer's units [20–23].

The important steps in the ANN methodology include the building of the model, training, and testing of data points, and authentication of the model. These steps are to be performed very specifically to obtain precise results. The characterization of properties is considered for the shrinkage of stretch fabrics. The input parameters, such as yarn count, construction parameters, weave structure, and on-loom width were measured. Spandex% was also taken into consideration. Shrinkage% is measured using the ANN model. These five input parameters and one output parameter data were stored in MS Excel and were fed to the model accordingly. As weave structure cannot be represented in a numerical value, weave factor is used as a numerical representation of the weave structure. It is the number of interlacements of warp and weft in each repeat. The formula used for weave factor is:

$$M = E/I$$

E = number of threads per repeat;
I = number of interactions per repeat.

In the case of irregular weave patterns, the weave interlacements of warp and weft yarns may be different; hence, two weave factors, M_1 and M_2, are considered:

$$M_1 = E_1/I_2 \qquad M_2 = E_2/I_1$$

where E_1 and I_2 can be measured by observing a pick in fabric, and E_2 and I_1 can be measured by observing ends in the repeat.

2.2.3. Significance of Backpropagation

Backpropagation is a learning technique used by artificial neural networks to generate a gradient descent for weights. A comparison is made between the desired and realized system outputs, and then connection weights are adjusted to reduce the discrepancy as much as is feasible. In this method, weights are changed in reverse order, from output to input. The different factors that need to be taken into consideration while training a machine learning model are very important, as they can significantly affect the prediction. A wide range of factors that could affect the shrink age behavior were considered, and then later ranked based on their affectability. The different input parameters selected here are, warp and weft count, ends/inch, and picks/inch, spandex percentage, and greige width of the fabrics. Following that, the fabric was cleaned and preprocessed using several scikit-learn data pre-processing packages. Preprocessing data is an important step in machine learning since it enhances the quality of the data and makes it easier to extract meaningful information. Data preprocessing in machine learning is the process of preparing raw data for use in the building and training of machine learning models. Data pre-processing in machine learning is essentially a data mining strategy that converts raw data into a more comprehensible and legible form. Standardization and normalization are used to ensure the accuracy of a dataset. As the precursor to many machine learning models, standardization is an important pre-processing step that is used to equalize the range of features in the input data set.

2.2.4. Building a Machine Learning Model to Predict Shrinkage Behavior Using ANN

Traditionally, neural networks have only three layers, which are the input layers, hidden layers, and output layers. Input layers are fed external data and, output layers feed their data to an external output, which is used in prediction. All these three layers are commonly referred to as dense layers. All the neurons are connected to each neuron in the next layers, and the output neuron has no forward connection to another layer, as shown in Figure 2.

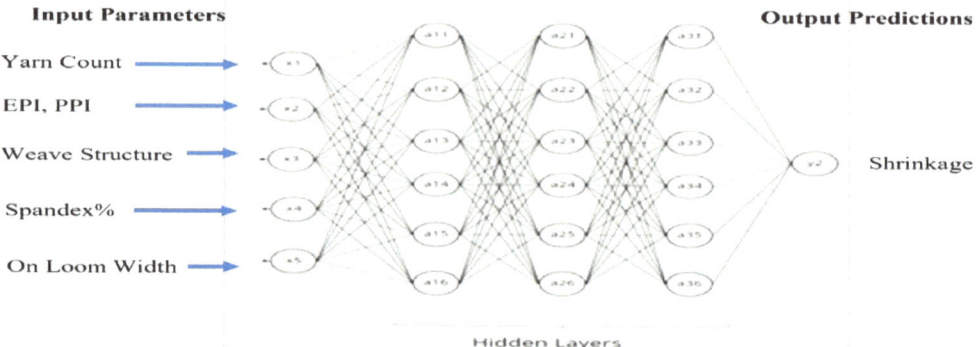

Figure 2. Model architecture for shrinkage prediction.

The architecture of the model is very important for the prediction of numerical output using machine learning. This architecture depends on how much depth the input data are related to the output data. The number of interrelated parameters might be the best choice for the number of deep layers in the neural network. The deeper the layer, the greater the number of weights required to manage a deeper neural network, which can result in predicting a good result. Four layers of neural networks were chosen, three of which will be hidden. Five neurons are chosen for each layer, and the output neuron will have shrinkage as an output. The number of epochs will be taken as four initially and will be changed as the model becomes accurate.

2.2.5. Training the Artificial Neural Network Model

Training a model for predictive analysis gives the model an intuition about how the output parameters will depend on input parameters. Data points are divided into two sets: one is a train set, and the other is a test set. Train sets were used to train the model, and a test set to test the model for unseen data and prediction. The number of times this process is repeated is the number of epochs. A higher number of epochs will increase time but also increase accuracy for current data. Training a model takes most of the time, and models are generally deployed on cloud storage, such as AWS and Google Cloud. These cloud storages are also equipped with TPU (tensor processing unit), which is very good for matrix multiplication and other operations. Nvidia GeForce GTX 1050 was used in this study.

2.2.6. Testing the Model on a Test Data Set

After building and training the model, the model accuracy was tested. The data set was divided into a test set and a train set. Twenty percent of the data were used for the test set. The model will predict the accuracy of model prediction on test data for each epoch. If low accuracy is obtained for the train set, then the time hyperparameters are changed accordingly, and if both the train and test set have low accuracy, then a different architecture for the model is to be developed. This will determine how the model predicts the test set data.

2.2.7. Authentication and Using the Model for the Prediction of a New Data Set

Experimental testing data are used for evaluating the accuracy of the model. This will give an idea of the error of the model concerning real-world scenarios, and corrections will be incorporated according to the type of error. These errors are mainly there if there is an unknown factor that exists, but is not in our inputs. Generally, these errors are negligible, but in case of very high accuracy, this can be added to the model as an input. The development of stretch fabric is an important factor as the prediction of the model's performance is based on these results. The error in the development of the fabric will lead to an error in the evaluation of the model. The error measurement of the machine learning model was used with respect to experimental results as a factor in this evaluation. If there are no significant results in the model, then the model architecture and parameters are reshaped. Evaluating the model is the final step in this study. Statistical results, such as the p-test, were used to test our model's accuracy and significance. The evaluation gives an idea about how much the model is reliable and precise in a real-world scenario.

3. Results and Discussion

3.1. Error Backpropagation Algorithm for Shrinkage Percentage

The algorithm related to the error backpropagation method to determine shrinkage percentage is given below.

The error signal at the output neuron j at the iteration n is defined by $e_j(n) = d_j(n) - y_j(n)$.

The internal activity level $v_j(n)$ produced at the input $v_j(n) = \sum_{i=0}^{p} w_{ji}(n) y_i(n)$.

The output of the neuron j at iteration n is $y_j(n) = \varphi_j(v_j(n))$.

The local gradient is $\delta j(n) = [d_j(n) - o_j(n)] o_j(n) [1 - o_j(n)]$.

For the hidden neuron j, the local gradient is $\delta j(n) = y_I(n) [1 - y_I(n)] \sum_k \delta_k(n) w_{kj}(n)$.

The correction $\Delta w_{ji}(n)$ applied to the synaptic weight is $\Delta w_{ji}(n) = \eta \times \delta j(n) \times y_i(n)$.

In forward pass, the synaptic weights remain unaltered throughout the network. In nonlinear functions, the below functions were included:

The log function is $y_I(n) = 1/(1 + \exp(-v_j(n))$.

The hyperbolic tangent function is $\varphi_j(v) = 2a/(1 - \exp(bv)) - a$.

The momentum constant is $\Delta w_{ji}(n) = \alpha \Delta w_{ji}(n-1) + \eta \, \delta j(n) \, y_i(n)$.

3.2. Correlation of Factors with Respect to Boil-Off Width

The correlation of factors with respect to boil-off width was established. The factors include warp count, weft count, greige PPI, greige EPI, and greige width. The R-value and p-value of fabrics were found with respect to boil-off width, and are given in Table 2. The R-value donates the correlation between the boil-off width and the different construction parameters of the fabric. The error matrices obtained from the machine learning model with respect to experimental shrinkage and shrinkage predicted by using Equations (1)–(4) are given in Table 3. The graphs depicting the correlation of warp and weft count with respect to boil-off width are shown in Figure 3. A low degree of negative correlation was observed between the boil-off width and warp count of 100% cotton. This means that as the warp count increases and the fineness of the yarn increases, the boil-off width decreases. A low degree of negative correlation was also observed between the boil-off width and weft count of cotton-spandex. As the weft count increases and the fineness of the yarn increases, the boil-off width decreases, and it can be seen from Figure 3 that the points are scattered over the region, depicting low correlation. The graphs depicting the correlation of greige EPI and PPI with respect to boil-off width are shown in Figure 4. A high degree of negative correlation was found between EPI and boil-off width. As the greige EPI increases, the boil-off width of the fabric decreases. A high degree of positive correlation was found between PPI and boil-off width. As the greige PPI increases, the boil-off width of the fabric also increases. The graph depicting the correlation of spandex percentage with boil-off width is shown in Figure 5. A low degree of negative correlation was observed between boil-off width and spandex percentage in weft core. This means that as the percentage of spandex in the weft yarn increases, the boil-off width of the fabric decreases. The graph depicting the correlation of greige width with boil-off width is shown in Figure 6. A high degree of positive correlation was observed between boil-off width and greige width. As the greige width of the fabric increases, the boil-off width of the fabric also increases. It can be witnessed from Figure 6 that a cluster of dense points occurs in the graph, depicting high correlation.

Table 2. R-value and *p*-value of fabrics with respect to boil-off width.

Properties	R-Value	*p*-Value
Warp count	−0.089	0.002
Weft count	−0.048	0.088
Spandex%	−0.119	2.136
Ends/inch	−0.040	0.150
Picks/inch	0.071	0.012
Greige width	0.635	1.184
M1	−0.168	2.042
M2	−0.119	2.265

Table 3. The error values for experimental and predicted shrinkage.

Errors	Values
Mean squared error	48.64
Mean absolute error (%)	4.21
Mean standard error	48.60
Root mean squared error	6.97

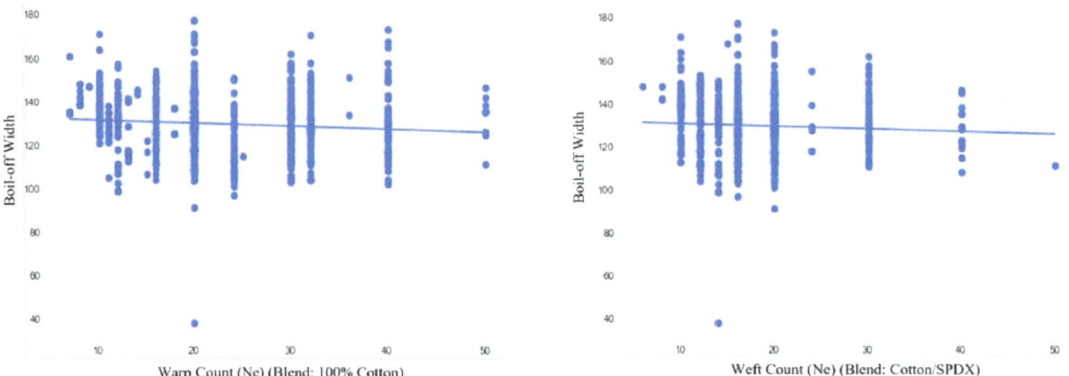

Figure 3. Warp and weft count with respect to boil-off width.

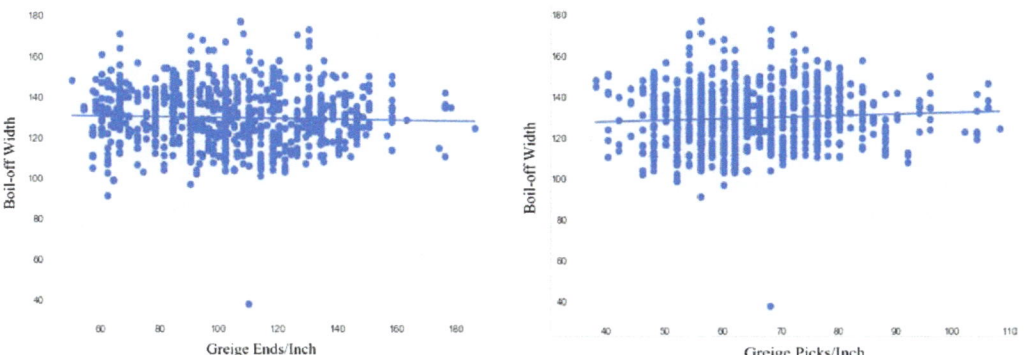

Figure 4. Greige EPI and PPI with respect to boil-off width.

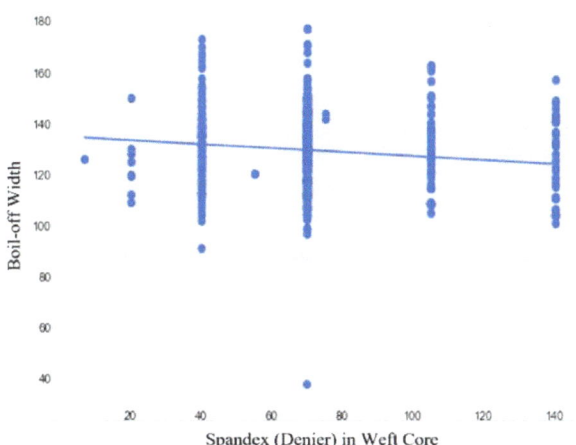

Figure 5. Spandex% with respect to boil-off width.

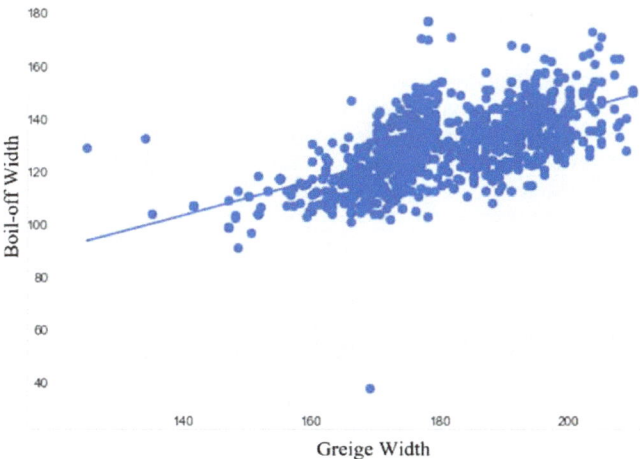

Figure 6. Greige width with respect to boil-off width.

Using an error measure, one may determine the forecasting model error. These are used to objectively compare the performance of opposing models. The error metrics that are employed here include MSE, RMSE, MAE, and MSE root mean squared error (RMSE) (SE). The correlation coefficient "r" is a unit-free value between -1 and 1. The statistical significance is indicated with a p-value.

Mean squared error (MSE) (average of the squares of the errors):

$$MSE = \frac{1}{n}\sum_{i=1}^{n}(Yi - \overline{Y}i)^2 \quad (1)$$

where n is the number of data points, Yi is the observed value, and $\overline{Y}i$ is the predicted value.

Mean absolute error (MAE) (a measure of errors between paired observations expressing the same phenomenon):

$$MAE = \frac{\sum_{i=1}^{n}|y_i - x_i|}{n} \quad (2)$$

where n is the number of data points, y_i is the predicted value, and x_i is the true value.

Mean standard error (SE) (standard deviation of its sampling distribution):

$$SE = \frac{\sigma}{\sqrt{n}} \quad (3)$$

where n is the number of observations, and σ is the standard deviation.

Root mean squared deviation (RMSD) is a measure of the differences between values (sample or population values) predicted by a model:

$$RMSD = \sqrt{\frac{\sum_{i=1}^{N}(xi - \overline{x}i)^2}{N}} \quad (4)$$

where i is a variable, n is the number of data points, xi is the observed value, and $\overline{x}i$ is the predicted value.

The mean absolute error percentage was found to be 4.21, which is within the acceptable range; the root mean squared error was also within the acceptable bounds. As a result, this model can accurately forecast the shrinkage behavior of stretch fabrics, allowing industries to use this tool to predict shrinkage behavior prior to the beginning of actual manufacturing. This is an advantageous tool for product development, quality assurance, and advanced fabric manufacturing.

4. Conclusions

It was determined that the artificial neural network can accurately forecast stretch fabric qualities based on basic fiber characteristics and fabric constructional parameters. The main input parameters that were considered to influence fabric shrinkage percentage were yarn count, construction of fabric (EPI, PPI), weave structure, the percentage content of spandex in the fabric, and the on-loom width. These factors have a significant effect on shrinkage percentage, which is predicted through the ANN model. The network prediction correlates well with the actual experimental results. The prediction of fabric qualities from constructional parameters has some uncertainty. The shrinkage properties are highly correlated with the weave structure of the fabric. The network projected opposing trends in a few cases, which are difficult to explain. However, the model prediction accuracy on training data reached 99.2%; therefore, using ANN is effective for numerical prediction of fabric parameters and determination of shrinkage percentage.

Author Contributions: Conceptualization, methodology, software, validation, formal analysis, writing, M.A.; visualization, supervision, project administration, B.K.B. All authors have read and agreed to the published version of the manuscript.

Funding: This research received no external funding.

Institutional Review Board Statement: Not applicable.

Informed Consent Statement: Not applicable.

Data Availability Statement: No data linked to this research.

Conflicts of Interest: The authors declare no conflict of interest.

References

1. Pannu, S.; Ahirwar, M.; Jamdigni, R.; Behera, B.K. Effect of Spandex Denier of Weft Core Spun Yarn on Properties of Finished Stretch Woven Fabric. *Int. J. Eng. Technol. Manag. Res.* **2020**, *7*, 21–32. [CrossRef]
2. Ahirwar, M.; Behera, B.K. Objective Hand Evaluation of Stretch Fabrics Using Artificial Neural Network and Computational Model. *J. Nat. Fibers* **2022**, *19*, 13640–13652. [CrossRef]
3. Aratani, Y.; Kojima, T. Stretch Fabrics. *Sen'i Gakkaishi* **1984**, *40*, 352–355. [CrossRef]
4. Varghese, N.; Thilagavathi, G. Development of woven stretch fabrics and analysis on handle, stretch, and pressure comfort. *J. Text. Inst.* **2015**, *106*, 242–252.
5. Ishimaru, S.; Isogai, Y.; Matsui, M.; Furuichi, K.; Nonomura, C.; Yokoyama, A. Prediction method for clothing pressure distribution by the numerical approach: Attention to deformation by the extension of knitted fabric. *Text. Res. J.* **2011**, *81*, 1851–1870. [CrossRef]
6. Kim, H.A.; Ryu, H.S. Hand and Mechanical Properties of Stretch Fabrics. *Fibers Polym.* **2008**, *9*, 574–582. [CrossRef]
7. Gurarda, A.; Meric, B. The Effects of Silicone and Pre-Fixation Temperature on the Elastic Properties of Cotton/Elastane Woven Fabrics. *AATCC Rev.* **2005**, *5*, 53–56.
8. Choi, M.S.; Ashdown, S.P. Effect of Changes in Knit Structure and Density on the Mechanical and Hand Properties of Weft-Knitted Fabrics for Outerwear. *Text. Res. J.* **2000**, *70*, 1033–1045. [CrossRef]
9. Kyzymchuk, O.; Melnyk, L. Stretch properties of elastic knitted fabric with pillar stitch. *J. Eng. Fiber Fabr.* **2018**, *13*, 1558925018820722. [CrossRef]
10. Varghese, N.; Thilagavathi, G. Handle, fit and pressure comfort of silk/hybrid yarn woven stretch fabrics. *Fibers Polym.* **2016**, *17*, 484–494. [CrossRef]
11. Ogulata, S.N.; Sahin, C.; Ogulata, R.T.; Balci, O. The prediction of elongation and recovery of woven bi-stretch fabric using artificial neural network and linear regression models. *Fibres Text. East. Eur.* **2006**, *14*, 46–49.
12. Kaynak, H.K. Optimization of stretch and recovery properties of woven stretch fabrics. *Text. Res. J.* **2017**, *87*, 582–592. [CrossRef]
13. Majumdar, P.K.; Majumdar, A. Predicting the breaking elongation of ring spun cotton yarns using mathematical, statistical, and artificial neural network models. *Text. Res. J.* **2004**, *74*, 652–655. [CrossRef]
14. *ASTM D3107-07*; Standard Method of Test for Stretch Properties of Fabrics Woven from Stretch Yarns. ASTM: Philadelphia, PA, USA, 1975.
15. Senthilkumar, M.; Anbumani, N.; Hayavadana, J. Elastane fabrics—A tool for stretch applications in sports. *Indian J. Fibre Text. Res.* **2011**, *36*, 300.
16. Behera, B.K.; Militky, J.; Mishra, R.; Kremenakova, D. Modeling of Woven Fabrics Geometry and Properties. In *Woven Fabrics*; IntechOpen: Rijeka, Croatia, 2012.
17. Tsai, Y.J. The influence of woven stretch fabric properties on garment and pattern design. *Taiwan Text. Res. J.* **2006**, *16*, 55–61. [CrossRef]
18. Park, S.W.; Hwang, Y.G.; Kang, B.C.; Yeo, S.W. Applying Fuzzy Logic and Neural Networks to Total Hand Evaluation of Knitted Fabrics. *Text. Res. J.* **2000**, *70*, 675–681. [CrossRef]
19. Noor, A.; Saeed, M.A.; Ullah, T.; Uddin, Z.; Ullah Khan, R.M.W. A review of artificial intelligence applications in apparel industry. *J. Text. Inst.* **2022**, *113*, 505–514.
20. Nwobi-Okoye, C.C.; Anyichie, M.K.; Atuanya, C.U. RSM and ANN Modeling for Production of Newbouldia Laevies Fibre and Recycled High Density Polyethylene Composite: Multi Objective Optimization Using Genetic Algorithm. *Fibers Polym.* **2020**, *21*, 898–909.
21. Raheel, M.; Liu, J. An Empirical Model for Fabric Hand: Part I: Objective Assessment of Light Weight Fabrics. *Text. Res. J.* **1991**, *61*, 31–38. [CrossRef]
22. Park, S.W.; Hwang, Y.G.; Kang, B.C.; Yeo, S.W. Total handle evaluation from selected mechanical properties of knitted fabrics using neutral network. *Int. J. Cloth. Sci. Technol.* **2001**, *13*, 106–114. [CrossRef]
23. Raheel, M.; Jiang, L. An Empirical Model for Fabric Hand. *Text. Res. J.* **1991**, *61*, 79–82. [CrossRef]

Disclaimer/Publisher's Note: The statements, opinions and data contained in all publications are solely those of the individual author(s) and contributor(s) and not of MDPI and/or the editor(s). MDPI and/or the editor(s) disclaim responsibility for any injury to people or property resulting from any ideas, methods, instructions or products referred to in the content.

Article

Design Elements That Increase the Willingness to Pay for Denim Fabric Products

Ryoga Miyauchi [1], Xiaoxiao Zhou [1] and Yuki Inoue [2,*]

[1] Graduate School of Humanities and Social Sciences, Hiroshima University, Hiroshima 730-0053, Japan
[2] Faculty of Economics, Hosei University, Tokyo 194-0298, Japan
* Correspondence: yuki.inoue@hosei.ac.jp; Tel.: +81-42-783-2552

Abstract: This study analyzed what design elements are attractive to consumers of denim fabric products. A questionnaire survey was used to investigate the brands and design elements that consumers prefer. Subsequently, the degree to which participating consumers liked the five design elements (traditional, transformative, pattern, multi-material, and decorative designs), fast fashion brands, and luxury brands were used as explanatory variables to determine the consumers' willingness to pay. A multiple regression analysis was performed on these variables. The results indicated that consumers who preferred traditional and transformative designs showed a positive effect on their willingness to pay for denim fabric products. Therefore, these elements could be attractive design elements that may command a high price point in new product planning proposals. Moreover, depending on the type of brand preferred by consumers, the impact of design elements on their purchase intention of denim fabric products has different consequences. This study analyzes the design elements preferred by consumers and contributes to the creation of design proposals by designers and apparel firms.

Keywords: denim fabric products; product design; luxury brand; fast fashion brand; fashion design

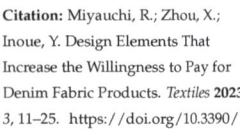

Citation: Miyauchi, R.; Zhou, X.; Inoue, Y. Design Elements That Increase the Willingness to Pay for Denim Fabric Products. *Textiles* **2023**, *3*, 11–25. https://doi.org/10.3390/textiles3010002

Academic Editor: Laurent Dufossé

Received: 22 November 2022
Revised: 28 December 2022
Accepted: 2 January 2023
Published: 5 January 2023

Copyright: © 2023 by the authors. Licensee MDPI, Basel, Switzerland. This article is an open access article distributed under the terms and conditions of the Creative Commons Attribution (CC BY) license (https://creativecommons.org/licenses/by/4.0/).

1. Introduction

Textiles are constantly evolving, with new types of textiles being developed recently [1–3]. The materials used in such textiles have also evolved [4], but traditional fabrics, such as denim, still endure. In the denim fabric product market, design is constantly evolving. Denim fabric products that were originally worn as work clothes have now been transformed into fashion products worn by a variety of consumers [5]. The determinants of consumers' intention to purchase denim fabric products are diversifying [6]. In addition to the quality aspects that consumers value [7], the aspects of design and aesthetic appeal are also complex [6]. The innovative and stylish features of denim fabric products underscore the importance of creative excellence in their development [8], and the competitiveness in this market has steadily increased over the last 50 years [5].

The denim fabric product market has established itself as a fashion product market, emphasizing the importance of creativity, although appropriate design elements for consumers are unclear [8]. Design is defined by three dimensions: product aesthetics, functionality, and symbolism [9,10]. Consumers are increasingly making brand choices based on products' aesthetic and symbolic value [11]. Functionality is also important in design [12]. Based on these three dimensions, we selected five elements: (1) traditional design (functionality), (2) transformative design (aesthetic), (3) pattern design (aesthetic), (4) multi-material design (aesthetic), and (5) decorative design (symbolism).

In addition to these five design elements (traditional, transformative, pattern, multi-material, and decorative designs), we focused on two brand types: fast fashion and luxury brands which have contrasting price ranges. We aimed to verify the influence of these five design elements and two brand types on consumers' willingness to pay (WTP) for denim products. WTP is effective in measuring the direct product value of the consumer market [13]. We attempted to answer the following research question: "What kind of product design is attractive to what kind of consumers in the denim fabric product market, or can be attractive?" Thus, the purpose of this study is to clarify to consumers the design elements suitable for denim fabric products.

The novelty of this study lies in its examination of the impact of design elements on consumers' WTP. Regarding consumers' WTP, Rahman's [14] study examined the impact of garment fit, body image, and appropriateness on consumers' purchase desire for jeans. In addition, Card et al. [15] examined the impact of laundering on the physical properties of denim to facilitate the design of denim garments to meet consumer needs. However, the impact of design elements on purchase desire has not been studied. Thus, this study fills a gap in research on the impact of design elements on consumers' purchase desires.

1.1. Hypothesis

We hypothesized the impact of consumers' preference for the five elements on their WTP for denim fabric products. In addition, consumers' favorite brands may affect their purchasing intention. Thus, we hypothesized the impact of consumers' preference for fast fashion or luxury brands and the five design elements mentioned above on their WTP for denim fabric products.

1.1.1. Traditional Design

Traditional design refers to maintaining the original design. Since the latter half of the 20th century, we have witnessed a vintage boom of denim fabric products [16]. Other factors in these vintage trends are changing values and the inclusion of vintage inspiration used by fashion designers in current designs [17]. These factors allow consumers to purchase the original reproductions of denim fabric products [5]. Thus, denim fabric products may influence consumers' purchasing intentions by adopting traditional design elements. Therefore, we proposed the following hypothesis:

H1a: *Consumers who prefer traditional design elements show high WTP for denim fabric products.*

1.1.2. Transformative Design

Transformative design refers to an unconventional design structure. "Transformative" refers to changing the shape or structure to something else without losing the substance [18]. Traditionally, denim's toughness and comfort were incompatible, but breakthrough innovations, such as washing and developing technology for commercialization, have managed to change the equation [15,19]. In addition, designers of luxury fashion brands have made developments to bring out the aesthetic appeal of unique fabrics [5]. Therefore, denim fabric products may influence consumers' purchasing intentions by adopting transformative design elements. Therefore, we proposed the following hypothesis:

H1b: *Consumers who prefer transformative design elements show high WTP for denim fabric products.*

1.1.3. Pattern Design

Pattern design refers to the application of a pattern. The use of certain patterns improves the appearance of a product for its aesthetic appeal [20,21]. With the improvement in aesthetic appeal and the icon of a brand, pattern design has come to be evaluated as a major factor influencing the purchasing motivation of current consumers [21]. Therefore, the adoption of pattern design in denim fabric products may influence consumers' purchasing intentions. Therefore, we proposed the following hypothesis:

H1c: *Consumers who prefer pattern design elements show high WTP for denim fabric products.*

1.1.4. Multi-Material Design

Multi-material design refers to the combination of various materials. The multi-material design of denim fabric products incorporates fabrics other than denim. In the fashion context, multi-material is positioned as a means to develop new designs [22]. Denim fabric products maintain their original uniqueness and expand the range of fashion designs by embodying combinations with new fabrics [23]. The adoption of a multi-material design in denim fabric products may influence consumers' purchasing intentions. Therefore, we proposed the following hypothesis:

H1d: *Consumers who prefer multi-material design elements show high WTP for denim fabric products.*

1.1.5. Decorative Design

Decorative design refers to adding decorations, such as embroidery and prints. Consumers' self-expression and personal fashion appeal became more important as designers began creating denim fabric products as fashion products [14]. Therefore, consumers add decorative design elements to their denim fabric products that are also worn as a form of self-expression. As a result, decorative design has come to be positioned by various designers as a means to expand and disrupt the market. Denim fabric products have been developed by consumers and designers through the addition of decorative designs. Therefore, the adoption of decorative designs in denim fabric products may enhance the product value to consumers. Therefore, we proposed the following hypothesis:

H1e: *Consumers who prefer decorative design elements show high WTP for denim fabric products.*

1.1.6. Fast Fashion Brands

Fast fashion brands capture the latest consumer trends by combining rapid response production capabilities with enhanced product design capabilities [24]. Trendy design, consistent quality, quick delivery, and fast arrival to the market have been made possible by the following features: low cost due to mass production, an increased number of fashion seasons, and structural changes [25]. The design, abundant choices, and affordability offered by fast fashion brands are factors that drive consumers to buy their products [26]. Hence, fast fashion brands propose attractive products to consumers. Moreover, their products are reasonably priced. When purchasing denim fabric products, consumers' preferences for fast fashion brands may affect their WTP. Therefore, the following hypotheses were proposed:

H2a: *Consumers who prefer traditional design elements and fast fashion brands decrease their WTP for denim fabric products.*

H2b: *Consumers who prefer transformative design elements and fast fashion brands decrease their WTP for denim fabric products.*

H2c: *Consumers who prefer pattern design elements and fast fashion brands decrease their WTP for denim fabric products.*

H2d: *Consumers who prefer multi-material design elements and fast fashion brands decrease their WTP for denim fabric products.*

H2e: *Consumers who prefer decorative design elements and fast fashion brands decrease their WTP for denim fabric products.*

1.1.7. Luxury Brands

Luxury brands are the opposite of fast fashion brands in terms of their price, creative design, and product value. Luxury brands propose designer-driven designs, unlike the consumer-driven designs of fast fashion brands [27]. Luxury brands propose to consumers the styles of modernity, eccentricity, luxury, elitism, and strength through their unique design, variety of materials, and advanced technology [28]. Such styles proposed by luxury brands motivate consumers to purchase their products [29,30]. In addition, luxury brands have a significant impact on consumers' WTP in terms of practical fashion lifestyle and perceived practical value [31]. Hence, by proposing original and creative designs from designers to consumers at a high price point, luxury fashion brands will increase consumers' WTP. When purchasing denim fabric products, consumers' preferences for luxury brands may affect their WTP. Therefore, the following hypotheses were proposed:

H3a: *Consumers who prefer traditional design elements and luxury brands increase their WTP for denim fabric products.*

H3b: *Consumers who prefer transformative design elements and luxury brands increase their WTP for denim fabric products.*

H3c: *Consumers who prefer pattern design elements and luxury brands increase their WTP for denim fabric products.*

H3d: *Consumers who prefer multi-material design elements and luxury brands increase their WTP for denim fabric products.*

H3e: *Consumers who prefer decorative design elements and luxury brands increase their WTP for denim fabric products.*

2. Materials and Methods

In this study, we conducted a multiple regression analysis using the dataset of an original questionnaire survey. The dependent variable is the extent of WTP for denim fabric products. The explanatory variables are the design elements and brands preferred by consumers and their interaction terms. The analytical model is shown in Figure 1.

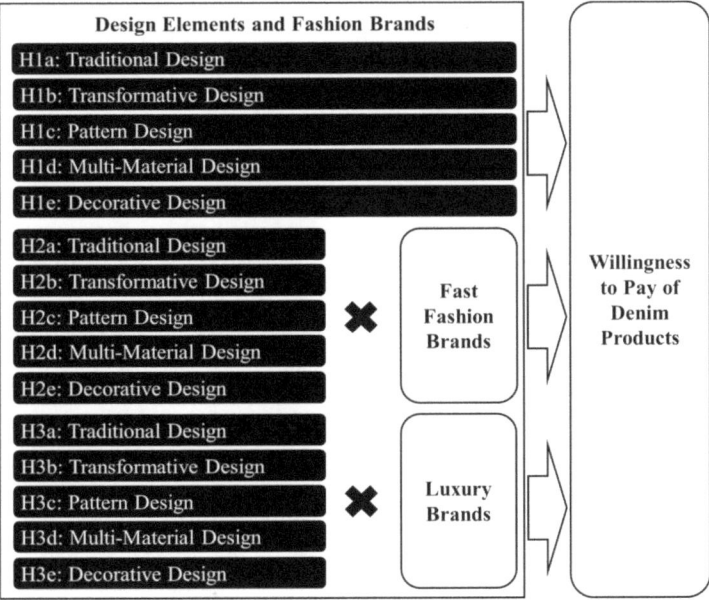

Figure 1. Analytical Model.

2.1. Questionnaire Summary

In this study, we conducted an online questionnaire survey. After creating the questionnaire, we asked Macromill, one of Japan's largest survey companies, to collect responses to the questionnaire. The sampling was random, with one condition that participants had to be 20 years or older. A screening process was developed, as described in Section 2.2. This survey was conducted over a period of four days, starting from 16 November 2021. While there are studies in the fashion industry that are specific to one country, or even one city [32], we selected the Japanese market for this study. In addition to consumer product value, technological innovations related to denim fabric products have become highly valued in Japan. Furthermore, various products have become increasingly widespread. Thus, we believe that the Japanese market is an appropriate research context.

Cochran's sample size formula was used to calculate the minimum sample data to ensure the reliability of the results [33]. The calculations were performed as follows: Based on Japan's Bureau of Statistics, the estimated number of people over 20 years of age in Japan in November 2021 was approximately 104.75 million [34]. Thereafter, reliability was set at 95%, the allowable error was set to be 5%, and the response rate was set to be 50%. Calculating all these factors suggested that a sample size of 384 people was required. To ensure sufficient sample size, the desired sample size was set to be 800, which was approximately twice as large as 384. Finally, 1107 participant responses were collected. Therefore, the sample size of this study was sufficient.

2.2. Questionnaire Design

2.2.1. Sampling

As this study is an analysis of the denim fabric product market, the desired respondents should be consumers who prefer buying denim fabric products. Therefore, the following screening questions were developed. For the first screening question, we asked about the total amount of money spent annually on clothing to confirm the degree of fashion consciousness in everyday life. To target fashion-loving consumers, we recruited consumers who chose "50,000 JPY" or more, which is higher than the average annual consumption value of clothing in Japan at 46,709 JPY (2019) [35]. The second screening question concerned how much consumers liked denim fabric products. Using a 7-point Likert scale, ranging from "extremely dislike" to "extremely like," we selected consumers who chose "moderately like," "like," and "extremely like." In the third screening question, we asked how strongly consumers' fashion consciousness was. Using a 7-point Likert scale, ranging from "extremely weak" to " extremely strong," we selected consumers who chose "moderately strong," "strong," and "extremely strong." Only respondents who passed the three screening questions were allowed to proceed with the questionnaire.

From the 1107 respondents, 407 (36.8%) were males and 700 (63.2%) were females. The respondents were divided into three categories according to their annual household income. A total of 427 respondents (38.6% of all participating consumers) had an annual income of less than 6 million JPY. A total of 393 respondents (35.5%) had an annual income of 6 million JPY or more but less than 12 million JPY, 126 (11.3%) respondents had an annual income of 12 million JPY or more, and 161 respondents (14.5%) did not answer the income question. The missing data were not a problem as annual income data were not used in the analysis. In addition, 508 participants were single (45.9%) and 599 were married (54.1%). This information was collected from the registration information in Macromill's questionnaire monitor.

After removing ineligible responses, the number of eligible respondents was reduced from 1107 to 1077. Therefore, the number of respondents whose responses were used in this study was 1077. The excluded conditions were as follows:

- (As shown in the Section on Survey Contents) The data on the purchase price per denim fabric product, as the dependent variable, were collected as a multiple-choice question. Respondents who selected "more," which could not be quantified, were excluded. Additionally, after confirming the distribution of responses after the loga-

rithmic conversion, we confirmed that the distribution was biased. Since this bias in the distribution would cause a bias in the statistical results, respondents who chose higher than 100,000 JPY were excluded as outliers.
- A Likert scale and binary variables were used in the questionnaire. Respondents who answered with the same number to all questionnaire items were excluded.
- No age limit was set for the respondents in the questionnaire; however, on one questionnaire, the participant's age was stated as 99 years old. We assumed that leaving such response in the dataset would impact the reliability of the answers. Therefore, we set the limit as 65 years for the respondents because it is the general retirement age in Japan. Respondents above this age were excluded.

2.2.2. Major Survey Contents

The respondents who passed the screening were asked the following questions: First, regarding the WTP for denim fabric products, the respondents were asked the average amount of money spent per piece of denim fabric product. The WTP for denim fabric products was obtained as the average purchase price per denim fabric product. The choices were as follows:

1. 1000; 2. 2000; 3. 3000; 4. 5000; 5. 7000; 6. 10,000; 7. 20,000; 8. 30,000; 9. 50,000; 10. 100,000; 11. 200,000; 12. 500,000; 13. 1,000,000; 14. More.

Subsequently, a multiple-choice question on what brand or maker the respondents would like to purchase products that use denim fabric from was asked. Specific brand names were presented along with the brand type to determine whether consumers preferred each brand type. We asked about each brand type with two choices, "agree (like)" or "disagree (dislike)". In addition to fast fashion brands and luxury brands, we asked questions on other brand types which were treated as the control variables in the analysis. Table 1 shows the questions regarding preference on the two types of fashion brands (fast fashion and luxury brands).

Table 1. Questions Regarding Preference on Fashion Brands.

Fast Fashion Brands	You like fast fashion brands (H&M, ZARA, UNIQLO, etc.) and often buy their products.
Luxury Brands	You like luxury brands (Gucci, Louis Vuitton, etc.) and often buy their products.

Finally, six questions were asked regarding clothing design preferences (the five design elements, and not limited to denim fabric products). Questions pertaining to all five design elements were answered using a five-point Likert scale (not applicable at all, not very applicable, neither, a little applicable, or very applicable). The specific contents of the questions are listed in Table 2.

Table 2. Questions Regarding the Five Design Elements.

Questionnaire Items	Questionnaire Contents
Traditional 1	You often buy products that maintain the old design.
Traditional 2	You like to buy replicas and reprinted brands.
Traditional 3	You share reprinted or old designs with others.
Traditional 4	You feel an attraction to products that the old design is maintained.
Traditional 5	You are someone who has knowledge of the old design.
Traditional 6	You like to search old designs in magazines and books.
Transformative 1	You prefer products that are not the same shape as others.
Transformative 2	You often buy unusual products or products that have never been seen before.
Transformative 3	You like eccentric products rather than simple ones.
Transformative 4	You find seemingly unusual products attractive.
Transformative 5	You prefer to wear something that does not match with that worn by others.
Transformative 6	When you see a seemingly unusual design in a collection or a magazine, you find it attractive.

Table 2. *Cont.*

Questionnaire Items	Questionnaire Contents
Pattern 1	You usually use patterned clothes.
Pattern 2	You like to buy products that have ethnic patterns.
Pattern 3	You prefer patterned products over simple products.
Pattern 4	You find products with original patterns and unique patterns attractive.
Pattern 5	You feel that the products that have patterns in every detail are highly valued.
Pattern 6	You like to put patterns in daily coordination.
Multi-Material 1	You like products that use multiple fabrics in one product.
Multi-Material 2	You often buy products with multiple different colors depending on the fabric.
Multi-Material 3	You like products that look different depending on the angle.
Multi-Material 4	You find a design created via a combination of fabrics attractive.
Multi-Material 5	You pay attention to the products that combine different fabrics.
Multi-Material 6	Many products that you chose for everyday wear use multiple fabrics.
Decorative 1	You like clothes with print designs and often buy them.
Decorative 2	You often wear products with an emblem.
Decorative 3	You prefer products with decorations (prints, patches, etc.) over simple designs.
Decorative 4	You often buy products that show the brand logo.
Decorative 5	You like to observe the decoration of clothing.
Decorative 6	The clothes you wear are often decorated.

2.3. Statistical Analysis

2.3.1. Dependent Variable

The dependent variable used was the average purchase price of denim fabric products, which was collected as a multiple-choice question, as described in the Section above. However, because the distribution did not have a bell shape, a natural logarithm conversion was used in the analysis following the correction of the distribution.

2.3.2. Explanatory Variables

Based on the factor analysis, we extracted the factors related to the degree to which the consumers liked the five design elements. The results are shown in Table 3. The average variance extracted via factor analysis confirmed the values of 0.5 or higher for the five design elements [36]. Regarding composite reliability, values of 0.6 or higher were confirmed for the five design elements [37]. The Cronbach's α was also confirmed to be 0.7 or higher for the five design elements [38]. Therefore, all the factors were reliable.

Table 3. Factor Analysis Results.

Questionnaire Items	Traditional Design	Transformative Design	Pattern Design	Multi-Material Design	Decorative Design
Traditional 1	0.76				
Traditional 2	0.86				
Traditional 3	0.83				
Traditional 4	0.80				
Traditional 5	0.82				
Traditional 6	0.85				
Transformative 1		0.73			
Transformative 2		0.84			
Transformative 3		0.78			
Transformative 4		0.86			
Transformative 5		0.78			
Transformative 6		0.79			

Table 3. Cont.

Questionnaire Items	Traditional Design	Transformative Design	Pattern Design	Multi-Material Design	Decorative Design
Pattern 1			0.66		
Pattern 2			0.65		
Pattern 3			0.82		
Pattern 4			0.85		
Pattern 5			0.79		
Pattern 6			0.83		
Multi-Material 1				0.84	
Multi-Material 2				0.81	
Multi-Material 3				0.81	
Multi-Material 4				0.82	
Multi-Material 5				0.87	
Multi-Material 6				0.83	
Decorative 1					0.78
Decorative 2					0.79
Decorative 3					0.87
Decorative 4					0.65
Decorative 5					0.61
Decorative 6					0.82
Average Variance Extracted	0.67	0.64	0.59	0.69	0.58
Composite Reliability	0.93	0.91	0.90	0.93	0.89
Cronbach's α	0.93	0.91	0.89	0.93	0.89

For H2a to H3e, the degree of preference for the five design elements and the interaction term for the consumers who preferred fast fashion and luxury brands were used as the explanatory variables.

2.3.3. Control Variables

In addition to the explanatory variables, control variables that might contribute to the dependent variable were selected.

Selected Shop Brands: The analytical results could be distorted if consumers were more likely to prefer selected shop brands. Therefore, we created a dummy variable related to selected shop brands with a binary response: 1 if respondents liked selected shop brands and 0 if they did not.

Casual Brands: The analytical results could be distorted if consumers were more likely to prefer casual brands. Therefore, we created a dummy variable related to casual brands with a binary response: 1 if respondents liked casual brands and 0 if they did not.

Gender: Gender carries different values for denim fabric products, which could affect the results of the analysis. We created a binary gender variable with 1 for men and 0 for women.

Age: Age has different values for denim fabric products, which could affect the results of the analysis. Age was calculated and answered numerically. After conversion to a natural logarithm, it was used as a control variable.

Annual Purchase of Clothes: The annual purchase of clothes by consumers might distort the results of the analysis. Therefore, we created a variable related to annual purchase of clothes, and the respondents answered the question numerically. After conversion to a natural logarithm, this was used as a control variable.

2.3.4. Empirical Specifications

In this study, multiple regression analysis was conducted. The analytical model is shown in Figure 1. In the interaction related to fast fashion and luxury brands, the problem of multicollinearity appeared when the analysis was conducted in parallel. Therefore, the analysis was conducted separately. In separate models, the mean variance inflation factor (VIF) in H1a to H1e was 1.36, with the maximum VIF being 1.85; the mean VIF in H2a to H2e was 2.40, with the maximum VIF being 4.51; and the mean VIF in H3a to H3e was 1.73, with the maximum VIF being 2.99 (the VIF empirical rule indicate that VIF values above 5 or 10 are multicollinear) [39]. Therefore, we confirmed that there was no multicollinearity problem. In addition, the Breusch–Pagan test was conducted to confirm the presence or absence of heterogeneity. All models had p-values of less than 0.05. Therefore, the Newey–West test was performed and modified.

3. Results

Table 4 shows the results of the analysis of Hypotheses H1a to H1e. Table 5 shows the analytical results, including the interaction terms between the variables related to the five design elements and that of fast fashion brands, in accordance with H2a to H2e. Table 6 shows the analytical results, including the interaction terms between the variables related to the five design elements and that of luxury brands, in accordance with H3a to H3e.

Table 4. Results of the Basic Analysis without Interaction Terms.

	Full Model		Stepwise	
	Coefficient	Standard Error	Coefficient	Standard Error
Traditional Design	0.07 **	0.02	0.07 **	0.02
Transformative Design	0.05 *	0.03	0.06 *	0.02
Pattern Design	−0.04 †	0.03	−0.05 *	0.02
Multi-Material Design	0.02	0.03		
Decorative Design	−0.03	0.03		
Gender	0.20 **	0.04	0.20 **	0.04
Age	0.00	0.06		
Annual Purchase of Clothes	0.30 **	0.03	0.30 **	0.03
Fast Fashion Brands	−0.42 **	0.04	−0.40 **	0.04
Selected Shop Brands	0.22 **	0.04	0.23 **	0.04
Casual Brands	0.02	0.04		
Luxury Brands	0.29 **	0.07	0.29 **	0.09
Constant	5.68 **	0.40	5.67 **	0.33
Adjusted R-squared	0.32		0.33	

** $p < 0.01$, * $p < 0.05$, † $p < 0.1$.

Table 5. Results of the Analysis with the Interaction Terms of Fast Fashion.

	Full Model		Stepwise	
	Coefficient	Standard Error	Coefficient	Standard Error
Traditional Design	0.11 **	0.03	0.12 **	0.04
Traditional Design × Fast Fashion Brands	−0.07 †	0.04	−0.08 *	0.04
Transformative Design	0.07 †	0.04	0.06 *	0.02
Transformative Design × Fast Fashion Brands	−0.03	0.05		
Pattern Design	−0.07	0.05	−0.04	0.03
Pattern Design × Fast Fashion Brands	0.04	0.05		
Multi-Material Design	0.05	0.05		
Multi-Material Design × Fast Fashion Brands	−0.05	0.06		
Decorative Design	−0.09 *	0.04	−0.08 *	0.04
Decorative Design × Fast Fashion Brands	0.11 *	0.05	0.10 *	0.04
Gender	0.20 **	0.04	0.20 **	0.04
Age	0.00	0.06		
Annual Purchase of Clothes	0.29 **	0.03	0.30 **	0.03
Fast Fashion Brands	−0.41 **	0.04	−0.41 **	0.04
Selected Shop Brands	0.23 **	0.04	0.23 **	0.04
Casual Brands	0.01	0.04		
Luxury Brands	0.29 **	0.07	0.30 **	0.07
Constant	5.66 **	0.40	5.65 **	0.33
Adjusted R-squared	0.33		0.33	

** $p < 0.01$, * $p < 0.05$, † $p < 0.10$.

Table 6. Results of the Analysis with the Interaction Terms of Luxury Brands.

	Full Model		Stepwise	
	Coefficient	Standard Error	Coefficient	Standard Error
Traditional Design	0.09 **	0.02	0.09 **	0.02
Traditional Design × Luxury Brands	−0.37 **	0.09	−0.34 **	0.08
Transformative Design	0.05 †	0.03	0.05 †	0.02
Transformative Design × Luxury Brands	0.22 †	0.12	0.27 **	0.10
Pattern Design	−0.05 †	0.03	−0.05 *	0.02
Pattern Design × Luxury Brands	0.10	0.12		
Multi-Material Design	0.02	0.03		
Multi-Material Design × Luxury Brands	0.00	0.11		
Decorative Design	−0.03	0.03		
Decorative Design × Luxury Brands	−0.02	0.10		
Gender	0.20 **	0.04	0.19 **	0.04
Age	0.00	0.06		
Annual Purchase of Clothes	0.29 **	0.03	0.29 **	0.03
Fast Fashion Brands	−0.41 **	0.04	−0.40 **	0.04
Selected Shop Brands	0.23 **	0.04	0.23 **	0.04
Casual Brands	0.03	0.04		
Luxury Brands	0.41 **	0.09	0.40 **	0.08
Constant	5.73 **	0.39	5.70 **	0.32
Adjusted R-squared	0.33		0.34	

** $p < 0.01$, * $p < 0.05$, † $p < 0.10$.

3.1. Results of the Basic Analysis without Interaction

Table 4 shows the analytical results for H1a to H1e. The stepwise determination coefficient was 0.33. Consequently, significant positive effects were confirmed for the two variables of traditional design ($p < 0.01$) and transformative design ($p < 0.05$). Therefore, H1a and H1b were supported. H1c was not supported because pattern design ($p < 0.05$) was confirmed to have a significant negative effect. Multi-material and decorative designs were removed as non-significant variables. Thus, H1d and H1e were not supported.

3.2. Results of the Analysis with the Interaction Term of Fast Fashion

Table 5 presents the analytical results for H2a to H2e. The coefficient of determination for the stepwise method was 0.33. H2a was supported because the interaction term between traditional design and fast fashion brands ($p < 0.05$) was confirmed to have a significant negative effect. Other hypotheses (H2b, H2c, H2d, and H2e) were not supported.

3.3. Results of the Analysis with the Interaction Term of Luxury Brands

Table 6 presents the analysis results for H3a to H3e. The determination coefficient for the stepwise method was 0.34. The interaction term between transformative design and luxury brands ($p < 0.01$) confirmed a significant positive effect. Consequently, H3b was supported. Other hypotheses (H3a, H3c, H3d, and H3e) were not supported.

4. Discussion

In this study, we suggested that the individuality and characteristics of each brand might affect the design elements. We analyzed two brand categories and set the interaction terms. Based on the basic analysis shown in Table 4, we compared the analytical results of the interaction terms for each brand. In addition, we considered the impact of preferred design elements on consumers. Finally, we interpreted the results of the hypotheses.

4.1. Discussion of the Basic Analysis

In the basic analysis (Table 4), traditional and transformative designs were confirmed to have significant positive effects. However, pattern design showed a significant negative effect. Denim fabric products are transformed into something of value in the market by designers who understand and interpret products related to tradition [5]. When it comes to denim fabric products, paying attention to history and tradition is necessary, even when devising a new design. Consumers who prefer traditional design have a high WTP for denim fabric products. Traditional design may contribute to the development of effective design elements for new product planning proposals.

The innovative features of transformative design depend on a well-formed silhouette. Hence, this does not become an obstacle for denim fabric, which is regarded as attractive. Design elements are valued because they provide consumers with attractive product value [8,40,41]. However, denim fabric, which is attractive for its originality, may be hindered by "color," "processing," and "decoration." Accordingly, consumers who prefer transformative design have a high WTP for denim fabric products. Thus, designers can propose attractive products by including innovative designs in new product planning.

Denim fabric products have a wide range of designs in the fabric itself [8]. Therefore, the direction in which the original fabric design improves the appearance of the product without applying any additional design has been considered. Pattern design aims to improve the attractiveness of a product by adding decoration to the fabric [21]. However, as mentioned, denim fabric products depend on the attractiveness of the fabric because of its aesthetic appeal. Fabric processing can bring out denim's decadence and "atmosphere", improving the design. Hence, a preference for pattern design has a negative impact on consumers' WTP for denim fabric products.

4.2. Discussion of the Analysis with Fast Fashion Brands

The two variables that provided significant results in the analysis of the interaction term of fast fashion brands were traditional and decorative designs.

Consumers who prefer traditional design demonstrate a negative effect on their WTP for denim fabric products when they prefer fast fashion brands. The price range of a traditional design element is high because the design fully utilizes the advanced technology required by consumers [16]. Traditional design is a source of new insights on fashion today [42,43], and fashion-loving consumers also influence the visual perception of these designs [21,44]. Thus, consumers who prefer fast fashion brands may be aware of superficial functional design; however, they do not have an attractive perception of the high quality required for traditional design elements.

Consumers who prefer fast fashion brands and decorative design show a positive effect on their WTP for denim fabric products. Consumers can quickly recognize the value of new design elements proposed by fast fashion brands [45]. In addition, they can purchase products at an affordable price [45]. This may be due to the fact that young consumers, who have a relatively low WTP, tend to prefer decorative design, which follows the trend. Therefore, decorative denim fabric products proposed by fast fashion brands, which can be purchased at an affordable price, may be effective as a means to increase their WTP. Fast fashion brands also recognize trends and continuously respond to consumer demand [46]. Therefore, a decorative design for denim fabric products could increase consumers' WTP by reflecting current fashion design preferred by consumers.

4.3. Discussion of the Analysis with Luxury Brands

The two variables that provided significant results in the analysis of the interaction term of luxury brands were traditional and transformative designs.

Consumers who prefer traditional design and luxury brands show a negative effect on their WTP for denim fabric products. The role of creativity in the luxury brand industry is to explore the product designs that disrupt existing value over time [47,48]. Consumers who prefer luxury brands are attracted to the creativity of designers, influencing their willingness to buy denim fabric products [32]. Traditional design tends to demand high-quality and faithful reproductions that consumers demand in contrast to the creative design of luxury brands [5,16]. Consumers who prefer luxury brands may discover the attractive value of creative and new products and choose to purchase them. Therefore, traditional design may differ from the needs of consumers who prefer luxury brands, which deal with attractive products driven by designers.

Consumers who prefer transformative design and luxury brands have more WTP for denim fabric products than those who do not. The transformative design element is evaluated as an element that increases the WTP for denim fabric products. However, it could increase the WTP for denim fabric products when targeting consumers who prefer luxury brands. Transformative design is a means of creative design invention, adding "new" value to products [18]. Consumers are influenced by the visual appeal to the shape of a new form of fashion devised by luxury brands [41,49]. Therefore, transformative design, as devised by designers of denim fabric products, could provide attractive value to consumers. Regarding luxury brands, a transformative design element is important to increase the WTP for denim fabric products.

5. Conclusions

Several studies have examined how to increase customers' purchasing intentions of denim products [14,15]. The novelty of this study lies in its examination of the impact of design elements on consumers' WTP for denim fashion products. This study fills a gap in research on the impact of the design elements of fashion products on consumers' WTP. This study targets the denim fabric market and confirms that consumers who prefer traditional and transformative designs positively show a high WTP for denim fabric products. Therefore, these elements may be attractive design elements that can be highly priced in new product planning proposals.

We also analyzed how the WTP for denim fabric products is affected by the preference for fast fashion brands and luxury brands. Consumers who prefer fast fashion brands demonstrate a negative effect on their WTP for denim fabric products if they prefer traditional designs. However, if they prefer a decorative design, it has a positive effect on their WTP. Consumers who prefer luxury brands show a negative effect on their WTP for denim fabric products if they prefer traditional designs. However, if they prefer a transformative design, it has a more positive effect on their WTP. Thus, it is possible to increase consumers' WTP for denim fabric products by appropriately judging the characteristics preferred by consumers.

5.1. Theoretical Implications

Consumers' own self-expression and personal fashion appeal are emphasized in denim fabric products [8,50–52]. In addition, as the determinants of consumers' intention to purchase denim jeans diversify, design items have also become complex [6]. In this study, we examined the design elements that consumers like, analyzed what kind of denim fabric products would be attractive, and contributed to research on product designs for denim fabric products. In addition, consumer interests could be related to the choice of brand category, and the contribution to purchase motivation depends on the characteristics of fashion brands that consumers like.

5.2. Practical Implications

Consumers who are targeted by the denim fabric product market and prefer traditional and transformative designs could show a high WTP for denim fabric products. Therefore, these elements may be attractive design elements that can be highly priced in new product planning proposals. Consumers who prefer pattern design may have a lower WTP for denim fabric products. The results of this study suggest that designers can propose attractive denim fabric products that are highly priced by incorporating traditional and transformative designs. However, with regard to low-priced brands, such as fast fashion brands—which appeal to consumer needs—and high-priced brands, such as luxury brands—where designers devise new designs—traditional design could not influence consumers' WTP. Transformative design can lead to a high WTP among consumers who are visually attracted to the creation of new designs, such as luxury brand designs. Decorative design could contribute to the WTP for denim fabric products by devising designs that appeal to the needs of consumers.

5.3. Limitations and Future Research

This study has some limitations. First, not all apparel products might have similar results, as we only analyzed the denim fabric product market. Therefore, different suggestions may be confirmed through a market analysis of different products.

Second, the questionnaire data for this study were obtained from the Japanese market. Although there are previous studies in the fashion industry that are specific to one country, or even one city [16,32], we must note the limitation in the acquired results from the data sample. Similar to the cultural differences among countries, consumers' preferences for design elements in fashion products could be different. Therefore, different findings and implications could be obtained if future studies conduct further investigations in other countries.

Third, other design elements in the denim fabric market could be suggested. Since the theory of design is extremely broad, it is possible to investigate new design innovations by exploring possible design elements other than those addressed in this study.

Finally, this study focused on two styles of brands: fast fashion and luxury brands. Since there are brands other than fast fashion and luxury brands that handle denim fabric products, conducting a survey that includes other brand categories could be made available.

Author Contributions: Conceptualization, R.M., X.Z. and Y.I.; methodology, R.M. and Y.I.; software, R.M.; validation, R.M.; formal analysis, R.M. and Y.I.; investigation, R.M. and Y.I.; resources, R.M. and Y.I.; data curation, R.M.; writing—original draft preparation, R.M., X.Z. and Y.I.; writing—review and editing, R.M., X.Z. and Y.I.; visualization, R.M., X.Z. and Y.I.; supervision, Y.I.; project administration, Y.I.; funding acquisition, Y.I. All authors have read and agreed to the published version of the manuscript.

Funding: This research received no external funding.

Institutional Review Board Statement: Not applicable.

Informed Consent Statement: Not applicable.

Data Availability Statement: The data presented in this study are available from the corresponding author upon request.

Conflicts of Interest: The authors declare no conflict of interest.

References

1. Júnior, H.L.O.; Neves, R.M.; Monticeli, F.M.; Dall Agnol, L. Smart Fabric Textiles: Recent Advances and Challenges. *Textiles* **2022**, *2*, 582–605. [CrossRef]
2. Tabassum, M.; Zia, Q.; Zhou, Y.; Wang, Y.; Reece, M.J.; Su, L. A Review of Recent Developments in Smart Textiles Based on Perovskite Materials. *Textiles* **2022**, *2*, 447–463. [CrossRef]
3. Krifa, M. Electrically conductive textile materials—Application in flexible sensors and antennas. *Textiles* **2021**, *1*, 239–257. [CrossRef]
4. Su, Y.; Li, W.; Cheng, X.; Zhou, Y.; Yang, S.; Zhang, X.; Chen, C.; Yang, T.; Pan, H.; Xie, G.; et al. High-performance piezoelectric composites via β phase programming. *Nat. Commun.* **2022**, *13*, 1–12. [CrossRef]
5. Fujioka, R.; Wuds, B. *Competitiveness of the Japanese Denim and Jeans Industry: The Cases of Kaihara and Japan Blue, 1970–2015*; European Fashion: Manchester, UK, 2020.
6. Rahman, O. Understanding Consumer's Perceptions and Buying Behaviors: Implications for Denim Jeans Design. *Wilson Coll. Text.* **2011**, *7*, 1–16.
7. Seninde, D.R.; Chambers IV, E.; Chambers, D.H.; Chambers, V.E. Development of a Consumer-Based Quality Scale for Artisan Textiles: A Study with Scarves/Shawls. *Textiles* **2021**, *1*, 483–503. [CrossRef]
8. Su, J.; Tong, X. Brand Personality, Consumer Satisfaction, and Loyalty: A Perspective from Denim Jeans Brands. *Fam. Consum. Sci.* **2016**, *44*, 427–446. [CrossRef]
9. Candi, M.; Haeran, J.; Suzanne, M. and Mayoor, M. Consumer responses to functional, aesthetic and symbolic product design in online reviews. *J. Bus. Res.* **2017**, *81*, 31–39. [CrossRef]

10. Homburg, C.; Martin, S. and Christina, K. New product design: Concept, measurement, and consequences. *J. Mark.* **2015**, *79*, 41–56. [CrossRef]
11. Simonson, A.; Schmitt, B.H. *Marketing Aesthetics: The Strategic Management of Brands, Identity, and Image*; Simon & Schuster: New York, NY, USA, 1997.
12. Papanek, V.J. *The Green Imperative: Natural Design for the Real World*; Thames and Hudson: New York, NY, USA, 1995.
13. Shogren, J.F.; Shin, S.Y.; Hayes, D.J.; Kliebenstein, J.B. Resolving Differences in Willingness to Pay and Willingness to Accept. *Am. Econ. Rev.* **1994**, *84*, 255–270.
14. Rahman, O. Denim Jeans: A Qualitative Study of Product Cues, Body Type, and Appropriateness of Use. *Fash. Pract. J. Des. Creat. Process Fash. Ind.* **2015**, *7*, 53–74.
15. Card, A.; Moore, M.A.; Ankeny, M. Garment Washed Jeans: Impact of Launderings on Physical Properties. *Int. J. Cloth. Sci. Technol.* **2006**, *18*, 43–52. [CrossRef]
16. Abrego, S. Cone Mills Denim: An Investigation into Fabrication, Tradition, and Quality. *Fash. Theory* **2019**, *23*, 515–530. [CrossRef]
17. Cassidy, T.D.; Bennett, H.R. The Rise of Vintage Fashion and the Vintage Consumer. *Fash. Pract. J. Des. Creat. Process Fash. Ind.* **2012**, *4*, 239–261. [CrossRef]
18. Moon, H.; Miller, D.R.; Kim, S.H. Product Design Innovation and Customer Value: Cross-Cultural Research in the United States and Korea. *J. Prod. Innov. Manag.* **2013**, *30*, 31–43. [CrossRef]
19. Miller, D.; Woodward, S. Manifesto for a Study of Denim, Social Anthropology. *Eur. Assoc. Soc. Anthropol.* **2008**, *15*, 335–351. [CrossRef]
20. Feijis, L.M.G.; Toeters, M.G. Cellular Automata-Based Generative Design of Pied-De-Poule Patterns Using Emergent Behavior: Case Study of How Fashion Pieces Can Help to Understand Modern Complexity. *Int. J. Des.* **2018**, *12*, 127–144.
21. Micheli, P.; Gemser, G. Signaling Strategies for Innovative Design: A Study on Design Tradition and Expert Attention. *J. Prod. Innov. Manag.* **2016**, *33*, 613–627. [CrossRef]
22. Loschek, I. *When Clothes Become Fashion: Design and Innovation Systems*; Berg Publishers: Oxford, UK, 2009.
23. Townsend, K. The Denim Garment as Canvas: Exploring the Notion of Wear as a Fashion and Textile Narrative. *Text. Cloth Cult.* **2015**, *9*, 90–107. [CrossRef]
24. Cachon, G.P.; Swinney, R. The Value of Fast Fashion: Quick Response, Enhanced Design, and Strategic Consumer Behavior. *Manag. Sci.* **2011**, *57*, 778–795. [CrossRef]
25. Bhardwaj, V.; Fairhurst, A. Fast Fashion: Response to Changes in the Fashion Industry. *Int. Rev. Retail. Distrib. Consum. Res.* **2010**, *20*, 165–173. [CrossRef]
26. Rahman, O.; Gong, M. Sustainable Practices and Transformable Fashion Design—Chinese Professional and Consumer Perspectives. *Int. J. Fash. Des. Technol. Educ.* **2016**, *9*, 233–247. [CrossRef]
27. Fuchs, C.; Prandelli, E.; Schreier, M.; Dahl, D.W. All that is users might not be gold: How labeling products as user designed backfires in the context of luxury fashion brands. *J. Mark.* **2013**, *77*, 75–91. [CrossRef]
28. Heine, K. The Personality of Luxury Fashion Brands. *J. Glob. Fash. Mark.* **2010**, *1*, 154–163. [CrossRef]
29. Jain, S.; Khan, M.N.; Mishra, S. Understanding Consumer Behavior regarding Luxury Fashion Goods in India Based on the Theory of Planned Behavior. *J. Asia Bus. Stud.* **2017**, *11*, 4–21. [CrossRef]
30. Wu, M.-S.S.; Chaney, I.; Chen, C.-H.S.; Nguyen, B.; Melewar, T.C. Luxury fashion brands: Factors influencing young female consumers' luxury fashion purchasing in Taiwan. *Qual. Mark. Res.* **2015**, *18*, 298–319. [CrossRef]
31. Li, G.; Li, G.; Kambele, Z. Luxury Fashion Brand Consumers in China: Perceived Value, Fashion Lifestyle, and Willingness to Pay. *J. Bus. Res.* **2012**, *65*, 1516–1522. [CrossRef]
32. Bye, E.; Sohn, M. Technology, Tradition, and Creativity in Apparel Designers: A Study of Designers in Tree US Companies. *Fash. Pract.: J. Des. Creat. Process Fash. Ind.* **2010**, *2*, 199–222. [CrossRef]
33. Cochran, W.G. *Sampling Techniques*, 3rd ed.; John Wiley & Sons: Hoboken, NJ, USA, 1977.
34. Statistics Bureau of Japan. Result of the Population Estimates. Available online: https://www.stat.go.jp/english/data/jinsui/tsuki/index.html (accessed on 9 October 2021).
35. Household Survey (Ministry of Internal Affairs and Communications). Available online: https://www.stat.go.jp/data/kakei/longtime/soutan.html (accessed on 10 January 2022).
36. Hair, J.F.; Black, B.; Babin, B.; Anderson, R.E.; Tatham, R.L. *Multivariate Data Analysis*; Pearson Prentice Hall, Pearson Education: Upper Saddle River, NJ, USA, 2006; pp. 1–816.
37. Bacon, D.R.; Sauer, P.L.; Young, M. Composite Reliability in Structural Equations Modeling. *Educ. Psychol. Meas.* **1995**, *55*, 394–406. [CrossRef]
38. Tavakol, M.; Dennick, R. Making Sense of Cronbach's Alpha. *Int. J. Med. Educ.* **2011**, *2*, 53–55. [CrossRef]
39. James, G.; Witten, D.; Hastie, T.; Tibshirani, R. *An Introduction to Statistical Learning: With Applications in R*, 1st ed.; Springer: Berlin/Heidelberg, Germany, 2014; p. 112.
40. Dunne, L.; Stivoric, J. Fashioning Bodily Knowledge: BodyMedia's Pervasive Body-Monitoring Portal. *Fash. Pract. J. Des. Creat. Process Fash. Ind.* **2013**, *5*, 107–116. [CrossRef]
41. Quinn, G.P.; Queen, J.P. *Experimental Design and Data Analysis for Biologists*; The Press Syndicate of the University of Cambridge: Cambridge, UK, 2002.

42. Cirella, S. Organizational Variables for Developing Collective Creativity in Business: A Case from an Italian Fashion Design Company. *Creat. Innov. Manag.* **2016**, *25*, 331–343. [CrossRef]
43. Gronow, J. Taste and Fashion: The Social Function of Fashion and Style. *Acta Sociol.* **1993**, *36*, 89–100. [CrossRef]
44. Mete, F. The Creative Role of Sources of Inspiration in Clothing Design. *Int. J. Cloth. Sci. Technol.* **2006**, *18*, 278–293. [CrossRef]
45. Bruce, M.; Daly, L. Buyer Behaviour for Fast Fashion. *J. Fash. Mark. Manag.* **2006**, *10*, 329–344.
46. Crofton, S.O.; Dopico, L.G. ZARA-Inditex and the Growth of Fast Fashion. *Essays Econ. Bus. Hist.* **2007**, *25*, 41–54.
47. Giacomin, J. What Is Human Centred Design? *Des. J.* **2014**, *17*, 606–623. [CrossRef]
48. Jones, G.G.; Pouillard, V. *Christian Dior: A New Look for Haute Couture*; Harvard Business School Entrepreneurial Management: Boston, MA, USA, 2009.
49. Chiva, R.; Alegre, J. Investment in Design and Firm Performance: The Mediating Role of Design Management. *J. Prod. Innov. Manag.* **2009**, *26*, 424–440. [CrossRef]
50. Louis, D.; Lombart, C. Impact of Brand Personality on Three Major Relational Consequences (Trust, Attachment, and Commitment to the Brand). *J. Prod. Brand Manag.* **2010**, *19*, 114–130. [CrossRef]
51. Rahman, O.; Jiang, Y.; Liu, W.S. Evaluative Criteria of Denim Jeans: A Cross-National Study of Functional and Aesthetic Aspects. *Des. J.* **2010**, *13*, 291–311. [CrossRef]
52. Sung, Y.; Kim, J. Effects of brand personality on brand trust and brand affect. *Phycol. Mark.* **2010**, *27*, 639–661. [CrossRef]

Disclaimer/Publisher's Note: The statements, opinions and data contained in all publications are solely those of the individual author(s) and contributor(s) and not of MDPI and/or the editor(s). MDPI and/or the editor(s) disclaim responsibility for any injury to people or property resulting from any ideas, methods, instructions or products referred to in the content.

MDPI AG
Grosspeteranlage 5
4052 Basel
Switzerland
Tel.: +41 61 683 77 34

Textiles Editorial Office
E-mail: textiles@mdpi.com
www.mdpi.com/journal/textiles

Disclaimer/Publisher's Note: The statements, opinions and data contained in all publications are solely those of the individual author(s) and contributor(s) and not of MDPI and/or the editor(s). MDPI and/or the editor(s) disclaim responsibility for any injury to people or property resulting from any ideas, methods, instructions or products referred to in the content.